Gene Delivery
to Mammalian Cells

METHODS IN MOLECULAR BIOLOGY™

John M. Walker, SERIES EDITOR

Gene Delivery to Mammalian Cells

Volume 1: Nonviral Gene Transfer Techniques

Edited by

William C. Heiser

Bio-Rad Laboratories
Hercules, CA

Humana Press Totowa, New Jersey

© 2004 Humana Press Inc.
999 Riverview Drive, Suite 208
Totowa, New Jersey 07512

www.humanapress.com

This publication is printed on acid-free paper. ∞
ANSI Z39.48-1984 (American Standards Institute)

Permanence of Paper for Printed Library Materials.

Cover illustration: Foreground graphic from Fig. 1A,B in Chapter 14 (Volume 1) "Delivery of DNA to Skin by Particle Bombardment," by Shixia Wang, Swati Joshi, and Shan Lu. Background graphic from Fig. 3 in Chapter 33 (Volume 2) "Retrovirus-Mediated Gene Transfer to Tumors: Utilizing the Replicative Power of Viruses to Achieve Highly Efficient Tumor Transduction In Vivo," by Christopher R. Logg and Noriyuki Kasahara.

Production Editor: Robin B. Weisberg.
Cover design by Patricia F. Cleary.

For additional copies, pricing for bulk purchases, and/or information about other Humana titles, contact Humana at the above address or at any of the following numbers: Tel.: 973-256-1699; Fax: 973-256-8341; E-mail: humana@humanapr.com; or visit our Website: www.humanapress.com

Photocopy Authorization Policy:

Printed in the United States of America. 10 9 8 7 6 5 4 3 2 1

1-59259-649-5 (e-ISBN)

Library of Congress Cataloging in Publication Data
Gene delivery to mammalian cells / edited by William C. Heiser.
    p. ; cm. -- (Methods in molecular biology ; 245-246)
    Includes bibliographical references and indexes.
    Contents: v. 1. Nonviral gene transfer techniques -- v. 2. Viral gene transfer techniques.
    ISBN 1-58829-086-7 (v. 1 : alk. paper) -- ISBN 1-58829-095-6 (v. 2 : alk. paper)
    ISSN 1064-3745
    1. Transfection. 2. Animal cell biotechnology. I. Heiser, William C. II. Series: Methods in molecular biology (Totowa, N.J.) ; 245-246.
    [DNLM: 1. Cloning, Molecular. 2. Gene Transfer Techniques. 3. Cells, Cultured--cytology. 4. Gene Targeting--methods. 5. Mammals--genetics. QH 442.2 G327 2004]
    QH448.4.G42 2004
    571.9'64819--dc21

                                                                        2003006980

Preface

The efficiency of delivering DNA into mammalian cells has increased tremendously since DEAE dextran was first shown to be capable of enhancing transfer of RNA into mammalian cells in culture. Not only have other chemical methods been developed and refined, but also very efficient physical and viral delivery methods have been established. The technique of introducing DNA into cells has developed from transfecting tissue culture cells to delivering DNA to specific cell types and organs in vivo. Moreover, two important areas of biology—assessment of gene function and gene therapy—require successful DNA delivery to cells, driving the practical need to increase the efficiency and efficacy of gene transfer both in vitro and in vivo.

These two volumes of the *Methods in Molecular Biology*™ series, *Gene Delivery to Mammalian Cells,* are designed as a compendium of those techniques that have proven most useful in the expanding field of gene transfer in mammalian cells. It is intended that these volumes will provide a thorough background on chemical, physical, and viral methods of gene delivery, a synopsis of the myriad techniques currently available to introduce genes into mammalian cells, as well as a practical guide on how to accomplish this. It is my expectation that it will be useful to the novice in the field as well as to the scientist with expertise in gene delivery.

Volume 1: Nonviral Gene Transfer Techniques discusses delivery of DNA into cells by nonviral means, specifically chemical and physical methods. *Volume 2: Viral Gene Transfer Techniques* details procedures for delivering genes into cells using viral vectors. Each volume is divided into sections; each section begins with a chapter that provides an overview of the basis behind the delivery system(s) described in that section. The succeeding chapters provide detailed protocols for using these techniques to deliver genes to cells in vitro and in vivo. Many of these techniques have only been in practice for a few years and are still being refined and updated. Some are being used not only in basic science, but also in gene therapy applications.

I wish to express my thanks to all of the authors who made *Gene Delivery to Mammalian Cells: Volume 1: Nonviral Gene Transfer Techniques* and *Volume 2: Viral Gene Transfer Techniques* possible. I would especially like to thank those who contributed the overview chapter to each section. They provided invaluable discussions, suggestions, and assistance on organizing those sections. I would particularly like to mention Joanne Douglas, Tom Daly, and Bill Goins for their suggestions on topics and authors, Dexi Liu and Shan Lu for their helpful discussions, and Mark Jaroszeski for his suggestions on organizing the entire editing process.

William C. Heiser

Contents

CONTENTS OF THE COMPANION VOLUME
Volume 2: Viral Gene Transfer Techniques

ix

Contributors

NICHOLAS AMBULOS • *Department of Microbiology, University of Maryland, Baltimore, MD*

JAMES R. BAKER, JR. • *Department of Internal Medicine and Center for Biologic Nanotechnology, The University of Michigan Health System, Ann Arbor, MI*

SHIKHA P. BARMAN • *ZYCOS Inc., Lexington, MA*

PETER A. BARRY • *Center for Comparative Medicine, University of California, Davis, CA*

LALITHA R. BELUR • *Institute of Human Genetics, University of Minnesota, Minneapolis, MN*

ANNA U. BIELINSKA • *Department of Internal Medicine and Center for Biologic Nanotechnology, The University of Michigan Health System, Ann Arbor, MI*

SUBHABRATA BISWAS • *Department of Medicine, University of Massachusetts Medical School, Worcester, MA*

VLADIMIR G. BUDKER • *Departments of Pediatrics and Medical Genetics, University of Wisconsin, Madison, WI*

SENG H. CHENG • *Genzyme Corporation, Framingham, MA*

EVELYN F. CHIAO • *Department of Pharmaceutical Sciences, School of Pharmacy, University of Pittsburgh, Pittsburgh, PA*

TE-HUI W. CHOU • *Department of Medicine, University of Massachusetts Medical School, Worcester, MA*

NATHALIE DUJARDIN • *Unité de Pharmacie galénique, Université Catholique de Louvain, Brussels, Belgium*

RICHARD GILBERT • *Department of Chemical Engineering, College of Engineering, University of South Florida, Tampa, FL*

TODD D. GIORGIO • *Departments of Biomedical Engineering and Chemical Engineering, Vanderbilt University, Nashville, TN*

MARY LYNNE HEDLEY • *ZYCOS Inc., Lexington, MA*

WILLIAM C. HEISER • *Life Science Group, Bio-Rad Laboratories, Hercules, CA*

LOREE C. HELLER • *Department of Surgery, College of Medicine, University of South Florida, Tampa, FL*

RICHARD HELLER • *Department of Surgery, College of Medicine, University of South Florida, Tampa, FL*

MIEN-CHIE HUNG • *Department of Molecular and Cellular Oncology, University of Texas, M. D. Anderson Cancer Center, Houston, TX*

MARK J. JAROSZESKI • *Department of Chemical Engineering, College of Engineering, University of South Florida, Tampa, FL*

SWATI JOSHI • *Department of Medicine, University of Massachusetts Medical School, Worcester, MA*

ROBERT KING • *Life Science Group, Bio-Rad Laboratories, Hercules, CA*

JOSEPH E. KNAPP • *Department of Pharmaceutical Sciences, School of Pharmacy, University of Pittsburgh, Pittsburgh, PA*

JOLANTA F. KUKOWSKA-LATALLO • *Department of Internal Medicine and Center for Biologic Nanotechnology, The University of Michigan Health System, Ann Arbor, MI*

DEXI LIU • *Department of Pharmaceutical Sciences, School of Pharmacy, University of Pittsburgh, Pittsburgh, PA*

SHAN LU • *Department of Medicine, University of Massachusetts Medical School, Worcester, MA*

JAMES J. LUDTKE • *Waisman Center, University of Wisconsin, Madison, WI*

JOHN MARSHALL • *Genzyme Corporation, Framingham, MA*

A. KIMBERLEY MCALLISTER • *Center for Neuroscience and Department of Neurology, University of California, Davis, CA*

THOMAS P. MCCREERY • *ImaRx Therapeutics Inc., Tucson, AZ*

R. SCOTT MCIVOR • *Institute of Human Genetics, University of Minnesota, Minneapolis, MN*

LLUIS M. MIR • *C.N.R.S., Institut Gustave-Roussy, Villejuif, France*

A. JAMES MIXSON • *Department of Pathology, University of Maryland, Baltimore, MD*

GREGORY S. PARI • *Department of Microbiology, University of Nevada, Reno, NV*

VÉRONIQUE PRÉAT • *Unité de Pharmacie galénique, Université Catholique de Louvain, Brussels, Belgium*

GUENHAËL SANZ • *Institut National de la Recherche Agronomique, Jouy-en-Josas, France*

SAULIUS ŠATKAUSKAS • *Department of Biology, Vytautas Magnus University, Kaunas, Lithuania*

YOUNG K. SONG • *Department of Pharmacology, School of Medicine, University of Pittsburgh, Pittsburgh, PA*

BILL SPOHN • *Department of Molecular and Cellular Oncology, The University of Texas, M.D. Anderson Center, Houston, TX*

SEAN SULLIVAN • *Department of Pharmceutics, University of Florida, Gainsville, FL*

ROBERT H. SWEITZER • *ImaRx Therapeutics Inc., Tucson, AZ*
HUI TIAN • *Department of Pharmaceutical Sciences, School of Pharmacy, University of Pittsburgh, Pittsburgh, PA*
EVAN C. UNGER • *Department of Radiology, University of Arizona Health Sciences Center, Tucson, AZ, and ImaRx Therapeutics Inc., Tucson, AZ*
SHIXIA WANG • *Department of Medicine, University of Massachusetts Medical School, Worcester, MA*
JON A. WOLFF • *Departments of Pediatrics and Medical Genetics and Waisman Center, University of Wisconsin, Madison, WI*
SHELBY K. WYATT • *Department of Biomedical Engineering, Vanderbilt University, Nashville, TN*
YIYANG XU • *Department of Microbiology, University of Nevada, Reno, NV*
DUEN-HWA YAN • *Department of Molecular and Cellular Oncology, The University of Texas, M.D. Anderson Center, Houston, TX*
NELSON S. YEW • *Genzyme Corporation, Framingham, MA*
GUISHENG ZHANG • *Department of Pharmaceutical Sciences, School of Pharmacy, University of Pittsburgh, Pittsburgh, PA*
GUOFENG ZHANG • *Waisman Center, University of Wisconsin, Madison, WI*
LEI ZHANG • *Department of Pathology, University of Maryland, Baltimore, MD*

I

DELIVERY USING CHEMICAL METHODS

1

Chemical Methods for DNA Delivery

An Overview

Dexi Liu, Evelyn F. Chiao, and Hui Tian

1. Introduction

Introduction of DNA into mammalian cells is a powerful tool for studying the function of various DNA sequences, and for gene therapy. The process of introducing DNA into cells for the purpose of gene expression is called transfection or gene delivery. Synthetic compounds used to facilitate DNA transfer are often named synthetic vectors or transfection reagents. Compared with biological (viral vectors) and physical methods that are covered elsewhere in this volume and in the next volume, the major advantages of synthetic vectors (or chemical methods) are their simplicity, ease of production, and relatively low toxicity. Many synthetic compounds have been developed since DEAE-dextran was first used in transfection experiments more than 35 years ago. Rapid progress in developing more efficient synthetic vectors has led to successful DNA delivery into a variety of cell types in vitro and in vivo. More importantly, in the last few years, we have witnessed significant efforts and progress in elucidating the mechanisms underlying synthetic vector-mediated DNA delivery. With the continuous effort to meet the need for safe and efficient gene-delivery methods for human gene therapy, it is foreseeable that significant advances will be made in the future. This article concentrates on four major types of chemical reagents that are available to most investigators: calcium phosphate, DEAE-dextran, cationic lipid, and cationic polymer. Each of these types of reagents has its advantages and disadvantages, some of which we briefly outline in this overview chapter.

From: *Methods in Molecular Biology, vol. 245:*
Gene Delivery to Mammalian Cells: Vol. 1: Nonviral Gene Transfer Techniques
Edited by: W. C. Heiser © Humana Press Inc., Totowa, NJ

One type of compound that is not covered in this section is biopolymers. These compounds, including polylysine *(1–7)*, histone *(8,9)*, chitosan *(10,11)*, and peptides *(12–17)*, have shown relatively low transfection efficiency when used alone. Although they might become relatively more important in the future, we feel their utility has not, at present, been demonstrated to be broad enough to recommend their routine use for transfection. For those who are interested in obtaining more information about the properties and activity of this group of polymers in DNA delivery, relevant information can be found in the references cited at the end of this chapter or in relevant chapters in this volume.

2. For What Purpose Are Synthetic Vectors Needed?

Cell membranes are sheetlike assemblies of amphipathic molecules that separate cells from their environment and form the boundaries of different organelles inside the cells. However, these physical structures, allow only the controlled exchange of materials among the different parts of a cell and with its immediate surroundings. DNA, on the other hand, is an anionic polymer, large in molecular weight, hydrophilic, and sensitive to nucleases in biological matrices. Unless special means are used, DNA molecules are not able to cross the physical barrier of membrane and enter the cells. In theory, for a successful introduction of DNA into cells and expression of the encoded gene, one needs to overcome three major hurdles. These include: (1) transfer of DNA from the site of DNA administration to the surface of target cells; (2) transfer of DNA across the plasma membrane into the cytoplasm; and (3) transfer of the DNA across the nuclear membrane into the nucleus to initiate gene expression. Thus, the corresponding properties for an ideal synthetic vector should include protecting DNA against nuclease degradation; transporting DNA to the target cells; facilitating transport of DNA across the plasma membrane; and finally promoting the import of DNA into the nucleus.

3. Properties of the Synthetic Compounds Most Commonly Used for DNA Delivery

3.1. Calcium Phosphate

Transfection with calcium phosphate was developed by Graham and Van der Eb in 1973 *(18)* and has been one of the most commonly used methods for DNA delivery into mammalian cells. This method takes advantage of the formation of small, insoluble, calcium-phosphate-DNA precipitates that can be adsorbed onto the cell surface and be taken up by cells through endocytosis. The procedure requires mixing of DNA with calcium ions, subsequent addition of phosphate to the mixture, and presentation of the final solution to cells in culture. Various types of cells have been transfected using this procedure. Transfection efficiency can be as high as 50%, depending on the cell type and the size and quality of the precipitate. The variation in the composition and particle size of the calcium-phosphate-DNA precipitates results in poor reproducibility. This method does not seem to work on cells in primary culture or in animals.

3.2. DEAE-Dextran

DEAE-dextran was the very first chemical reagent used for DNA delivery. It was initially reported by Vaheri and Pagano in 1965 *(19)* for enhancing the viral infectivity of cells. Similar to cationic polymers (*see* **Subheading 3.4**), DNA and DEAE-dextran form aggregates through electrostatic interaction. A slight excess of DEAE-dextran in the mixture results in net positive charge in the DEAE-dextran/DNA complexes formed. These complexes, when added to cells, bind to the negatively charged plasma membrane and then are internalized through endocytosis. Compared to calcium phosphate, the transfection efficiency of DEAE-dextran is much higher, although it varies with the type of cells and other experimental conditions. Transfection efficiency as high as 80% has been reported with DEAE-dextran/DNA complexes. The method is relatively simple and reproducible, but requires low or no serum during transfection. DEAE-dextran is toxic to cells at high concentration.

3.3. Cationic Lipids

The use of cationic lipids for DNA delivery was pioneered by Felgner and colleagues *(20)*. The first cationic lipid synthesized for the purpose of DNA delivery was *N*-(2,3-dioleyloxypropyl) *N, N, N*-trimethylamonium chloride (DOTMA). When hydrated, DOTMA forms liposomes with or without additional lipid components. When mixed with DNA, the positive charges at the liposome surface electrostatically interact with the negative charges on the phosphate backbone of the DNA to form DNA/liposome complexes (lipoplexes). Addition of lipoplexes to cells in culture normally results in significant levels of gene expression, with an efficiency ranging from 5% to more than 90% depending on the type of cell line used. A broad spectrum of transfection activity among many cell types, low toxicity, and high efficiency are the major advantages of cationic lipids.

Major efforts have been made over the past decade to optimize the transfection activity of cationic lipids. Results from in vitro and in vivo studies have revealed that, among the many physicochemical properties that affect transfection activity of cationic liposomes, the most important one is the cationic lipid structure. The common features shared by all transfection-active cationic lipids developed so far include three structural domains: (1) a positively charged head group, (2) a hydrophobic anchor, and (3) a linker connecting the head group and the hydrophobic anchor. The structure of the head group varies from one to four ammonium groups, which can be primary, secondary, tertiary, or quaternary for multivalent cationic lipids *(21,22)*. The most common structure for the hydrophobic anchor consists of either two hydrocarbon chains or cholesterol. With a few exceptions, glycerol is the linker in cationic lipids with a double-chain anchor, whereas a variety of linkers have been used in cationic lipids with a cholesterol anchor *(21,22)*. Our studies on structure–function relationships for intravenous transfection revealed that the optimal lipid structure should include the following: (1) a cationic head group and its neighboring hydrocarbon chain being in a 1,2-relationship on the backbone, (2) an ether bond to link the hydrocarbon chain to the backbone, and (3) paired oleyl chains as the hydrophobic anchor *(25)*. Cationic lipids without these structural features had lower intravenous transfection activity in mice.

For in vitro transfection, structure-based rules that allow for prediction of transfection activity of cationic lipids have yet to be established. The transfection activity of cationic lipids varies significantly with the type of cells used. Inclusion of specific lipids (helper lipids) such as dioleoylphosphatidylethanolamine (DOPE) into cationic liposomes enhances transfection activity in some cell types but not others *(23)*. For transfection of epithelial cells by airway administration, Lee et al. have shown that cholesterol-based cationic lipids (lipid-67) exhibit much higher activity than their noncholesterol-based counterparts *(24)*. The mechanism of lipid-67-based gene transfer into lung epithelial cells is not known.

Although many complex structures between DNA and cationic liposomes have been identified *(26–29)*, the most important parameters affecting the transfection activity of cationic liposomes appear to be the particle size of the complexes *(30)* and the charge ratio (amines to DNA phosphates ratio, $+/-$) *(20–25)*. Complexes with larger particle size (>200 nm) and with charge ratio of slightly greater than 1 appear to be optimal for in vitro transfection. For intravenous transfection, however, a charge ratio ($+/-$) of greater than 12 is required for an optimal transfection activity *(31,32)*. Cholesterol generally functions as a better helper lipid than DOPE for systemic transfection *(31,33,34)*.

3.4. Cationic Polymers

Different from the hydrophobic cationic lipids, cationic polymers are a group of highly water-soluble molecules. Three general types of cationic polymers have been used for transfection: linear (polylysine, spermine, and histone), branched, and spherical. The most extensively studied cationic polymers for DNA delivery are polyethyleneimine (PEI) *(35,36)* and dendrimers *(37–39)*. PEIs are highly branched organic polymers produced by polymerizing aziridine *(40)*. In principle, PEI, with every third atom in the polymer being amino nitrogen, is a type of organic compound that contains the highest density of potential positive charges. It has been estimated that about 25, 50, and 25% of nitrogens in branched PEIs are primary, secondary, and tertiary amines, respectively. The enriched nitrogen atoms and the diversified nitrogen forms provide considerable buffer capacity for the PEIs over a wide pH range *(41,42)*. Both branched and linear types of PEIs have been used for transfection. The in vitro transfection efficiency of PEI 800 KDa on a large variety of cell lines and primary culture is comparable to that of cationic lipids.

Dendrimers are a new class of branched, spherical, and starburst molecules. Dendrimers differ in their initiator structure and in the number of layers of the building blocks in each molecule. (The number of layers is also called the number of generations.) The common initiators include ammonia (NH_3) as trivalent initiator and ethylenediamine as a tetravalent initiator. Polymerization takes place in a geometrically outward fashion, resulting in branched polymer with spherical geometry and containing interior tertiary and exterior primary amines. The defined structure and large number of surface amino groups of dendrimers have led to these polymers being employed as a carrier for DNA delivery. Different forms of dendrimers have been shown to be active in transfection *(37–39)*. Their precisely controlled size and shape provide them with the potential advantage of forming more homogeneous and highly reproducible DNA complexes.

When mixed with DNA, cationic polymers readily self-assemble with DNA and generate small tortoidal or spherical structures of approx 40–100 nm *(42)*, depending on polymer size, structure, DNA to polymer ratio, and the type and concentration of ions in the buffer. The complexes formed between DNA and cationic polymers are called polyplexes. When added to cells in culture, polyplexes are taken up by the cells. Similar to that of cationic lipids, transfection activity of cationic polymers varies with cell type, structure, and size of the polymer, and polymer to DNA ratio. Compared to cationic lipids, the major drawback of cationic polymers is their relatively high toxicity.

3.5. Combined Systems

Transfection activity of combined synthetic compounds has been explored. Depending on the transfection reagents selected, a significant enhancement in transfection activity can be achieved. In fact, much higher transfection activity of cationic liposomes was reported when mixed with polylysine *(43)*, protamine sulfate *(44)*, peptides *(45,46)*, or PNA (Chapter 6 in this volume). The mechanisms for such synergistic effect between cationic liposomes and polymers are not known, but it is believed that the structure of DNA complexes in the combined system may be more effective in escaping the endosomal degradation and/or more efficient in facilitating DNA transfer into the nucleus.

4. Mechanisms of DNA Delivery by Synthetic Vectors

A central tenet of DNA delivery by synthetic compounds is that DNA molecules can only be delivered into a cell when they are converted into a particulate form. This appears to be true for all of the synthetic compounds that have been developed for transfection thus far. Obviously, synthetic compounds such as calcium phosphate, DEAE-dextran, cationic lipids, and cationic polymers are all capable of forming particles with DNA. With the exception of the calcium phosphate method that forms calcium-phosphate-DNA precipitates, other synthetic compounds form complexes with DNA through electrostatic interaction between DNA and the synthetic vectors. Once complexed with a synthetic vector, DNA molecules are protected against nuclease-mediated degradation. To accomplish DNA delivery, these complexes need to (1) bind to the cell surface, (2) cross the plasma membrane, (3) release DNA into the cytoplasm, and finally, (4) transport the DNA into the nucleus. The following is a brief summary of the current understanding on these events as they are involved in synthetic compound-mediated DNA delivery.

4.1. Binding to the Cell Surface

Without exception, binding of DNA complexes to the cell surface with the most commonly used synthetic compounds is accomplished by electrostatic interaction between the positively charged complexes and the negatively charged cell surface. It has been shown that cell surface binding of DNA complexes can be significantly enhanced by centrifugation of the transfection reagents onto cell surface *(47)*. Another approach to enhance complex binding to the cell surface is to use DNA complexes of larger size *(30)*. Enhanced binding usually correlates with higher transfection efficiency.

4.2. Crossing the Plasma Membrane and Entering the Cytoplasm

The fact that particle structure of DNA complexes is required for DNA delivery into cells stimulates the strong notion that endocytosis is the major pathway, through which DNA molecules are internalized *(48–51)*. Because the endocytotic process ends at the lysosome where DNA is degraded, escape of DNA from the endosome at an early stage of endocytosis is believed to be critical for cytosolic DNA delivery and is considered to be one of the most important rate-limiting steps that determine overall transfection efficiency. The mechanism of how DNA molecules leave the endosome and enter the cytoplasm is unknown. However, a number of strategies have been explored to enhance endosomal release. The first of these involves the use of DOPE as a helper lipid for liposome-based DNA delivery. It is believed that DOPE is capable of inducing membrane fusion between the endosome and the liposome through the formation of an inverted hexagonal structure, which, in turn, results in membrane destabilization and release of DNA into the cytoplasm *(52)*. The second strategy is inclusion of membrane fusion proteins or peptides into DNA complexes *(53)*. The mechanism of action of these compounds is similar to that of DOPE: low pH in the early endosome induces the proteins or peptides to fuse with membrane, releasing the DNA into the cytoplasm *(53)*. Finally, buffering capacity of charged groups such as the primary, secondary, and tertiary amines in PEIs and polyamidoamine in dendrimers may play an important role in endosomal DNA release in polyplex-mediated DNA delivery. Protonation of these groups in the endosome may create sufficient osmotic pressure to induce endosomal lysis *(35,36)*.

4.3. DNA Release from Complexes

Although formation of complexes between DNA and synthetic compounds is essential for cell entry, dissociation of DNA from the complexes after the complex has entered the cell appears to be essential for successful transfection. The necessity for DNA release from the complexes was demonstrated by Zanber et al. *(54)*. In their experiments, lipoplexes or free DNA was injected directly into the nucleus of oocytes and the level of gene expression in these injected oocytes was analyzed at a later time. No gene expression was detected in oocytes injected with lipoplexes as compared to a significant level of gene expression in those injected with free DNA. With respect to cationic liposome-based transfection, work by Xu and Szoka *(55)* suggested that DNA release from the complexes result from fusion of liposomes and cell membranes. Neutralization of positive charges of synthetic vector in DNA complexes by negatively charged cellular lipids and other components has been proposed as the mechanism of cytosolic DNA release from the complexes *(55)*.

4.4. Nuclear Transport of DNA

Once DNA molecules are in the cytosol, they must still enter the nucleus. How this occurs is largely unknown, but the transport of DNA from the cytosol into the nucleus does seem to occur because transgene expression is achieved. Microinjection of plasmid DNA directly into cytoplasm revealed that transport of DNA from the cytoplasm into the nucleus was of extremely low efficiency *(54,56)*. An additional factor affecting the transfer of DNA into the nucleus is nuclease activity *(56)*. Cytoplasmic nuclease activity can reduce the amount of intact DNA molecules available for nuclear import and gene expression, supporting the notion that movement of DNA from cytoplasm to the nucleus is one of the most important limitations to a successful gene transfer. Interestingly, studies by Pollard et al. *(57)* demonstrated that PEI of 25KDa is able to promote gene delivery from the cytoplasm into the nucleus. Such activity, however, is cell type dependent. Different from cationic liposomes, dissociation of PEI/DNA complexes to allow transcription does not seem to be a problem because injection of these complexes into the nucleus produces a level of gene product comparable to injection of plasmid DNA alone *(57)*.

Dean et al. *(58)* showed that the efficiency of nuclear transport of DNA is dependent on the DNA sequence. Cytosolic injection of DNA containing replication origin and SV40 promoter sequences resulted in a significantly higher level of DNA importation into the nucleus than that observed with plasmids without these specific sequences *(58)*. Inclusion of a nuclear localization signal, normally a short lysine- and arginine-rich peptide, as part of synthetic vector has been considered as one of the promising strategies to enhance DNA import into the nucleus *(59)*.

5. Optimizing Transfection

In general, expression of the genetically coded information in DNA in a particular type of cell presents its own particular set of problems that must be overcome to achieve a high level of expression. Unfortunately, selection of synthetic compounds for a given type of cell is still largely empirical. There is not a set of hard-and-fast rules to follow. In fact, a particular synthetic compound is almost as likely to be the exception as it is to follow any set of rules. Keeping this caveat in mind, we will make some general comments that we hope will aid readers in selecting an initial transfection reagent.

Selecting an appropriate synthetic vector for DNA delivery may require some homework on the researcher's part. The first piece of advice is to consult with the technical service department of commercial resources (**Table 1**) for an established protocol, because many companies have already optimized the experimental condition for their products in selected cell types. Researchers could adopt these conditions if the types of cells to be used in their experiments are the same or similar. This may significantly reduce the effort required to optimize the experimental conditions.

In cases where there is not enough information available to make a selection, researchers may wish to obtain transfection reagents from a number of commercial sources to identify the one that gives the best results. Many companies sell kits that contain multiple reagents. Alternatively, if researchers are new to this type of study or are dealing with new types of cells, they will have to optimize the experimental conditions themselves following the protocols and suggestions described in the following chapters of this section.

As a standard procedure, the most convenient way to do optimization work is to use a reporter system. The most commonly used reporter genes for this purpose include those that code for luciferase (Luc), green fluorescence protein (GFP), β-galactosidase (β-gal), chloramphenicol acetyltransferase (CAT), or secreted alkaline phosphatase (SEAP). Most reporter gene-containing plasmids and related reagents for gene-expression analysis are commercially available from many resources (**Table 2**). The following is a brief description of the advantages and disadvantages of the most commonly used reporter assay systems.

Table 1
Commercial Sources for Transfection Reagents[a]

Company	Product	Composition
Amersham-Biosciences 800-526-3593 *www.amershambiosciences.com*	CellPhect Transfection Kit	CaPO4 or DEAE-Dextran
Bio-Rad Labs 800-424-6723 *www.bio-rad.com*	CytoFectene Transfection Reagent	Cationic lipid
BD Biosciences–CLONTECH 800-662-2566 *www.clontech.com*	CLONfectin™ CalPhos™	Cationic lipid Calcium phosphate
CPG Inc. 800-362-2740 *www.cpg-biotech.com*	GeneLimo™ Transfection Reagent Plus GeneLimo™ Transfection Reagent Super	Polycationic lipids + lipid compound Polycationic lipids + lipid compound
Gene Therapy Systems 888-428-0558 *www.genetherapysystems.com*	GenePORTER™ GenePORTER™ 2 BoosterExpress™ Reagent Kit PGene Grip™ Vector/Transfection Systems	DOPE + Proprietary compounds Proprietary material Proprietary material GenePORTER + plasmid vector
Glen Research 703-437-6191 *www.glenres.com*	Cytofectin GS	Cationic lipid
Invitrogen 800-955-6288 *www.invitrogen.com*	LipofectAMINE 2000™ Reagent LipofectAMINE PLUS™ Reagent Lipofectin® Reagent Transfection Reagent Optimization System	Polycationic lipid Polycationic lipid (DOSPA:DOPE) Cationic lipid (DOTMA:DOPE) LipofectAMINE PLUS™ + Lipofectin® + CellFectin® +DMRIE
	CellFectin® Reagent OligofectAMINE™ Reagent	Cationic lipopolyamine Proprietary

Table 1
Continued

Company	Product	Composition
InvivoGen 888-457-5873 *www.invivogen.com*	LipoVec™ LipoGen™	Cationic phosphonolipids Non-liposomal lipid
MBI Fermentas 800-340-9026 *www.fermentas.com*	ExGen 500 ExGen 500 in vivo delivery agent	Cationic polymer Cationic polymer
Novagen 800-526-7319 *www.novagen.com*	GeneJuice™ Transfection Reagent	Proprietary polyamine
PanVera Corporation 800-791-1400 *www.panvera.com*	TransIT®-TKO Reagents TransIT®-LT-1 and LT-2 TransIT® Express Transfection Reagent TransIT® PanPack TransIT®-Insecta TransIT®-293 TransIT®-Keratinocyte	Polyamine® Polyamine® Polyamine® Polyamine + cationic lipid® Cationic lipid® Polyamine® Polyamine
Promega 800-356-9526 *www.promega.com*	TransFAST™ Transfection Reagent Tfx™-10, -20, and -50 Reagents Tfx™ Reagents Transfection Trio Transfectam® Reagents ProFection® Mammalian Transfection system	Cationic lipid Cationic lipid Cationic lipid Cationic lipid DEAE-Dextran or calcium phosphate
Qbiogene 800-424-6101 *www.qbiogene.com*	GeneSHUTTLE™-20 and -40 Reagents In vivo GeneSHUTTLE™ Reagent DuoFect™ Reagent System Calcium Phosphate transfection kit	Polycationic lipids Liposome-mediated DOTAP:Chol. Receptor-mediated endocytosis Calcium phosphate

Company	Product	Type
Qiagen 800-426-8157 www.qiagen.com	TransMessenger™ Transfection Reagent	Proprietary lipid
	SuperFect® Transfection Reagent	Activated dendrimer
	Effectene™ Transfection Reagent	Nonliposomal lipid
	Transfection Reagent Selector Kit	Dendrimer + nonliposomal lipid
	PolyFect® Transfection Reagent	Activated dendrimer
Roche Applied Science 800-262-1640 www.biochem.roche.com	FuGENE 6 Transfection Reagent	Nonliposomal proprietary lipid
	X-tremeGENE Ro-1539 Transfection Reagent	Proprietary lipid
	X-tremeGENE Q2 Transfection Reagent	Proprietary lipid
	DOSPAR	Polycationic lipid
	DOTAP	Cationic lipid
Sigma-Aldrich Corporation 800-325-3010 www.sigma-aldrich.com	DEAE-Dextran Transfection Reagent	DEAE-Dextran
	Calcium Phosphate Transfection Kit	Calcium phosphate
	Escort™, -II, -III, and -V	Cationic lipid
	In vivo Liposome Transfection Reagents	Cationic lipid
Cell & Molecular Technologies 800-543-6029 www.specialtymedia.com	Mammalian Cell Transfection Kit	Calcium phosphate
	Transient Expression Kit	DEAE-Dextran
Stratagene 800-424-5444 www.stratagene.com	LipoTaxi®	Cationic lipid
	Mammalian Transfection Kit	Calcium phosphate + DEAE-Dextran
	MBS Mammalian Transfection Kit	Modified calcium phosphate
	Primary Enhancer™ Reagent	Lipid + calcium phophate
	GeneJammer™	Proprietary polyamine
	Genetransfer®	Cationic liposomes
Wako USA 800-992-9256 www.wakousa.com		

[a]This table is assembled in part by the information provided in **ref. 67.**

Table 2
Commercial Resources for Reporter Systems[a]

Company	Product	Reporter system
Applied Biosystems	Galacto-Light™	β-galactosidase
800-345-5224	Galacto-Light™ Plus	β-galactosidase
www.appliedbiosystems.com	Luc-Screen®	luciferase
	Phospho-Light™	SEAP
BioVision Inc.	Luciferase Reporter Assay Kit	Luciferase
800-891-9699		
www.biovisionlabs.com		
BD Biosciences—CLONTECH	Great EscAPe™	SEAP
800-662-2566	pSEAP2 Reporter Vector	SEAP
www.clontech.com	pCMVβ	β-galactosidase
	Luminescent β-gal Reporter System3	β-galactosidase
	pβgal Reporter Vectors	β-galactosidase
	Luciferase Reporter Assay Kit	Luciferase
	pCMS-EGFP Vector	GFP
	pGFP Variant Vectors	GFP
	pIRES2-EGFP Vector	GFP
	EGFP Cotransfection Marker Vector	GFP
	pd2EGFP Vector	GFP
	Living Colors™ Fluorescent Timer	DsRed
	Living Colors™ DsRed2 Vectors	DsRed
	pDsRed Vector	DsRed
	Promoterless Enhanced Fluorescent Protein Vectors	ECFP, EGFP, or EYFP
	N-terminal Enhanced Fluorescent Protein Vectors	ECFP, EGFP, or EYFP
	C-terminal Enhanced Fluorescent Protein Vectors	ECFP, EGFP, or EYFP
	Promoterless Destabilized Vectors	D2ECFP, d2EGFP, or d2EYFP

Company	Product	Reporter
Gene Therapy Systems 888-428-0558 *www.genetherapysystems.com*	β-galactosidase assay kits X-galactosidase assay kit	β-galactosidase β-galactosidase
ICN Biomedicals 800-854-0530 *www.icnbiomed.com*	CAT Assay Aurora GAL-XE Aurora AP SEAP Aurora GUS	CAT β-galactosidase β-glucuronidase
Intergen 800-431-4505 *www.intergen.com*	CATalyse™ Assay ONGP lac Z Reporter Gene Assay	CAT β-galactosidase
Invitrogen 800-955-6288 *www.invitrogen.com*	pTracer™-CMV2 pTracer™-EF pTracer™-SV40 LacZ Reporter Assay Kit PLAP Reporter Assay Kit	GFP Super GFP GFP β-galactosidase plap gene
InvivoGen 888-457-5873 *www.invivogen.com*		
Marker Gene Tech Inc. 541-342-3760 *www.markergene.com*	Luciferase Assay Kit FACS lacZ β-galactosidase Detector Kit In vivo lacZ β-galactosidase Intracellular Detector Kit	Luciferase β-galactosidase β-galactosidase
Molecular Devices 800-635-5577 *www.moleculardevices.com*	Fluorescent di-β-galactopyranoside CLIPR™ Luciferase Assay Kit	β-galactosidase Luciferase
Novagen 800-526-7319 *www.novagen.com*	BetaBlue™ Staining Kit BetaRed™ Staining Kit BetaRed™-β-galactosidase assay kit BetaFluor™-β-galactosidase assay kit	β-galactosidase β-galactosidase β-galactosidase β-galactosidase

Table 2
Continued

Company	Product	Reporter system
PerkinElmer Life Sciences	LucLite®	Luciferase
800-323-1891	LucLite® Plus	Luciferase
www.perkinelmer.com	GeneLux Enhanced Luciferase Assay Kit	Phosphotinus-luciferase
Promega	Bright-Glo™ Luciferase Assay System	Luciferase
800-356-9526	Dual-Luciferase® Reporter Assay System	Luciferase
www.promega.com	Steady-Glo® Luciferase Assay System	Luciferase
Pierce Intergen	D-Luciferin	Luciferase
800-874-3723		
www.piercenet.com		
Roche Molecular Biochemicals	β-Gal Reporter Gene Assay	β-galactosidase
800-262-1640	Luciferase Reporter Gene Assay	Luciferase
www.biochem.roche.com	SEAP Reporter Gene Assay	SEAP
Stratagene	β-Galactosidase Assay Kit	β-galactosidase
800-424-5444	In situ β-Galactosidase Assay Kit	β-galactosidase
www.stratagene.com		
Thermo Labsystems	GenGlow100 Kit	Luciferase
800-522-7763	GenGlow 1000 Kit	Luciferase
www.therm.com	Luciferase Substrate	D-luciferin

*ᵃThis table is assembled in part by the information provided in **ref. 60**.*

For more information about the reporter assay systems, readers are referred to an excellent product report by Hillary Sussman *(60)*.

The luciferase gene is cloned from the North American firefly (*Photinus pyralis*) *(61)*. Its product, luciferase, catalyzes the oxidative carboxylation of luciferin in the presence of ATP and acetyl-CoA, resulting in the release of photons that can be measured in a luminometer or scintillation counter. The major advantage of the luciferase assay is that it is rapid, convenient, and has a broad linear range covering seven or eight orders of magnitude. The detection limit can be as low as 10^{-20} mol of luciferase. In addition, luciferase activity detected is closely coupled to protein synthesis because this enzyme is not posttranslationally modified.

Use of GFP as reporter has gained significant popularity in recent years *(62)*. The GFP gene is derived from the sea jellyfish (*Aequorea victoria*). There are no substrates or cofactors needed for GFP detection because it fluoresces upon ultraviolet irradiation. The main advantage of using GFP as a reporter is that gene expression can be directly observed under a fluorescence microscope in individual living cells. Stability of GFP in presence of heat, denaturants, detergents, and most proteases is another major advantage. Because there is no amplification involved in GFP measurement, the sensitivity of GFP assay is significantly lower than that of the luciferase assay. Combination of GFP and fluorescence-activated cell sorting (FACS) provides a useful tool for separation of transfected from untransfected cells.

β-gal, encoded by the LacZ gene, hydrolyzes sugar molecules. Although a background level exists in mammalian cells and animal tissues, β-gal is still frequently used in transfection experiments. Depending on the substrates used, detection of β-gal activity can be colorimetric, fluorescent, or chemiluminescent. Hydrolysis of O-nitrophenyl-β-D-galactose or chlorophenol red β-galactoside yields colored products that can be quantified spectrophotometrically. β-gal expression can also be visualized histochemically through hydrolysis of 5-bromo-4-chloro-3-indoyl-β-D-galactopyranoside (X-Gal), which produces a blue precipitate. Several companies have developed products for the quantification of β-gal by employing fluorogenic substrate or chemiluminescent dioxetane chemistry that increase detection sensitivity over the colorimetric assay (**Table 2**).

The bacterial CAT enzyme catalyzes transfer of an acetyl group from acetyl-CoA to the antibiotic chloramphenicol. The quantitative analysis of CAT activity is carried out using radio-labeled substrates such as ^{3}H-acetyl-CoA or ^{14}C-chloramphenicol. The need to use radioactive materials and a narrow range for detection has made the CAT assay much less popular than other reporter systems.

Another reporter system for transfection optimization is SEAP. SEAP is a mutated form of human placental alkaline phosphatase. SEAP offers the great advantage of being secreted by the cell. Its activity can thus be measured repeatedly over time in a nondestructive fashion. Because there is no need to prepare cell extracts, the assay is rapid and allows the cells to be further studied. With the development of sensitive dioxetane substrate for alkaline phosphatase, the detection limit of SEAP is less than 10^{-18} mole of the enzyme.

6. Summary and Outlook

The use of synthetic vectors for DNA delivery into mammalian cells in vitro is now a well-established technique. There is also a rapid increase in use of synthetic vectors for in vivo DNA delivery. Cationic liposomes have already been used in clinical trials *(63–66)*. However, most of the in vivo applications are conducted through local regional administration. In addition, the level of gene expression reported from these in vivo attempts has not seemed to be sufficient to bring about a medical benefit. Therefore, significant efforts are needed to improve in vivo DNA-delivery efficiency of the synthetic vectors.

In the development of chemical reagents for DNA delivery, research and early activities have typically been driven by the need for better efficiency. In focusing their efforts in this way, researchers have sought to create efficiency by building on the knowledge of the biology underlying DNA delivery and by synthesis of many new compounds sharing certain chemical and physical properties. Although there is no doubt that this strategy has produced some synergies in developing various transfection reagents, future development will focus around further improvement in gene-delivery efficiency for human gene therapy.

In attempting to predict what surprises the next 5–10 years might hold for chemical methods for DNA delivery, we believe that there will be continuous improvements made in in vivo DNA delivery for those synthetic compounds that have already been developed. Perhaps new compounds will be developed that will be able to deliver DNA to all types of cells with equally high efficiency. It is even conceivable that new compounds or new formulations will be developed for gene delivery in a target-specific manner. Practically speaking, the driving force for future development is firmly rooted in public interest in gene therapy. The rapid progress of the past few years makes it intriguing to speculate that it is possible that synthetic compounds might become an important component in the repertoire for human gene therapy.

Acknowledgment

We wish to thank Drs. Joseph E. Knapp, Xiang Gao, and William C. Heiser for critical reading of this manuscript. We are also grateful to Dr. Todd Giorgio for sharing unpublished information on PNA-mediated gene transfer.

References

1. Wu, G. Y. and Wu, C. H. (1987) Receptor-mediated gene transformation by a soluble DNA carrier system. *J. Biol. Chem.* **262**, 4429–4432.
2. Ferkol, T., Zperales, J. C., Mularo, F., and Hanson, R. W. (1996) Recepter-mediated gene transfer into macrophages. *Proc. Natl. Acad. Sci. USA* **93**, 101–105.
3. Wagner, E., Cotton, M., Foisner, R., and Xbirnstiel, M. L. (1991) Transferrin-polycation-DNA complexes: the effect of polycations on the structure of the complex and DNA delivery to cells. *Proc. Natl. Acad. Sci. USA* **88**, 4255–4259.
4. Wolfert, M. A. and Seymour, L. W. (1996) Atomic force microscopic analysis of the influence of the molecular weight of poly-L-lysine on the size of polyelectrolyte complexes formed with DNA. *Gene Ther.* **3**, 269–273.
5. Perales, J., C., Ferkol, T., Beegen, H., Ratnoff, O. D., and Hanson, R. W. (1994) Gene transfer in vivo: sustained expression and regulation of genes introduced into the liver by receptor-targeted uptake. *Proc. Natl. Acad. Sci. USA* **91**, 4086–4090.
6. McKenzie, D. L., Collard, W. T., and Rice, K. G. (1999) Comparative gene transfer efficiency of low molecular weight polylysine DNA-condensing peptides. *J. Pept. Res.* **54**, 311–318.
7. Erbacher, P., Roche, A. C., Monsigny, M., and Midoux, P. (1995) Glycosylated polylysine/DNA complexes: gene transfer efficiency in relation with the size and the sugar substitution level of glycosylated polylysines and with the plasmid size. *Bioconjug. Chem.* **6**, 401–410.
8. Fritz, J. D., Herweijer, H., Zhang, G., and Wolff, J. A (1996) Gene transfer into mammalian cells using histone-condensed plasmid DNA *Hum. Gene. Ther.* **7**, 1395–1404.
9. Demirhan, I., Hasselmayer, O., Chandra, A., Ehemann, M., and Chandra, P. (1998) Histone-mediated transfer and expression of the HIV-1 tat gene in Jurkat cells. *J. Hum. Virol.* **1**, 430–440.
10. Richardson, S. C, Kolbe, H. V., and Duncan, R. (1999) Potential of low molecular mass chitosan as a DNA delivery system: biocompatibility, body distribution and ability to complex and protect DNA. *Int. J. Pharm.* **178**, 231–243.
11. Roy, K., Mao, H. Q., Huang, S. K., and Leong, K. W. (1999) Oral gene delivery with chitosan-DNA nanoparticles generates immunologic protection in a murine model of peanut allergy. *Nat. Med.* **5**, 387–391.
12. Wadhwa, M. S., Collard, W. T., Adami, R. C., McKenzie, D. L., and Rice, K. G. (1997) Peptide-mediated gene delivery: influence of peptide structure on gene expression. *Bioconj. Chem.* **8**, 81–88.
13. Duguid, J. G., Li, C., Shi, M., Logan, M. J., Alila, H., Rolland, A., et al. (1998) A physicochemical approach for predicting the effectiveness of peptide-based gene delivery systems for use in plasmid-based gene therapy. *Biophys. J.* **74**, 2802–2814.
14. Mahat, R. I., Monera, O. D., Smith, L. C., and Rolland, A. (1999) Peptide-based gene delivery. *Curr. Opin. Mol. Ther.* **1**, 226–243.
15. Plank, C., Tang, M. X., Wolfe, A. R., and Szoka, F. C. (1999) Branched cationic peptides for gene delivery: role of type and number of cationic residues in formation and in vitro activity of DNA polyplexes. *Hum. Gene Ther.* **10**, 319-332.

16. McKenzie, D. L., Kwok, K. Y., and Rice, K. G. (2000) A potent new class of reductively activated peptide gene delivery agents. *J. Biol. Chem.* **275,** 9970–9977.
17. Niidome, T., Takaji, K., Urakawa, M., Ohmori, N., Wada, A., Hirayama, T., and Aoyagi, H. (1999) Chain length of cationic alpha-helical peptide sufficient for gene delivery into cells. *Bioconjug. Chem.* **10,** 773–780.
18. Graham, F. L. and van der Eb, A. J. (1973) Transformation of rat cells by DNA of human adenovirus 5. *Virology* **54,** 536–539.
19. Pagano, J. S. and Vaheri, A. (1965) Enhancement of infectivity of poliovirus RNA with diethylaminoethyl-dextran (DEAE-D). *Arch. Gesamte Virusforsch.* **17,** 456–464.
20. Felgner, P. L., Gadek, T. R., Holm, M., Roman, R., Chan, H. W., Wenz, M., et al. (1987) Lipofection: a highly efficient, lipid-mediated DNA-transfection procedure. *Proc. Natl. Acad. Sci. USA* **84,** 7413–7417.
21. Miller, A. D. (1998) Cationic liposomes for gene therapy. *Angew Chem. Int. Ed. Engl.* **37,** 1768–1785.
22. Byk, G., Dubertret, C., Escriou, V., Frederic, M., Jaslin, G., Rangara, R., et al. (1998) Synthesis, activity, and structure-activity relationship studies of novel cationic lipids for DNA transfer. *J. Med. Chem.* 41, 224–235.
23. Hui S. W., Langner, M., Zhao, Y. L., Ross, P., Hurley, E., Chan, K. (1996) The role of helper lipids in cationic liposome-mediated gene transfer. *Biophys. J.* **71,** 590–599.
24. Lee, E. R., Marshall, J., Siegel, C. S., Jiang, C., Yew, N. S., Nichols, M. R., et al. (1996) Detailed analysis of structures and formulations of cationic lipids for efficient gene transfer to the lung. *Hum. Gene Ther.* **7,** 1701–1717.
25. Ren, T., Song, Y. K., Zhang, G., and Liu, D. (2000) Structural basis of DOTMA for its high intravenous transfection activity in mouse. *Gene Ther.* **7,** 764–768
26. Sternberg, B., Sorgi, F. L., and Huang, L. (1994) New structures in complex formation between DNA and cationic liposomes visualized by freeze-fracture electron microscopy. *FEBS Lett.* **356,** 361–366.
27. Räler, J. O., Koltover, I., Salditt, T., and Safinya, C. R. (1997) Structure of DNA-cationic liposome complexes: DNA intercalation in multilamellar membranes in distinct interhelical packing regimes. *Science* **275,** 810–814.
28. Lasic, D. D., Strey, H., Stuart, M., Podgornik, R., and Frederik, P. M. (1997) The structure of DNA-liposome complexes. *J. Am. Chem. Soc.* **119,** 832–833.
29. Templeton, N. S., Lasic, D. D., Frederik, P. M., Strey, H. H., Roberts, D. D., and Pavlakis, G. N. (1997) Improved DNA: liposome complexes for increased systemic delivery and gene expression. *Nat. Biotechnol.* **15,** 647–652.
30. Rose, P. C. and Hui, S. W. (1999) Lipoplex size is the major determinant of in vitro lipofection efficeincy. *Gene Ther.* **6,** 651–659.
31. Song, Y. K., Liu, F., Chu, S., and Liu, D. (1997) Characterization of cationic liposome-mediated gene transfer in vivo by intravenous administration. *Hum. Gene Ther.* **8,** 1585–1594.
32. Liu, F. Qi, H., Huang, L., and Liu, D. (1997) Factors controlling the efficeincy of cationic lipid-mediated transfection in vivo via intravenous administration. *Gene Ther.* **4,** 517–523.

33. Hong, K., Zheng, W., Baker, A., and Papahadjopoulos, D. (1997) Stabilization of cationic liposome-plasmid DNA complexes by polyamines and poly(ethylene glycol)-phospholipid conjugates for efficient in vivo gene delivery. *FEBS Lett.* **400,** 233–237.

34. Liu, Y., Mounkes, L. C., Liggitt, H. D., Brown, C. S., Solodin, I., Heath, T. D., and Debs, R. J. (1997) Factors influencing the efficiency of cationic liposome-mediated intravenous gene delivery. *Nat. Biotechnol.* **15,** 167–173.

35. Boussif, O., Lezoualch, F., Zanta, M. A., Mergny, M. D., Scherman, D., Demeneix, B., and Behr, J. P. (1995) A versatile vector for gene and oligonucleotide transfer into cells in culture and in vivo: polyethylenimine. *Proc. Natl. Acad. Sci. USA* **92,** 7297–7301.

36. Abdallah, B., Hassan, A., Benoist, C., Goula, D., Behr, J. P., and Demeneix, B. A. (1996) A powerful nonviral vector for in vivo gene transfer into the adult mammalian brain: polyethylenimine. *Hum. Gene Ther.* **7,** 1947–1954.

37. Tang, M., X. Redemann, C. T., and Szoka, F. C. Jr. (1996) In vitro gene delivery by degraded polyaminedoamine dendrimers. *Bioconj. Chem.* **7,** 703–714.

38. Kukowska-Latallo, J. F., Bielinska, A. U., Johnson, J., Spindler, R., Tomalia, D. A., and Baker, J. R. Jr. (1996) Efficient transfer of genetic material into mammalian cells using Starburst polyamidoamine dendrimers. *Proc. Natl. Acad. Sci. USA* **93,** 4897–4902.

39. Dunlap, D., D., Maggi, A., Soria, M., R., and Monaco, L. (1997) Nanoscopic structure of DNA condensed for gene delivery. *Nucleic Acids Res.* **25,** 3095–3101.

40. Dick, C. R. and Ham, G. E. *(1970)* Characterization of polyethyleimine. *J. Macromol. Sci. Chem.* **A4** 1301–1314.

41. Suh, J., Paik, H. J., and Hwang B. K. (1994) Ionization of poly(ethylenimine) and poly(allylamine) at various pH's. *Bioorg. Chem.* **22,** 318–327.

42. Tang, M.X. and Szoka, F.C. (1997) The influence of polymer structure on the interactions of cationic polymers with DNA and morphology of the resulting complexes. *Gene Ther.* **4,** 823–832.

43. Gao, X. and Huang, L. (1996) Potentiation of cationic liposome mediated gene delivery by polycations. *Biochemistry* **35,** 1027–1036.

44. Sorgi, F. L., Bhattacharya, S., and Huang, L. (1997) Protamine sulfate enhances lipid-mediated gene transfer. *Gene Ther.* **4,** 961–968.

45. Chen, Q. R., Zhang, L., Stass, S. A., and Mixson, A. J. (2001) Branched co-polymers of histidine and lysine are efficient carriers of plasmids. *Nucleic Acids Res.* **29,** 1334–1340.

46. Murphy E. A., Waring A. J., Murphy, J. C., Willson R. C., and Longmuir, K. J. (2001) Development of an effective gene delivery system: a study of complexes composed of a peptide-based amphiphilic DNA compaction agent and phospholipid. *Nucleic Acid Res.* **29,** 3694–3704.

47. Boussif, O., Zanta, M. A. and Behr, J. P. (1996) Optimized galenics improve in vitro gene transfer with cationic molecules up to 1000-fold. *Gene Ther.* **3,** 1074–1080.

48. Legendre, J. Y. and Szoka, F. C. (1992) Delivery of plasmid DNA into mammalian cell lines using pH-sensitive liposomes, comparison with cationic liposomes. *Pharm. Res.* **9**, 1235–1242.

49. Wrobel, I. and Collins, D. (1995) Fusion of cationic liposomes with mammalian cells occurs after endocytosis. *Biochim. Biophys. Acta* **1235**, 296–304.

50. Friend, D. S., Papahadjopoulos, D., and Debs, R. J. (1996) Endocytosis and intracellular processing accompanying transfection mediated by cationic liposomes. *Biochim. Biophys. Acta* **1278**, 41–50.

51. Zhou, X. and Huang, L. (1994) DNA transfection mediated by cationic liposomes containing lipopolylysine: characterization and mechanism of action. *Biochim. Biophys. Acta* **1189**, 195–203.

52. Farhood, H., Serbina, N., and Huang, L. (1995) The role of dioleoylphosphatidylethanol-amine in cationic liposome mediated gene transfer. *Biochim. Biophys. Acta* **1235**, 289–295.

53. Wagner, E., Plank, C., Zatloukal, K., Cotton, M., and Birnstiel, M. L. (1992) Influenza virus hemagglutinin HA-2 N-terminal fusogenic peptides augment gene transfer by transferrin-polylysine/DNA complexes: towards a synthetic virus-like gene transfer vehicle. *Proc. Natl. Acad. Sci. USA* **89**, 7934–7938.

54. Zabner, J., Fasbender, A. J., Moninger, T., Poellinger, K. A., and Welsh, M. J. (1995) Cellular and molecular barriers to gene transfer by a cationic lipid. *J. Biol. Chem.* **270**, 18997–19007.

55. Xu, Y. and Szoka, F. C. Jr. (1996) Mechanism of DNA release from cationic liposome/DNA complexes used in cell transfection. *Biochemistry* **35**, 5616–5623.

56. Lechardeur, D., Sohn, K. J., Haardt, M., Joshi, P. B., Monck, M., Graham, R. W., et al. (1999) Metabolic instability of plasmid DNA in the cytosol: a potential barrier to gene transfer. *Gene Ther.* **6**, 482–497.

57. Pollard, H., Remy, J. S., Loussouarn, G., Demolombe, S., Behr, J. P., and Escande, D. (1998) Polyethylenimine but not cationic lipids promotes transgene delivery to the nucleus in mammalian cells. *J. Biol. Chem.* **273**, 7507–7511.

58. Dean, D. A., Dean, B. S., Muller, S., and Smith, L. C. (1999) Sequence requirements for plasmid nuclear import. *Exp. Cell Res.* **253**, 713–722.

59. Subramanian, A., Ranganathan, P., and Diamond, S.L. (1999) Nuclear targeting peptide scaffolds for lipofection of nondividing mammalian cells. *Nat. Biotech.* **17**, 873–877.

60. Sussman, H. (2001) Choosing the best reporter assay. *The Scientist* **15**, 25–27.

61. de Wet, J. R., Wood, K. V., Helinski, D. R., and DeLuca, M. (1985). Cloning of firefly luciferase cDNA and the expression of active luciferase in Escherichia coli. *Proc. Natl. Acad. Sci. USA* **82**, 7870–7873.

62. Conn, P. M. (1999) Green flourescent protein. *Methods Enzymol.* **302**.

63. Caplen, N. J., Alton, E. W., Middleton, P. G., Dorin, J. R., Stevenson, B. J., Gao, X., et al. (1995) Liposome-mediated CFTR gene transfer to the nasal epithelium of patients with cystic fibrosis. *Nat. Med.* **1**, 39–46.

64. McLachlan, G., Ho, L. P., Davidson-Smith, H., Samways, J., Davidson, H., Stevenson, B. J., et al. (1996) Laboratory and clinical studies in support of cystic fibrosis gene therapy using pCMV-CFTR-DOTAP. *Gene Ther.* **3,** 1113–1123.
65. Gill, D. R., Southern, K. W., Mofford, K. A., Seddon, T., Huang, L., Sorgi, F., et al. (1997) A placebo-controlled study of liposome-mediated gene transfer to the nasal epithelium of patients with cystic fibrosis. *Gene Ther.* **4,** 199–209.
66. Nabel, G.J., Nabel, E. G., Yang, Z. Y., Fox, B. A., Plautz, G. E., Gao, X., et al. (1993) Direct gene transfer with DNA-liposome complexes in melanoma: expression, biologic activity, and lack of toxicity in humans. *Proc. Natl. Acad. Sci. USA* **90,** 11307–11311.
67. Ostresh, M. (1999) No barriers to entry, transfection tools get biomolecules in the door. *The Scientists* **11**, 21.

2

Gene Transfer into Mammalian Cells Using Calcium Phosphate and DEAE-Dextran

Gregory S. Pari and Yiyang Xu

1. Introduction

Simplicity and cost are just two of the factors that have sustained the popularity of calcium phosphate and, to a lesser extent, DEAE-dextran transfection methods. However, notwithstanding these factors, the calcium phosphate method, especially use of *N,N-bis*(2-hydroxyethyl)-2-aminoethanesulfonic acid (the BES variation), has proven to be the only method sufficient for the cotransfection of multiple plasmids into a wide variety of cell types. Although not as widely used today, DEAE-dextran-mediated transfection is a highly reproducible method for transient expression of a foreign gene.

The DEAE-dextran-mediated transfection method was widely used in the early to mid-1980s because of the simplicity, efficiency, and reproducibility of the procedure *(1–5)*. One major drawback of this method is the poor efficiency in forming stable cell lines. In addition, cellular toxicity is high because it is necessary to expose the cells to dimethyl sulfoxide (DMSO). Consequently, DEAE-dextran-mediated transfection has fallen out of favor with many investigators, giving way mostly to lipid-mediated transfection. However, because lipid-mediated transfection can be costly and inefficient in some cell types, many laboratories may want to consider the DEAE-dextran method. Some investigators have found DEAE-dextran-mediated transfection to be highly effective in certain cell lines. Several reports demonstrated that this is the method of choice for delivering DNA to primary cultured human macrophages *(6,7)*. In addition, it appears that DEAE-dextran enhances the transfection efficiency of mammalian cells when using electroporation *(8)*.

From: *Methods in Molecular Biology, vol. 245:*
Gene Delivery to Mammalian Cells: Vol. 1: Nonviral Gene Transfer Techniques
Edited by: W. C. Heiser © Humana Press Inc., Totowa, NJ

In contrast to DEAE-dextran, the calcium phosphate co-precipitation procedure has remained a popular method to efficiently deliver DNA to a wide variety of cell types. The main advantage of calcium phosphate DNA transfection is the high efficiency for the generation of constitutively expressing cell lines. Calcium phosphate is the method of choice for the simultaneous transfection of multiple plasmids. In our laboratory, we routinely co-transfect as many as 12 different plasmid constructs at the same time into mammalian cells. Plasmid DNA to be transfected must be of the highest purity, usually only double-banded CsCl DNA is used for transfection.

The original calcium phosphate method used a HEPES-based buffer system *(9)*. This method is simple to use, but is limited in the range of cell lines that can be efficiently transfected. Many variations of the HEPES-based system exist, and some have optimized this method for a particular cell type *(10)*. A variation of the original calcium phosphate transfection method, one that uses BES buffer, has emerged as a very versatile and highly efficient transfection method *(11)*. The BES method uses a different buffer system in which the pH is lower than the HEPES-based procedure. A lower pH, coupled with incubation in a reduced CO_2 atmosphere for 15 h, allows the DNA-calcium phosphate precipitate to form slowly on the cells. This results in a significant increase in the efficiency of transfection and a higher percentage of stably expressing cell lines than the HEPES-based procedure. Co-transfection efficiencies are also much higher using the BES method versus the original HEPES-based buffer transfection method. This feature is of particular importance to establish a co-transfection replication assay *(12,13)*. Because of these advantages, we present only the BES method in this chapter.

The following are representative protocols for DEAE-dextran transfection of adherent and suspension cells, as well as a protocol for the BES method for calcium phosphate-mediated transfection.

2. Materials

2.1. DEAE-Dextran Transfection

1. Tris-buffered saline (TBS): Prepare the following sterile solutions: Solution A: 80 g/L NaCl, 3.8 g/L KCl, 2 g/L Na_2HPO_4, 30 g/L Tris base. Adjust pH to 7.5. Solution B: 15 g/L $CaCl_2$, 10 g/L $MgCl_2$. Filter-sterilize both solutions and store at $-20°C$. To make 100 mL of working solution, add 10 mL of Solution A to 89 mL of water and then add 1 mL of Solution B, filter-sterilize and store at 4°C.
2. Suspension Tris-buffered saline (STBS): 25 mM Tris-HCl, pH 7.4, 137 mM NaCl, 5 mM KCl, 0.6 mM Na_2HPO_4, 0.7 mM $CaCl_2$, 0.5 mM $MgCl_2$. Dissolve in distilled H_2O and filter-sterilize.

3. Phosphate-buffered saline (PBS): 137 mM NaCl, 2.7 mM KCl, 4.3 mM Na$_2$HPO$_4$, 1.4 mM KH$_2$PO$_4$, pH 7.3.
4. Dulbecco's modified Eagle's medium (DMEM) supplemented with 10% fetal bovine serum (FBS).
5. DEAE-dextran: 10 mg/mL in TBS.
6. 10% DMSO.
7. For suspension cultures, cells are grown in RPMI 1640 medium supplemented with 10–20% FBS.

2.2. Calcium Phosphate Co-Precipitation

1. DMEM supplemented with 10% FBS.
2. CsCl-purified double-banded DNA.
3. 2.5 M CaCl$_2$ filter sterilized through a 0.45-μm filter.
4. 2X BES-buffered saline (BBS): 50 mM N,N-bis(2-hydroxyethyl)-2-aminoethanesulfonic acid, 1.5 mM Na$_2$HPO$_4$, 280 mM NaCl. Adjust pH to 6.95 with 1 N NaOH (*see* **Note 1**).
5. 35°C 3% CO$_2$ humidified incubator.

3. Methods
3.1. DEAE-Dextran Methods

Two DEAE-dextran methods are commonly used. The first is the basic protocol, which can be used on all anchorage-dependent cells. The second can be used on cells that normally grow in suspension or with anchorage-dependent cells that have been trypsinized and are in suspension. This procedure may increase transfection efficiency in some cells. For adherent cells, it is advisable to try the basic protocol first, then if transfection efficiency is low, try the suspension procedure.

3.1.1. Anchorage-Dependent Cells

1. Plate 5×10^5 cells in a 10 cm tissue culture dish (*see* **Note 2**). Cells should be plated at least 24 h before transfection and should be no more than 40–60% confluent.
2. For each plate of cells to be transfected, ethanol precipitate 5 μg of DNA in a 1.5-mL microcentrifuge tube and resuspend the pellet in 40 μL of TBS. If the same DNA is used for multiple plates, precipitate all the DNA in one tube. Ethanol precipitation sterilizes DNA.
3. Remove media from the cells and wash cells with 10 mL of PBS. After washing, add 4 mL (for a 10-cm dish) of DMEM supplemented with 10% FBS.
4. Aliquot 80 μL of DEAE-dextran into 1.5-mL microfuge tubes and warm to 37°C in a water bath. Add the resuspended DNA (5 μg of DNA per 80 μL of DEAE-dextran) slowly to the tube while vortexing gently.

5. Add 120 µL of the DNA/DEAE-dextran mixture to the plate in a dropwise fashion using a 200 µL pipet tip. Swirl the plate after each drop is applied to ensure that the mixture is distributed evenly (*see* **Note 3**).
6. Incubate the plates for 4 h in a 37°C incubator with a 5% CO_2 atmosphere. This incubation time can be shortened for some cell types.
7. Remove the medium. At this point, the cells may look a little sick but this is normal.
8. Add 5 mL of 10% DMSO in PBS. Incubate for 1 min at room temperature. Remove the DMSO and wash the cells with 5 mL of PBS. Replace the PBS with 10 mL of DMEM supplemented with 10% FBS.

3.1.2. Cells in Suspension

1. Ethanol precipitate DNA and resuspend the pellet in 500 µL of STBS (*see* **Subheading 2**). Use 10 µg of DNA per 2×10^7 cells. Cells can be either normally growing suspension cells, for example B-cells, or trypsinized anchorage-dependent cells.
2. Pellet cells in a 50-mL conical centrifuge tube.
3. Resuspend cells in 5 mL of STBS and re-pellet as in **step 2**.
4. Prepare a 2X solution of DEAE-dextran (200–1000 µg/mL) in STBS and add 500 µL of this solution to 500 µL of the DNA resuspended in 500 µL of STBS from **step 1**. Mix well. Resuspend pelleted cells with this DEAE-dextran/DNA solution. Use a final concentration of 100–500 µg/mL of DEAE-dextran.
5. Incubate cells in a CO_2 incubator for 30–90 min. Tap cells occasionally to keep them from clumping. Incubation times vary and should be determined experimentally.
6. Add DMSO to cells dropwise to a final concentration of 10%, mix well while adding.
7. Incubate cell with DMSO for 2–3 min. Add 15 mL of STBS to cell.
8. Pellet cells, wash with 10 mL of STBS and pellet again. Wash cells in medium with serum and pellet. After this centrifugation, resuspend cells in complete medium (RPMI plus 10–20% FBS). If cells are normally anchorage-dependent, re-plate on a 10 cm dish or in a 75-cm² flask. If cells are normally grown in suspension, incubate cells in normal growth media in 25-cm² flasks. The onset of expression from transfected plasmids varies depending on cell type. Usually expression begins between 24–48 h post-transfection.

3.2. Calcium Phosphate Co-Precipitation Method

Like DEAE-dextran transfection, two calcium phosphate transfection methods are routinely used: a HEPES-based method and a BES buffer method. Both are good for transient expression, but the BES-buffer procedure is much more efficient for making established constitutively expressing cell lines, in some cells 50% efficiency can be achieved. In addition, this procedure works better on a wider variety of cell types, is excellent for co-transfection, and is easier to perform than the traditional HEPES-based method. Because the BES method is so versatile and offers these advantages, this method is presented here.

1. Plate approx 5×10^5 cells on a 10-cm tissue culture dish 24 h before transfection. The cells should be no more than 50% confluent for making established cell lines and about 70% confluent for transient expression. Smaller plates (e.g., 6 cm) may be used and in some cases this is actually sufficient and easier.
2. Dilute the 2.5 M CaCl$_2$ solution to 0.25 M with sterile water.
3. Precipitate plasmid DNA in 17×100 mm polypropylene tubes as follows: Add 20–30 µg of plasmid DNA per tube (Falcon # 2058) or 10–20 µg for a 6-cm plate (*see* **Note 4**). For transfecting cells in a 100-mm dish or in two 60-mm dishes, add 20–30 µg of DNA to the tube followed, in order, by 500 µl of 0.25 M CaCl$_2$, then 500 µL of 2X BBS. Use one-half of this mixture on each plate when using 60-mm dishes. For transfecting cells in one 60-mm dish, add 10–20 µg of DNA followed, in order, by 250 µL of 0.25 M CaCl$_2$, then 250 µl of 2X BBS. In both cases, mix well and incubate at room temperature for 10–20 min (*see* **Note 5**).
4. Add the calcium phosphate/DNA mixture to cells in a dropwise fashion, swirling the plate after each drop. Incubate the cells overnight in a 35°C 3% CO$_2$ incubator (*see* **Note 6**).
5. Wash cells twice with 5 mL of PBS, then add 10 mL of DMEM with 10% FBS. Incubation of cells from this point on is done in a 5% CO$_2$ 37°C incubator.
6. For transient expression, harvest cells 48 h post-transfection. For selection of stably integrated expression clones, split cells (1:10) 48 h post-transfection into selection medium. For co-transfection *see* **Note 7**. Many investigators have used the BES method to elucidate genes required for viral DNA replication. The BES method is well-suited for this purpose because of the high co-transfection efficiency (*see* **Note 8**).

4. Notes

1. BBS pH is critical. Make three solutions ranging in pH from 6.93–6.98. Usually a visual inspection of cells after an overnight transfection will indicate which BBS mixture and DNA concentration works best. A coarse and clumpy precipitate will form when DNA concentrations are too low, a fine, almost invisible precipitate will form when DNA concentration are too high. An even granular precipitate forms when the concentration is just right. This usually correlates to the highest level of gene expression or formation of stably integrated cell lines.

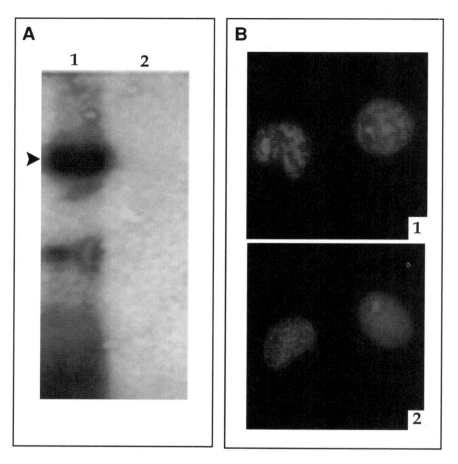

Fig. 1. Cotransfection of 11 plasmids using BES calcium phosphate coprecipitation.
(**A**) Replication assay. Human foreskin fibroblasts (HFF) were transfected with 10 plasmids encoding human cytomegalovirus (HCMV) replication genes along with a plasmid encoding the cloned origin of lytic replication (oriLyt). Total cellular DNA was harvested 5 d post-transfection and cleaved with *Eco*RI and *Dpn*I. DpnI, a four base-pair
recognition enzyme, will cleave input DNA (unreplicated DNA) multiple times and
*Eco*RI will linearize the HCMV oriLyt. Replicated plasmid is *Dpn*I-resistant and is indicated by the arrow. DNA is separated on an agarose gel and hybridized with the par-
ent plasmid vector. Lanes: 1, All required plasmids plus HCMV oriLyt; 2, omission of
one plasmid required for oriLyt replication. (**B**) HCMV replication compartment formation requires cotransfection of essential replication proteins. Cotransfections included a replication protein fused in frame with EGFP. HFF cells were transfected with
10 plasmids encoding HCMV replication proteins along with a plasmid encoding a
replication protein fused in-frame with EGFP. Transfected cells were fixed and visualized using a confocal microscope. Panel 1: cotransfections were performed the same as
in A sample number 1, cells were fixed and visualized using a confocal microscope.

2. Smaller dishes may be used; adjust the number of cells and DNA used proportionally. Higher densities of some cell types may be necessary to achieve good transfection efficiencies. If cell death is too high owing to the toxicity of DEAE, then try plating cells at a higher density.

3. It may be necessary to determine the optimal concentration of DEAE-dextran needed for good transfection efficiencies. Vary the volume of TBS used to resuspend DNA and the amount of DEAE-dextran. For example:

DNA in TBS, µL	DEAE-dextran, mg/mL
80	160
40	80
20	40

4. Only high-quality plasmid DNA will work. Use only double-banded CsCl purified DNA. Carrier DNA is not necessary and actually will decrease efficiency. Also, linear DNA does not transfect well.

5. At this point no precipitate should be visible. Use three different concentrations of DNA to help identify the DNA concentration necessary for optimal transfection.

6. CO_2 level is critical. Measure the level with a Fyrite gas analyzer. Temperature is somewhat less critical. A 37°C incubator can be used.

7. When performing cotransfections, vary the amount of effector plasmid in relation to the other plasmids in the mix keeping the total amount of DNA the same. The ratio of plasmids used in the mixture can be the difference between success and failure. We routinely find that a higher concentration of effector plasmid in the mix yields better results. We commonly use this cotransfection method to assay the level of promoter activity effected by certain vital transactivators.

8. The co-transfection-replication assay involves the co-transfection of several different plasmids (in the case of HCMV 11 different plasmids) each encoding a gene required for DNA replication. Depending on the number of plamids in the transfection mixture, vary the amount of transactivators and effector plasmid. As many as 11 plasmids can be tranfected at one time. Each plasmid can contain one or many genes required for replication of a cloned origin of DNA replication (**Fig. 1**).

Panel 2: cotransfections were the same as in (A), sample number 2, in which one plasmid encoding an essential protein was omitted from the transfection mixture. Inclusion of all of the essential proteins results in a more organized pattern of fluorescence typical of DNA replication compartments.

References

1. Fujita, T., Shubiya, H., Ohashi, T., Yamanishi, K., and Taniguchi, T. (1986) Regulation of human interleukin-2 gene: functional DNA sequences in the 5' flanking region for the gene expression in activated T lymphocytes. *Cell* **46**, 401–407.
2. Lopata, M. A., Cleveland, D. W., and Sollner-Webb, B. (1984) High-level expression of a chloramphenicol acetyltransferase gene by DEAE-dextran-mediated DNA transfection coupled with a dimethysulfoxide or glycerol shock treatment. *Nucleic Acids Res.* **12**, 5707–5711.
3. Reeves, R., Gorman, C., and Howard, B. (1985) Minichromosomes assembly of nonintegrated plasmid DNA transfected into mammalian cells. *Nucleic Acids Res.* **13**, 3599–3605.
4. Selden, R. F., Burke-Howie, K., Rowe, M. E., Goodman, H. M., and Moore, D. D. (1986) Human growth hormone as a reporter gene in regulation studies employing transient gene expression. *Mol. Cell. Biol.* **6**, 3173–3179.
5. Sussman, D. J. and Milman, G. (1984) Short-term, high-efficiency expression of transfected DNA. *Mol. Cell. Biol.* **4**, 1641–1646.
6. Mack, K. D., Wei, R., Elbagarri, A., Abbey, N., and McGrath, M. S.(1998) A novel method for DEAE-dextran mediated transfection of adherent primary cultured human macrophages. *J. Immunol. Methods* **211**, 79–86.
7. Rupprecht, A. P. and Coleman, D. L. (1991) Transfection of adherent murine peritoneal macrophages with a reporter gene using DEAE-dextran. *J. Immunol. Methods* **144**, 157–63.
8. Gauss, G. H. and Lieber, M. R. (1992) DEAE-dextran enhances electroporation of mammalian cells. *Nucleic Acids Res.* **20**, 6739–6740.
9. Graham, F. L. and Van der Eb, A. J. (1973) A new technique for the assay of infectivity of human adenovirus 5 DNA. *Virology* **52**, 456–460.
10. Segura, I., Gonzalez, M. A., Serrano, A., Abad, J. L., Bernad, A., and Riese, H. H. (2001) High transfection efficiency of human umbilical vein endothelial cells using an optimized calcium phosphate method. *Anal. Biochem.* **296**, 143–147.
11. Chen, C. and Okayama, H. (1987) High-efficiency transformation of mammalian cells by plasmid DNA. *Mol. Cell. Biol.* **7**, 2745–2752.
12. Pari, G. S. and Anders, D. G. (1993) Eleven loci encoding trans-acting factors are required for transient complementation of human cytomegalovirus oriLyt-dependent DNA replication. *J. Virol.* **67**, 6979–6988.
13. Pari, G. S., Kacica, M. A., and Anders, D. G. (1993) Open reading frames UL44, IRS1/TRS1, and UL36-38 are required for transient complementation of human cytomegalovirus oriLyt-dependent DNA synthesis. *J. Virol.* **67**, 2575–2582.

3

DNA Delivery to Cells in Culture Using Peptides

Lei Zhang, Nicholas Ambulos, and A. James Mixson

1. Introduction

There are now several cationic peptide carriers that efficiently import plasmids and oligonucleotides into cells. As a result, we anticipate that cationic peptides will play an increasingly important role with in vitro and in vivo gene delivery systems. Cationic peptides usually bind through ionic interactions to the negatively charged phosphate backbone of DNA, although additional noncovalent bonds may stabilize the interaction between the polymer and DNA (**Fig. 1**). Alternatively, cationic peptides may be covalently conjugated to DNA to promote entry into the cell. Regardless of the type of linkage between the peptide and DNA, peptide-mediated delivery can be characterized by the pathway of entry into cells: endosomolytic (*1–5*) or membrane-penetrating (*6–9*). Endosomolytic peptides enter cells through endocytosis, whereas membrane-penetrating peptides bypass the endocytotic pathway and may fuse directly with the cellular membranes. Although this chapter will discuss several methods for preparing peptide/DNA complexes, we will focus on the solid phase synthesis of peptides and the complexes that these peptides form with DNA.

1.1. Endosomolytic Peptides

Peptide-DNA complexes (polyplexes) enter cells most commonly by the endosomal pathway. Poly-L-lysine was one of the first peptides used for transfection (*10–13*), although it and related peptides (e.g., poly-L-ornithine) have low transfection efficiency (*14*). Nevertheless, this polymer along with its cargo is effectively taken up by many cells. The import of DNA by these cationic poly-

From: *Methods in Molecular Biology, vol. 245:*
Gene Delivery to Mammalian Cells: Vol. 1: Nonviral Gene Transfer Techniques
Edited by: W. C. Heiser © Humana Press Inc., Totowa, NJ

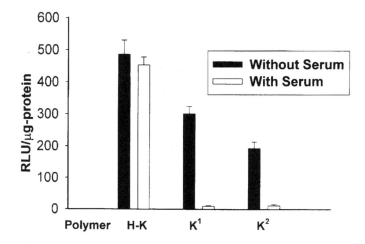

k1-poly-L-lysine composed of 18 lysines and 1 glycine- molecular weight of 2381
k2--poly-L-lysine (Sigma Co.) with molecular weight between 1,000 and 4,000

Fig. 1. Comparison of poly-L-lysine with HK polymer. In contrast to polylysine (K), serum has minimal effect on gene expression with the H-K triplex. Poly-L-lysine (K) or H-K polymers were first incubated with the PCI-Luc for 30 min followed by incubating with DOTAP liposomes. After incubating the triplex with MDA-MB-435 cells for 4 h, luciferase expression was measured 48 h later. Although K has nearly twice as many positive charges per molecule to interact with DNA compared to HK, the transfection complex containing the K-polymer is very sensitive to the presence of serum. In contrast, the transfection complex containing the HK polymer is resistant to serum and consequently transfection remains high. These results are consistent with the occurrence of noncovalent bonds other than ionic interactions.

mers into cells may be owing to their enhancement of DNA condensation and to the net negative charge on the surface of most eukaryotic cells. Nevertheless, disruption of endosomes by poly-L-lysine is inefficient and several strategies have been adopted to increase polylysine-mediated transfection. The simplest approach to augment endosomal disruption is to co-incubate pH-buffering agents with polylysine-DNA complexes; thus, lysosomotropic agents such as chloroquine have been found to increase significantly transfection of polylysine-mediated gene transfers *(14)*. Alternatively, gene expression may be enhanced by condensing polylysine-DNA complexes with high concentrations of NaCl *(15)*. Conjugation of transfection-enhancing domains to poly-L-lysine will be discussed in subsequent sections.

In addition to polylysine, several peptides have been discovered that are more effective in augmenting the cytosolic and nuclear delivery of DNA compared to polylysine (**Tables 1** and **2**); this is owing in part to their having com-

Table 1
Peptide Carriers

	References
Endosomolytic peptides	
Linear and branched cationic polyamino acids	Pouton et al. (*13*), Plank et al. (*14*), Singh et al. (*31*)
HA (GLFEAIAGFIENGWEGMIDGGGC)	Wagner et al. (*1*), Bailey et al. (*16*), Waelti et al. (*17*)
GALA (WEAALAEALAEALAEHLAEALAEALEALA)	Parente et al., (**19**)
KALA (WEAKLAKALAKALAKHLAKALAKALKACEA)	Wyman et al. (*2*)
Amphiphilic α-helical oligopeptides-Ac-(LARL)$_6$-NHCH3	Niidome et al. (*3*)
HK(KHKHKHKHKGKHKHKHKHK); Imidazole-containing polylysine	Chen et al. (*5*), Midoux et al. (*4*)
Membrane-penetrating proteins	
Tat-derived peptide (YGRKKRRQRRR)	Astriab-Fisher et al. (*34*)
Antennapedia-derived peptide (RQIKIWFQNRRMKWKK)	Astriab-Fisher et al. (*34*)
HIV-1 gp 41 (GALFLGFLGAAGSTMGA)	Morris et al. (*37*), Chaloin et al. (*29*), Morris et al. (*38*)
Transportan (GWTLNSAGYLLGKINLKALAALAKKIL)	Pooga et al. (*30*)
Nuclear Localization Peptides	
Simian virus 40 (PKKKRKV)	Kalderon et al.(*39*), Kalderon et al.(*40*)
NLS (PKKKRKVEDPYC)	Zanta et al. (*46*)
Nucleoplasm (KRPAATKKAGQAKKKK)	Robbins et al .(*41*)
Oncoprotein c-Myc (PAAKRVKL)	Makkerh et al. (*42*)
Nuclear ribonucleoprotein A1–(GNQSSNFGMKGGNFGGRSSGPYG-GGGQYFAKPRNQGGY)	Subramanian et al. (*44*), Bogerd et al. (*45*)

Table 2
Polymer–DNA Complexes

Polymer	Domains	Optimal peptide/DNA ratio	DNA	Marker/target	Entry Mechanism	References
Loligomer	DNA binding, NLS	3:1[a]	P	Luc	E	Singh et al. (*31*)
MPG	DNA binding, NLS	10:1[b]	P, AS	Luc, *Cdc25C* *Ca channel*	F	Chaloin et al. (*29*), Morris et al. (*37*), Morris et al. (*38*)
Galactosylated lysine	DNA binding, galactose	1:0/0.6[a]	P	Cat	E	Hashida et al. (*28*)
Histidylated lysine, HK	DNA binding, pH-buffering	2:1 to 65:1[b]	P, AS	Luc, *ICAM-1*	E	Midoux et al., (*4*), Pichon et al. (*55*), Chen et al. (*5*)
(LARL)$_6$	DNA binding, *hydrophobic*	2:1[b]	P	Luc	E	Niidome et al. (*3*)

Abbreviations:NLS, nuclear localization signal; P, plasmid; AS, antisense oligonucleotide; Luc or Cat; plasmid-encoding luciferase or chloramphenicol acetyl transferase; E and F represent endocytosis and fusogenic pathways, respectively. The target of the antisense oligonucleotides is underlined.

[a]Optimal weight: weight ratio, [b]Optimal positive:negative charge ratio.

ponents that may facilitate endosomal disruption. Examples of these endoso-molytic peptides include an N-terminal sequence from hemagglutinin-A2 virus (GLFEAIAGFIENGWEGMIDGGGC) *(1,16–18)*, KALA *(2)*, GALA *(19,20)*, (LARL)$_6$ *(3)*, and histidine (or imidazole)-lysine polymers *(4,5,21,22)*. Peptides such as GALA, KALA, and hemagglutinin-derived virus sequences form α-helices at acidic pH, thereby disrupting endosomes; HK polymers, akin to non-degradable polymers (polyethylinimine, polyamidomine, etc.), may disrupt endosomes through their buffering properties. These peptides can be further classified by their ability to bind to DNA. For example, KALA, histidylated polylysine, and (LARL)$_6$ α-helical peptide can interact directly with DNA; in contrast, GALA and hemagglutinin A2 virus-derived peptides must be further modified in order to bind with DNA (e.g., the addition of cationic peptide domain) *(23)*. Nevertheless, to achieve optimal transfection, peptides almost always need to be modified with cell-specific ligands and/or a nuclear localization signal *(23)*. We emphasize that endosomolytic peptides may be used alone or in combination with other carriers such as liposomes or viruses to increase transfection *(5,21,24–28)*.

1.2. Membrane-Penetrating Peptides

Compared to endosomolytic peptides, membrane-penetrating (fusogenic) peptides of DNA are relative newcomers in the transport of DNA molecules *(8,29–35)*. Membrane-penetrating peptides were used initially to transport proteins into cells *(36)*, but more recently some of these peptides have been shown to carry oligonucleotides directly into cells through their fusogenic properties (**Tables 1** and **3**). It has also been suggested, but not proven, that these peptides may be efficient carriers of plasmids. Delivery of DNA and/or proteins by these peptides occurs at 4°C, indicating that intracellular delivery of these peptide-DNA conjugates may be owing to the formation of "inverted micelles." Evidence that these peptide/DNA conjugates enter by a pathway other than endocy-

Table 3
Polymer–DNA Conjugates

Conjugate	Target/marker	DNA	Entry Mechanism	References
Tat-MDR conjugate	P-glycoprotein, galanin receptor	AS	F	Astriab-Fisher et al. *(34)*, Pooga et al. *(43)*
ANT-MDR conjugate	P-glycoprotein, galanin receptor	AS	F	Astriab-Fisher et al. *(34)*, Pooga et al. *(43)*

Abbreviations:AS, antisense oligonucleotide; F represents that the mechanism of entry of the conjugate is fusogenic.

tosis is further supported by lack of transfection enhancement with lysosomo-tropic agents (e.g., chloroquine); lysosomotropic agents usually increase trans-fection of non buffering transfection carriers mediated by endocytosis. However, the precise mechanism of entry of these "membrane-penetrating" peptides re-mains unclear and endocytosis has not definitively been ruled out. One distin-guishing characteristic of these fusogenic peptides is that they contain several arginines, in contrast to the lysine-rich cationic peptides associated with endo-cytosis; hydrogen bonding of the guanidinium side chain of arginine is a feature not shared by the lysine-containing cationic peptides. These arginine-rich pep-tides include Tat-derived (YGRKKRRQRRR), Antennapedia-derived peptides (RQIKIWFQNRRMKWKK), and polyarginine peptides *(35)*, and such peptides for in vitro experiments have been shown to be efficient carriers of oligonu-cleotides (**Tables 1** and **3**) *(34)*.

Recently, phage particles displaying Tat peptide on the surface have shown efficient gene transfer in vitro *(7)*. As a result, it has been suggested that mem-brane-penetrating peptides may be effective carriers of large molecular-weight DNA *(8)*. In one study with plasmids, however, gene expression with branched polyarginine polymers was low *(14)*. More studies are required to assess the efficacy of these arginine rich membrane-penetrating polymers on the delivery of plasmids. Other membrane-penetrating peptides with high hydrophobicity that do not contain arginine-rich sequences have been shown to be effective car-riers of therapeutic oligonucleotides and/or luciferase-encoding plasmids *(29, 30,37,38)*.

1.3. Nuclear Localization Signals

In addition to the membrane-penetrating, endosomolytic, or DNA-condens-ing properties, incorporation of nuclear localization signals (NLS) into peptides may increase transfection significantly *(30,39–46)* (**Tables 1–3**). For example, a branched peptide chimera containing multiple lysine pentapeptides and NLS (SV40 large T-antigen) significantly enhanced transfection of DNA compared to the unmodified branched polylysine *(31)*. In addition, one report has found that an oligonucleotide containing a lysine tail accumulated significantly in the nucleus, which the authors suggested might be owing to the NLS qualities of the lysine tail *(47)*. NLS are usually strongly cationic *(30,39–41,43,46)*, but may be weakly cationic *(42)* or neutral *(44,45)*. Obviously, cationic NLS sequences may interact and compete with other cationic sequences within the polymer. If this is problematic, a neutral or less cationic NLS can be incorporated into the peptide. The number of NLS per DNA molecule appears to be critical in opti-mizing transfection *(48)*.

1.4. Assembly of Complex

The challenge of peptide delivery systems is to define the precise formulations to increase delivery of DNA. The investigator must determine if self-assembly of several peptides, each of which contains one of these domains (DNA-condensing, membrane, destabilizing, and NLS), is preferable to a signal peptide containing all of these domains. The choice of how to proceed with this assembly will probably be determined by understanding the mechanisms involved rather than constraints on peptide synthesis. For example, the peptide MPG combines the putative membrane-penetrating domain of HIV GP-41 with the NLS domain of SV40 to transport oligonucleotides and plasmids efficiently into cells (**Table 2**) *(29,37,38)*. Nevertheless, if release of DNA from the polymer within the endosome is critical for disruption of endosomes *(49)*, then the NLS sequence should not be incorporated within the DNA condensing segment of the peptide-delivery agent, but rather should be associated directly with the DNA.

2. Materials

2.1. Sources of Media, Cells, and Chemicals

1. Dulbecco's Modified Eagle's medium (DMEM), fetal calf serum (FCS), Glutamine, Opti-MEM, and DH5α bacteria are available from Invitrogen (Carlsbad, CA).
2. EBM medium, EGM-2 medium, and bovine brain extracts (BBE) are available from Clonetics (San Diego, CA).
3. Superbroth is available from Advance Biotechnologies Incorporated (Columbia, MD).
4. Luciferase Assay System with Reporter Lysis Buffer System is available from Promega (Madison, WI).
5. Ion-exchange chromatography columns for DNA purification are available from Qiagen, Inc.(Chatsworth, CA).
6. Coomassie Plus Protein Assay Reagent is available from Pierce (Rockford, IL).
7. 1, 2-dioleoyl-3-trimethylammonium-propane (DOTAP) dissolved in chloroform and cholesterol are available from Avanti, Inc. (Birmingham, AL).
8. Cell lines: MDA-MB-435 was a gift from Dr. Erik Thompson, Lombardi Cancer Center, Washington, DC), MDA-MB-231, CRL-5800, and Chinese hamster ovary (CHO) cells are available from American Tissue Culture Collection (ATCC, Manassas, VA). Bovine aortic endothelial (BAE) cells are available from Coriell Cell Repositories (Camden, NJ); and normal human dermal fibroblasts (NHDF), human umbilical vein endothelial cells

(HUVEC), and human microvascular vein endothelial cells (HMVEC) are available from Clonetics. MDA-MB-435 and MDA-MB-231 cells are un-differentiated estrogen-independent breast cancer lines obtained from un-related human donors, whereas the CRL-5800 cell line was isolated from a human nonsmall cell lung cancer.

9. The resin, Fmoc PAL-PEG low load (0.1–0.2 g/mmol) resin, is available from Applied Biosystems (ABI, Foster City, CA).
10. Amino acids and chemicals for peptide synthesis.

    a. Piperidine and HATU (*N*-[(dimethylamino)-1H-1,2,3-triazolo[4,5-b] pyridin-1-ylmethylene]-*N*-methylmethanaminium hexafluorophosphate *N*-oxide) are available from ABI.

    b. *N*-Hydroxybenzotriazole (HOBt) is available from AnaSpec (San Jose, CA); all amino acids are available from AnaSpec, and side chains are protected as follows: Fmoc-lysine (Boc; tert-butyloxycarbonyl), Fmoc-aspartic acid and Fmoc-glutamic acid (OtBu; tert-butyl ester), Fmoc-arginine (Pbf; 2,2,4,6,7-pentamethyldihydrobenzofuran-5-sulfonyl), Fmoc-serine and Fmoc-threonine (tBu; tert-butyl), Fmoc-histidine, Fmoc-asparagine, Fmoc-glutamine, Fmoc-cysteine (Trt; trityl), and for the branched core Fmoc-lysine [Dde; 1-(4,4-dimethyl-2,6-dioxocyclo-hex-1-ylidine]ethyl].

    c. *N*,*N*-Diisopropylethylamine (DIEA), hydrazine, and triisopropylsilane are available from Sigma-Aldrich (St. Louis, MO).

    d. Acetonitrile, *N*,*N*-dimethylformamide (DMF), dichloromethane (DCM), trifluoroacetic acid (TFA), and ethyl ether are available from Burdick and Jackson (Muskegon, MI).

    e. Phenol is available from Invitrogen.

    f. Acetic anhydride is available from EM Scientific (Cherry Hill, NJ).

    g. Reagent K: 88% TFA, 5% water, 5% phenol, and 2% triisopropylsilane.

    h. Buffer A: 0.1% TFA; Buffer B: 0.1% TFA, 80% acetonitrile.

2.2. Plasmids

1. pCI-Luc, a plasmid encoding luciferase with a CMV promoter, is available from Promega.
2. Isolate and purify pCI-Luc from *Escherichia coli* (DH5α) using Qiagen Maxi Kits (**see** Note 1).

2.3. Maintaining Cells and Culture Medium

1. Maintain primary human umbilical vein endothelial cells (HUVEC, HMVEC) in EGM-2 Bullet Kit media. Perform experiments with these cell lines between passages 3 and 6.

2. Maintain MDA-MB-435, MDA-MB-435, and CHO, NHDF, and BAE (for gene expression) cells in DMEM containing 10% FCS and 2 m*M* glutamine. Perform experiments on NHDF between passages 2 and 6.
3. Maintain BAE cells, for antisense experiments, in EBM containing 5% FCS and 1.2 µg/mL of BBE.

2.4. Liposomes

1. Place 25 mg of DOTAP (2.5 mL) in a round-bottomed flask and dry lipids to a fine film with a Rotary Evaporator.
2. After hydrating lipids with water (1.6 mL), transfer liposome solution into an 18 mL glass-stoppered test tube (Kontes, Vineland, NJ).
3. Sonicate liposomes in the presence of argon until clear in a Branson 1210 bath sonicator.
4. Extrude the solution of liposomes 10 times through a 50 nm polycarbonate membrane with LipsoFast-Basic extruder.
5. Dilute liposome solution with 25 mL of water so that the final concentration is 1 µg/µL.
6. In addition to preparing cationic liposomes, several cationic liposomes are commercially available, including Lipofectamine (Invitrogen), Lipofectin (Invitrogen), DOTAP (Roche, Indianapolis, IN), and [1,3-di-oleoyloxy-2-(6-carboxy-spermyl)-propylamid] (DOSPER, Roche).

2.5. Peptides

1. Poly-L-lysine (K_2; molecular-weight range 1000–4000) is available from Sigma-Aldrich.
2. The biopolymer core facility at the University of Maryland synthesized the following polymers on a Symphony multiple peptide synthesizer (PTI, Tucson, AZ):
   a. HK (19 mer; molecular weight-2, 454)–[K-H-K-H-K-H-K-H-K-G-K-H-K-H-K-H-K-H-K];
   b. HHK4b (83 mer; molecular weight-10, 570)–[K-H-K-H-H-K-H-H-K-H-H-K-H-H-K-H-H-K-H-K]4 K-K-K; and
   c. K^1 (19 mer; molecular weight 2, 381)–[K-K-K-K-K-K-K-K-K-G-K-K-K-K-K-K-K-K-K].

   Details of their syntheses are discussed in **Subheading 3.1.**
3. Reconstitute polymers with water at a concentration of 30 mg/mL.

2.6. Equipment

1. Symphony multiple peptide synthesizer (Protein Technologies Inc [PTI], Tucson, AZ).

2. HPLC (Beckman, Fullerton, CA) with a Dynamax 21.4 × 250 mm C-18 reversed phase preparative column (Varian, Palo Alto, CA).
3. Voyager matrix-assisted laser desorption/ionization-time of flight (MALDI-TOF) mass spectrometry (Applied Biosystems).
4. Rotary Evaporator with water bath (Buchi, Switzerland).
5. Lipofast Basic extruder (Avestin Inc, Ottawa, ON, CAN).
6. Branson 1210 bath sonicator (Bransonic, Danbury, CT).
7. TD 20/20 luminometer (Turner Design, Sunnyvale, CA).

3. Methods

3.1. Solid-Phase Peptide Synthesis (SPPS)

The SPPS methods described here have been utilized by our group to synthesize histidine-rich peptides. Although they were designed to overcome difficulties encountered during the synthesis of these peptides, the methods should also be applicable to nonhistidine-containing peptide carriers. For an alternative strategy of preparing peptide carriers, *see* **Notes 2** and **3**.

3.1.1. General SPPS Methodology (50,51)

1. Synthesize peptides on a Symphony multiple peptide synthesizer at a 0.025-mM scale using Fmoc (9-fluorenylmethyloxycarbonyl) chemistry.
2. Synthesize the HK peptides with Fmoc-PAL-PEG-PS resin (ABI) at a substitution level of 0.1–0.2 g/mMol (*see* **Note 4**).
3. Remove Fmoc group (deprotection) by incubating three times for 20 min with 20% piperidine containing 0.1 M HOBt. (First remove Fmoc from resin, and then in successive cycles remove Fmoc from N-terminal amino acid of nascent peptide.)
4. Following deprotection, wash the resin six times for 1 min with DMF/DCM (1:1; primary solvent).
5. Catalyze amino acid coupling with HATU/DIEA as the activator. Mix Fmoc-Lysine(Boc), with HATU/ DIEA in a 1:1:1.5 molar ratio. (For subsequent couplings, similarly activate desired amino acid.)
6. Mix activated amino acid with resin for 45 min at room temperature (RT).
7. Wash three times for 1 min with the primary solvent.
8. Cap any uncoupled free N-terminal amino groups on the resin using 50% acetic anhydride in DMF. Wash resin with 50% acetic anhydride in DMF two times for 15 min.
9. Wash three times for 1 min with the primary solvent.
10. Repeat **steps 3–9** for each successive amino acid addition performed until the desired polymer is completed.

11. Cleave peptides from the resin and remove the side-chain protecting groups by treatment with reagent K for 4 h *(52)*.
12. Filter in-line peptides on the Symphony and deposit into collection tubes.
13. Precipitate peptides with three volumes of ice-cold ethyl ether. Wash three times in ice-cold ethyl ether and dry.
14. Purify peptides on a high-performance liquid chromatography (HPLC) column (Beckman-Coulter) with System Gold operating software, using a Dynamax 21-4 $\times$ 250mm C-18 reversed-phase preparative column with a binary solvent system, buffers A and B, with a gradient of 0–100% buffer B over 60 min at a flow rate of 20 mL/min. Detection is at 214 nm.
15. Analyze purified peptides by MALDI-TOF mass spectrometry to verify the predicted molecular mass (*see* **Note 5**).

3.1.2. Synthesis of HHK4b Branched Peptide (53)

1. Remove Fmoc group (deprotection) with 20% piperidine/0.1 *M* HOBt and wash with primary solvent as described in **steps 3** and **4** in **Subheading 3.1.1.**
2. Mix Fmoc-lysine (Dde) with HATU/DIEA in a 1:1:1.5 molar ratio.
3. Mix activated amino acid with resin for 45 min at RT.
4. Wash three times for 1 min with the primary solvent.
5. Cap any uncoupled free N-terminal amino groups on the resin using 50% acetic anhydride in DMF. Wash with capping reagent two times for 15 min.
6. Wash three times for 1 min with the primary solvent.
7. Repeat two cycles of **steps 1–6** to couple two additional Fmoc-Lys(Dde) residues. 8. Following the capping step after addition of the third Fmoc-Lys (Dde), remove Dde groups by incubation of the peptide resin in 2% hydrazine for 30 min.
9. Wash the peptide resin six times for 1 min with the primary solvent.
10. To synthesize branches, repeat **steps 3–10** of **Subheading 3.1.1.** as necessary for synthesis of the desired sequence (*see* **Note 6**).
11. Cleave, purify, and analyze HHK4b (*see* **steps 11–15** in **Subheading 3.1.1.**)

3.2. Preparation of Peptide/Plasmid Complexes

1. Inoculate MDA-MB-435 cells in wells of a 24-multiwell plate so that they are 40–80% confluent on the day of transfection (*see* **Note 7**).
2. Dilute 2.5 µL (7.5 nmol) of HK or 0.093 µL (0.375 nmol) of HHK4b with Opti-MEM to a total volume of 40 µL.
3. Dilute 1.5 µg of PCI-Luc with Opti-MEM to a total volume of 40 µL.
4. Add 40 µL of HK or HHK4b polymer to 40 µL of PCI-Luc with gentle vortexing (*see* **Notes 8–11**).

5. Let stand for 30 min at RT. Go to **step 8** if liposomes are not added.
6. Mix 1.5 µl of DOTAP stock solution (1 µg/µL) with 38.5 µL of Opti-MEM. Add this diluted DOTAP solution to the polymer:DNA complex to augment transfection. Gently swirl for 5 s (*see* **Notes 11–13**).
7. Let stand for 30 min at RT.
8. Dilute the liposome:polymer: DNA or polymer: DNA complex with either 200 µL or 240 µL, respectively, of Opti-MEM (± 15% serum).
9. Add 300 µL of complex per well.
10. Remove complexes 4 h after transfection and add DMEM with 10% serum (or appropriate medium for a particular cell as defined in **Subheading 2.3.**) (*see* **Notes 14–20**).

3.3. Luciferase Measurement

1. Aspirate media 24–48 h after transfection
2. Lyse cells with 100 µL of Reporter lysis buffer.
3. Gently rock the cells in the 24-multiwell plate for 10 min at RT. Transfer the lysates to microfuge tubes.
4. Centrifuge lysates in a microfuge at 10,000*g* for 5 min at 4°C, transfer the supernatant to a new tube, and place on ice.
5. Mix 10 µL of the supernatant with 60 µL of luciferase assay buffer.
6. Determine RLU levels of each sample with the direct current TD 20/20 luminometer (*see* **Note 21**).
7. Measure the amount of lysate protein with Coomassie Plus Protein Reagent to normalize RLU values.

4. Notes

1. Assay plasmids on an analytical agarose gel to ensure that there is no contamination with other nucleic acids, including bacterial chromosomal DNA.
2. Solution phase methods have been explored as an alternative to solid-phase chemistry. Solution-phase methods are free of some of the limitations of solid phase and lend themselves more easily to combinatorial chemistry. Disadvantages of solution phase methods include: (a) purification of these peptides and (b) the precise order or location of amino acids is not known. Nevertheless, several peptides have been synthesized with solution-phase chemistry for gene-therapy delivery. For instance, histidines and the closely related imidazole groups have been linked to backbones of poly-L-lysine *(4,22)*. Furthermore, co-polymers of amino acids have been prepared by solution synthesis and these approaches could easily be adapted to gene therapy *(54,55)*. Notably, synthesis of large numbers of synthetic polymers upon different diacrylate supports offers the possibility of rapidly discovering new vectors for gene delivery *(56)*.

3. We are not aware of any commercially available cationic peptide that has been formulated for in vitro or in vivo gene delivery systems. A limited number of cationic peptides/polymers (e.g., polylysine, polyarginine, etc.) are available from companies such as Sigma-Aldrich.

4. Although our experience has been primarily with histidine-containing polymers as transfection agents, we emphasize that these histidine-rich peptides are particularly difficult to synthesize. Thus, the methods described in **Subheading 3.1.** for histidine-rich peptides should be adequate for the synthesis of most peptides. Notably, low-density resin is preferable to high density resin for polymers that are difficult to synthesize (e.g., histidine-rich peptides).

5. For highly branched histidine-containing polymers (e.g., four branches emanating from a lysine core), determination of the molecular weight with MALDI mass spectroscopy may prove elusive. Electrospray mass spectroscopy provides a particularly useful alternative for determining the molecular weight of these highly branched histidine-containing polymers. Nonetheless, these methods may sometimes fail, possibly owing to poor ionization, and in a few cases, amino acid analysis and/or the use of sodium dodecyl sulfate (SDS) gels may be helpful in determining the approximate weight of an HPLC-purified polymer.

6. The four branches of HHK4b emanate from the α-amino and three ε amino groups of the lysine tripeptide core.

7. This method can easily be adapted to plasmids encoding other reporter proteins, such as chloramphenicol acetyl transferase.

8. Although our polymer/DNA experience is based on experiments with HK polymers, we believe that this is a useful paradigm for preparation of other peptide/DNA complexes. Nevertheless, the transfection time and optimal amount and ratio of polymer and DNA for in vitro transfection experiments must be determined for each cell line. To determine the optimal transfection conditions, we suggest mixing varying amounts of the polymer (0.06 nmol to 7.5 nmol), DNA (0.5 and 1.5 μg), and liposomes (0.5, 1, and 1.5 μg).

9. Solubility of polymer:DNA complexes is dependent on their concentration, but at the tissue-culture concentrations noted in **Subheading 3.2.**, solubility is not a problem with histidine/lysine co-polymers and DNA complexes.

10. The optimal stoichiometric (+/−) ratio is usually between 1 and 10 for the cationic peptide and DNA. When the mechanism of entry is through endocytosis, the optimal +/− stoichiometric ratio for transfections is markedly decreased with the use of branched carriers compared to linear carriers. For calculating the optimal stoichiometric ratio of HK polymer/DNA complexes, 4% of imidazole side chains of histidine carry a positive charge at pH 7.4, assuming that the pKa of imidazole is 6.0. All ε-groups of lysine in the HK polymers carry a positive charge except for the lysines in the core

of branched HK polymers. For example, each HHK4b polymer containing 83 amino acids (32 K, 48 H, 3 core K) carries 33.92 positive charges at physiologic pH. Because 0.375 nmol of HHK4B in complex with 0.228 pmol of PCI-Luc DNA (5 kbp or 10 kb) was the optimal transfection amount, the optimal stoichiometric ratio of HHK4b to plasmid DNA is 5.57 ($0.375 \text{ nmol} \times 33.92/0.228 \text{ pmol} \times 10^4 = 0.375 \text{ nmol} \times 33.92/2.28 \text{ nmol} = 12.78 \text{ nmol}/2.28 \text{ nmol} = 5.57$)

11. For transfecting plasmids, branched polymers alone are more effective carriers than the linear polymers. In combination with liposomes however, the optimal HK polymer (linear vs branched) for transfection of plasmids depends on the cell line. For example, the liposome:linear HK carrier is the optimal carrier for endothelial (BAEC, HUVEC, and HMVEC) and NHDF cells, whereas the liposome:branched HHK4b carrier is the optimal carrier for MDA-MB-435, MDA-MB-231, CHO, and CRL-5800 cells *(49)*.

12. Although the liposome/HK (or HHK4b) combination enhances more than polymer alone *(5)*, it is not clear if liposomes enhance other endosomolytic peptide carriers.

13. In the absence of serum, all liposome preparations (Lipofectamine, Lipofectin, DOTAP, DOTAP/cholesterol, and DOSPER) in combination with the HK or HHK4b polymers enhance gene expression between 5 and 40-fold compared to the liposome carrier alone *(5,21)*. In the presence of serum, HK or HHK4b/liposome combinations increased transfection more than a 100-fold compared to liposome carriers alone *(5,21)* (with the exception of Lipofectin). For MDA-MB-435 cells, the most effective liposome preparations are Lipofectamine and DOSPER.

14. For transfecting antisense oligonucleotides, the branched HHK4b in combination with liposomes is more effective than the linear HK/liposome or branched HHK4b carrier alone in endothelial cells (unpublished results). Although we believe that the HHK4b/liposome combination will be a more effective carrier of oligonucleotides compared with linear HK/liposome combination, these results are based on three endothelial cell lines.

15. In vitro transfection experiments with antisense oligonucleotides are easily adapted from the plasmid transfection experiments. Dependent on the target of the oligonucleotide, assays (cell number, apoptosis, Northern) can be done to assess the efficacy of the therapy *(34,38,57)*.

16. Although there are several examples of enhanced in vitro transfection by addition of cell-specific ligands, these experiments are usually done as preludes for in vivo experiments. Thus, the addition of cell-specific ligands to peptides for only in vitro experiments, in our opinion, will be quite limited, and other transfection agents or methods are probably preferable in most cases. Nonetheless, in cells that prove very difficult to transfect for in vitro

or ex vivo experiments, addition of these cell-specific ligands may prove useful.

17. Addition of NLS to endosomolytic peptides should allow more efficient transfection in many cell lines. Although technically easier to incorporate the NLS and the endosomolytic domains within the same peptide, these domains may be more effective on separate components of the transfection complex.

18. Several reports describe the use of peptide-oligonucleotide conjugates to deliver antisense molecules to their intracellular targets (**Table 3**). To prepare these peptide-oligonucleotide conjugates, there are two methods utilizing separate synthesis of the peptide and the oligonucleotide with subsequent post-assembly conjugation. First, peptide-oligonucleotide conjugates have been prepared via disulfide formation *(29,34,58,59)*. The oligonucleotide containing a 5′-thiol group is activated with a 2-pyrididylthiol moiety, purified by HPLC, and then conjugated to the peptide containing a C-terminal cysteine group. Second, the N-terminal thioester-functionalized peptide is conjugated to a 5′-cysteinyl oligonucleotide in aqueous/organic solutions *(60,61)*. (This latter technology has been licensed to Glenn Technologies [Sterling, VA].) With both methods, a range of peptide-oligonucleotide conjugates have been prepared and purified by reversed-phase HPLC (RP = HPLC).

19. Submicromolar levels of the peptide-oligonucleotide conjugate targeting the P-glycoprotein receptor were incubated with cells for 6 h in the presence or absence of serum. Significant inhibition of this surface receptor was observed *(34)*. Translocation across the membrane has been seen in less than 5 min with peptide-DNA conjugates *(29)*.

20. After determining the optimal transfection ratio for the polymer and DNA complex, several other properties may be useful in optimizing the transfection complex *(14,62)*. These characteristics (with assay) are the following: (a) size (N4 Plus Submicron Particle Sizer [Coulter] and electron microscopy); (b) surface charge (Delsa 440SX [Coulter]; gel retardation assay), (c) aggregation (spectroscopy); and (d) integrity of the complex in the presence of serum (ethidium bromide and DNAse sensitivity assay). Clearly, the utility of these assays depends on the type of experiment. For example, Ni-iodome et al. *(3)* found a direct correlation between aggregation of polymer and the ability to transfect cells. Notably, we have been unable to measure the size of the HK/DNA complexes with the N4 plus particle sizer, but after adding liposomes, we are able to measure the ternary complex with the particle sizer *(5,21)*. Other investigators have had difficulty determining the size of peptide/DNA complexes (at transfection concentrations) *(14)*. We

suspect that it will be difficult to quantify the size of many peptide/DNA complexes with the N4 plus, unless a limited amount of aggregation occurs.

21. Although the TD 20/20 luminometer is as sensitive as a photon luminometer, the relative light units obtained from the direct current TD luminometer are significantly different from the values obtained by a photon luminometer. The relative light units of a photon counting Lab Monolight 2010 model (Analytical Luminescence Laboratory, San Diego, CA) can be approximated by multiplying the relative light units determined with the TD 20/20 model by 1.03×10^3.

References

1. Wagner, E., Plank, C., Zatloukal, K., Cotten, M., and Birnstiel, M. L. (1992) Influenza virus hemagglutinin HA-2 N-terminal fusogenic peptides augment gene transfer by transferrin-polylysine-DNA complexes: toward a synthetic virus-like gene-transfer vehicle. *Proc. Natl. Acad. Sci. USA* **89**, 7934–7938.

2. Wyman, T. B., Nicol, F., Zelphati, O., Scaria, P. V., Plank, C., and Szoka, F. C., Jr. (1997) Design, synthesis, and characterization of a cationic peptide that binds to nucleic acids and permeabilizes bilayers. *Biochemistry* **36**, 3008–3017.

3. Niidome, T., Ohmori, N., Ichinose, A., Wada, A., Mihara, H., Hirayama, T., and Aoyagi, H. (1997) Binding of cationic-helical peptides to plasmid DNA and their gene transfer abilities into cells. *J. Biol. Chem.* **272**, 15307–15312.

4. Midoux, P. and Monsigny, M. (1999) Efficient gene transfer by histidylated polylysine/pDNA complexes. *Bioconjug. Chem.* **10**, 406–411.

5. Chen, Q. R., Zhang, L., Stass, S. A., and Mixson, A. J. (2001) Branched co-polymers of histidine and lysine are efficient carriers of plasmids. *Nucleic Acids Res.* **29**, 1334–1340.

6. Xia, H., Mao, Q., and Davidson, B. L. (2001) The HIV Tat protein transduction domain improves the biodistribution of beta-glucuronidase expressed from recombinant viral vectors. *Nat. Biotechnol.* **19**, 640–644.

7. Eguchi, A., Akuta, T., Okuyama, H., Senda, T., Yokoi, H., Inokuchi, H., et al. (2001) Protein transduction domain of HIV-1 Tat protein promotes efficient delivery of DNA into mammalian cells. *J. Biol. Chem.* **276**, 26204–26210.

8. Snyder, E. L. and Dowdy, S. F. (2001) Protein/peptide transduction domains: potential to deliver large DNA molecules into cells. *Curr. Opin. Mol. Ther.* **3**, 147–152.

9. Kokunai, T., Urui, S., Tomita, H., and Tamaki, N. (2001) Overcoming of radioresistance in human gliomas by p21WAF1/CIP1 antisense oligonucleotide. *J. Neurooncol.* **51**, 111–119.

10. Leonetti, J. P., Rayner, B., Lemaitre, M., Gagnor, C., Milhaud, P. G., Imbach, J. L., and Lebleu, B. (1988) Antiviral activity of conjugates between poly(L-lysine) and synthetic oligodeoxyribonucleotides. *Gene* **72**, 323–332.

11. Leonetti, J. P., Degols, G., and Lebleu, B. (1990) Biological activity of oligonucleotide-poly-L-lysine) conjugates: mechanism of cell uptake. *Bioconjug. Chem.* **1**, 149–153.

12. Degols, G., Leonetti, J. P., Gagnor, C., Lemaitre, M., and Lebleu, B. (1989) Antiviral activity and possible mechanisms of action of oligonucleotides-poly(L-lysine) conjugates targeted to vesicular stomatitis virus mRNA and genomic RNA. *Nucleic Acids Res.* **17**, 9341–9350.

13. Pouton, C. W., Lucas, P., Thomas, B. J., Uduehi, A. N., Milroy, D. A., and Moss, S. H. (1998) Polycation-DNA complexes for gene delivery: a comparison of the biopharmaceutical properties of cationic polypeptides and cationic lipids. *J. Control Release* **53**, 289–99.

14. Plank, C., Tang, M. X., Wolfe, A. R., and Szoka, F. C., Jr. (1999) Branched cationic peptides for gene delivery: role of type and number of cationic residues in formation and in vitro activity of DNA polyplexes. *Hum. Gene Ther.* **10**, 319–332.

15. Ferkol, T., Perales, J. C., Mularo, F., and Hanson, R. W. (1996) Receptor-mediated gene transfer into macrophages. *Proc. Natl. Acad. Sci. USA* **93**, 101–105.

16. Bailey, A. L., Monck, M. A., and Cullis, P. R. (1997) pH-induced destabilization of lipid bilayers by a lipopeptide derived from influenza hemagglutinin. *Biochim. Biophys. Acta* **1324**, 232–244.

17. Waelti, E. R. and Gluck, R. (1998) Delivery to cancer cells of antisense L-myc oligonucleotides incorporated in fusogenic, cationic-lipid-reconstituted influenza-virus envelopes (cationic virosomes). *Int. J. Cancer* **77**, 728–733.

18. Schoen, P., Chonn, A., Cullis, P. R., Wilschut, J., and Scherrer, P. (1999) Gene transfer mediated by fusion protein hemagglutinin reconstituted in cationic lipid vesicles. *Gene Ther.* **6**, 823–832.

19. Parente, R. A., Nir, S., and Szoka, F. C., Jr. (1990) Mechanism of leakage of phospholipid vesicle contents induced by the peptide GALA. *Biochemistry* **29**, 8720–8728.

20. Simoes, S., Slepushkin, V., Pretzer, E., Dazin, P., Gaspar, R., Pedroso de Lima, M. C., and Duzgunes, N. (1999) Transfection of human macrophages by lipoplexes via the combined use of transferrin and pH-sensitive peptides. *J. Leukoc. Biol.* **65**, 270–279.

21. Chen, Q. R., Zhang, L., Stass, S. A., and Mixson, A. J. (2000) Co-polymer of histidine and lysine markedly enhances transfection of liposomes. *Gene Ther.* **7**, 698–704.

22. Putnam, D., Gentry, C. A., Pack, D. W., and Langer, R. (2001) Polymer-based gene delivery with low cytotoxicity by a unique balance of side-chain termini. *Proc. Natl. Acad. Sci. USA* **98**, 1200–1205.

23. Plank, C., Zatloukal, K., Cotten, M., Mechtler, K., and Wagner, E. (1992) Gene transfer into hepatocytes using asialoglycoprotein receptor mediated endocytosis of DNA complexed with an artificial tetraantennary galactose ligand. *Bioconjug. Chem.* **3**, 533–539.

24. Curiel, D. T., Wagner, E., Cotten, M., Birnstiel, M. L., Agarwal, S., Li, C. M., et al. (1992) High-efficiency gene transfer mediated by adenovirus coupled to DNA-polylysine complexes. *Hum. Gene Ther.* **3**, 147–154.

25. Ebbinghaus, S. W., Vigneswaran, N., Miller, C. R., Chee-Awai, R. A., Mayfield, C. A., Curiel, D. T. and Miller, D. M. (1996) Efficient delivery of triplex forming oligonucleotides to tumor cells by adenovirus-polylysine complexes. *Gene Ther.* **3**, 287–297.

26. Li, S. and Huang, L. (1997) In vivo gene transfer via intravenous administration of cationic lipid-protamine-DNA (LPD) complexes. *Gene Ther.* **4**,891-900.
27. Li, S., Rizzo, M. A., Bhattacharya, S. and Huang, L. (1998) Characterization of cationic lipid-protamine-DNA (LPD) complexes for intravenous gene delivery. *Gene Ther.* **5**, 930–937.
28. Hashida, M., Takemura, S., Nishikawa, M., and Takakura, Y. (1998) Targeted delivery of plasmid DNA complexed with galactosylated poly-L-lysine). *J. Control Release* **53**, 301–310.
29. Chaloin, L., Vidal, P., Lory, P., Mery, J., Lautredou, N., Divita, G., and Heitz, F. (1998) Design of carrier peptide-oligonucleotide conjugates with rapid membrane translocation and nuclear localization properties. *Biochem. Biophys. Res. Commun.* **243**, 601–608.
30. Pooga, M., Soomets, U., Hallbrink, M., Valkna, A., Saar, K., Rezaei, K., et al. (1998) Cell penetrating PNA constructs regulate galanin receptor levels and modify pain transmission in vivo. *Nat. Biotechnol.* **16**, 857–861.
31. Singh, D., Bisland, S. K., Kawamura, K., and Gariepy, J. (1999) Peptide-based intracellular shuttle able to facilitate gene transfer in mammalian cells. *Bioconjug. Chem.* **10**, 745–754.
32. Schwartz, J. J. and Zhang, S. (2000) Peptide-mediated cellular delivery. *Curr. Opin. Mol. Ther.* **2**, 162–167.
33. Uherek, C. and Wels, W. (2000) DNA-carrier proteins for targeted gene delivery. *Adv. Drug Deliv. Rev.* **44**, 153–166.
34. Astriab-Fisher, A., Sergueev, D. S., Fisher, M., Shaw, B. R., and Juliano, R. L. (2000) Antisense inhibition of P-glycoprotein expression using peptide-oligonucleotide conjugates. *Biochem. Pharmacol.* **60**, 83–90.
35. Futaki, S., Suzuki, T., Ohashi, W., Yagami, T., Tanaka, S., Ueda, K., and Sugiura, Y. (2001) Arginine-rich peptides. An abundant source of membrane-permeable peptides having potential as carriers for intracellular protein delivery. *J. Biol. Chem.* **276**, 5836–5840.
36. Schwarze, S. R., Ho, A., Vocero-Akbani, A., and Dowdy, S. F. (1999) In vivo protein transduction: delivery of a biologically active protein into the mouse. *Science* **285**,1569–1572.
37. Morris, M. C., Vidal, P., Chaloin, L., Heitz, F., and Divita, G. (1997) A new peptide vector for efficient delivery of oligonucleotides into mammalian cells. *Nucleic Acids Res.* **25**, 2730–2736.
38. Morris, M. C., Chaloin, L., Mery, J., Heitz, F., and Divita, G. (1999) A novel potent strategy for gene delivery using a single peptide vector as a carrier. *Nucleic Acids Res.* **27**, 3510–3517.
39. Kalderon, D., Roberts, B. L., Richardson, W. D., and Smith, A. E. (1984) A short amino acid sequence able to specify nuclear location. *Cell* **39**, 499–509.
40. Kalderon, D. and Smith, A. E. (1984) In vitro mutagenesis of a putative DNA binding domain of SV40 large-T. *Virology* **139**, 109–137.

41. Robbins, J., Dilworth, S. M., Laskey, R. A., and Dingwall, C. (1991) Two interdependent basic domains in nucleoplasmin nuclear targeting sequence: identification of a class of bipartite nuclear targeting sequence. *Cell* **64**, 615–623.
42. Makkerh, J. P., Dingwall, C., and Laskey, R. A. (1996) Comparative mutagenesis of nuclear localization signals reveals the importance of neutral and acidic amino acids. *Curr. Biol.* **6**, 1025–1027.
43. Pooga, M., Lindgren, M., Hallbrink, M., Brakenhielm, E., and Langel, U. (1998) Galanin-based peptides, galparan and transportan, with receptor-dependent and independent activities. *Ann. NY Acad. Sci.* **863**, 450–453.
44. Subramanian, A., Ranganathan, P., and Diamond, S. L. (1999) Nuclear targeting peptide scaffolds for lipofection of nondividing mammalian cells. *Nat. Biotechnol.* **17**, 873–877.
45. Bogerd, H. P., Benson, R. E., Truant, R., Herold, A., Phingbodhipakkiya, M., and Cullen, B. R. (1999) Definition of a consensus transportin-specific nucleocytoplasmic transport signal. *J. Biol. Chem.* **274**, 9771–9777.
46. Zanta, M. A., Belguise-Valladier, P., and Behr, J. P. (1999) Gene delivery: a single nuclear localization signal peptide is sufficient to carry DNA to the cell nucleus. *Proc. Natl. Acad. Sci. USA* **96**, 91–96.
47. Sazani, P., Kang, S. H., Maier, M. A., Wei, C., Dillman, J., Summerton, J., et al. (2001) Nuclear antisense effects of neutral, anionic and cationic oligonucleotide analogs. *Nucleic Acids Res.* **29**, 3965–3974.
48. Ciolina, C., Byk, G., Blanche, F., Thuillier, V., Scherman, D., and Wils, P. (1999) Coupling of nuclear localization signals to plasmid DNA and specific interaction of the conjugates with importin alpha. *Bioconjug. Chem.* **10**, 49–55.
49. Chen, Q. R., Zhang, L., Luther, P. W., and Mixson, A. J. (2002) Optimal transfection with the HK polymer depends on its degree of branching and the pH of endocytic vesicles. *Nucleic Acids Res.* **30**, 1338–1345.
50. Chang, C. D. and Meienhofer, J. (1978) Solid phase peptide synthesis using mild base cleavage of N-α-fluonenylmethyloxy-carbonyl amino acids, exemplified by a synthesis of dihydrosomatostatin. *Int. J. Pept. Protein Res.* **11**, 246–249.
51. Atherton, E., Fox, H., Harkiss, D., Logan, C. J., Sheppard, R. C., and Williams, B. J. (1978) A mild procedure for solid phase peptide synthesis: use of fluorenylmethoxycarbonylamine acids. *J. Chem. Soc. Chem. Commun.* **18**, 537–539.
52. King, D. S., Fields, C. G., and Fields, G. B. (1990) A cleavage method which minimizes side reactions following Fmoc solid phase peptide synthesis. *Int. J. Peptide Prot. Res.* **36**, 255–266.
53. Robey, F. A. (1994) Starting material for cyclic peptides, peptomers, and peptide conjugates. *Methods Mol. Biol.* **35**, 73–90.
54. Patchornik, A., Berger, A., and Katchalski, E. (1957) Poly-L-histidine. *J. Am. Chem. Soc.* **79**, 5227–5236.
55. Norland, K. S., Gasman, G. D., Katchalsky, E., and Blout, E. R. (1963) Some optical properties of poly-l-bennyl-L-histidine and poly-L-histidine. *Biopolymers* **1**, 277–294.

56. Lynn, D. M., Anderson, D. G., Putnam, D., and Langer, R. (2001) Accelerated discovery of synthetic transfection vectors: parallel synthesis and screening of a degradable polymer library. *J. Am. Chem. Soc.* **123**, 8155–8156.
57. Pichon, C., Roufai, M. B., Monsigny, M., and Midoux, P. (2000) Histidylated oligolysines increase the transmembrane passage and the biological activity of antisense oligonucleotides. *Nucleic Acids Res.* **28**, 504–512.
58. Zuckermann, R., Corey, D., and Schultz, P. (1987) Efficient methods for attachment of thiol specific probes to the 3'-ends of synthetic oligodeoxyribonucleotides. *Nucleic Acids Res.* **15**, 5305–5321.
59. Vives, E. and Lebleu, B. (1997) Selective coupling of a highly basic peptide to an oligonucleotide. *Tetrahedron Lett.* **38**, 1183–1186.
60. Stetsenko, D. A. and Gait, M. J. (2000) Efficient conjugation of peptides to oligonucleotides by "native ligation." *J. Org. Chem.* **65**, 4900–4908.
61. Stetsenko, D. A. and Gait, M. J. (2000) New phosphoramidite reagents for the synthesis of oligonucloetides containing a cysteine residue useful in peptide conjugiaton. *Nucl. Nucl. Acids* **19**, 1751–1764.
62. Li, S., Tseng, W. C., Stolz, D. B., Wu, S. P., Watkins, S. C., and Huang, L. (1999) Dynamic changes in the characteristics of cationic lipidic vectors after exposure to mouse serum: implications for intravenous lipofection. *Gene Ther.* **6**, 585–594.

4

DNA Delivery to Cells in Culture Using PNA Clamps

Todd D. Giorgio and Shelby K. Wyatt

1. Introduction

Peptide nucleic acid (PNA) is a DNA mimic in which the deoxyribose phosphate backbone has been replaced by N-(2-aminoethyl) glycine linkages. PNAs, first described by Nielsen et al. in 1991 (*1*), possess a number of useful properties including rapid and high-affinity binding to DNA, RNA, and PNA, resistance to degradation by nucleases and proteases and poor affinity for proteins that normally bind nucleic acids. For these reasons and others, PNA has been proposed for use as a therapeutic agent in controlling gene expression through either antisense or antigene activity.

The chemical properties of PNA oligomers that confer interesting bioactivity also limit transmembrane transport, resulting in negligible cell internalization (*2,3*). In addition, PNA lacks the polyanionic charges necessary for condensation and complexation with cationic liposomes through electrostatic interactions. PNA–DNA hybrids, however, possess a distributed negative charge contributed from the DNA. Condensed particles can be formed from the interaction of PNA–DNA hybrids with cationic lipids and these lipoplexes are rapidly incorporated into mammalian cells in culture.

The ratio of PNA to DNA in the hybrid modulates lipoplex formation with cationic lipids. Many interesting bioactive structures are composed primarily from DNA, with new functionality contributed by a relatively small amount of hybridized PNA.

PNA hybridization affinity with complementary DNA is much stronger than the affinity of the corresponding DNA–DNA hybrid. A "PNA clamp" is

From: *Methods in Molecular Biology, vol. 245:*
Gene Delivery to Mammalian Cells: Vol. 1: Nonviral Gene Transfer Techniques
Edited by: W. C. Heiser © Humana Press Inc., Totowa, NJ

an oligomer that has two identical PNA sequences joined by a flexible hairpin linker. When a PNA clamp is mixed with a complementary DNA target sequence, a highly stable PNA–DNA–PNA triplex hybrid is formed. Double-stranded DNA also forms the PNA–DNA–PNA triplex upon addition of a complimentary PNA clamp with the displacement of one DNA strand *(4)*. The resulting complex can be primarily double-stranded DNA (including circular plasmid DNA) that retains full biological function with little perturbation of native structure and conformation outside the PNA-clamped region.

Chemical modification of the PNA clamp allows a variety of characteristics to be imparted to the target DNA in an essentially irreversible fashion. Fluorescently labeled PNA clamps (fluorescein, rhodamine, Cy3) can be triplexed with plasmid DNA coding for reporter genes to allow direct, simultaneous measurements of gene delivery, intracellular localization, and gene expression inside live cells. PNA clamps possessing common ligation chemistries (biotin, maleimide, 3-(2-pyridyldithio)propionic acid *N*-hydroxysuccinimide ester [SPDP]) are available, making possible a wide range of functional PNA-clamped DNA hybrids. A transferrin-PNA clamp, for example, was shown to enhance receptor-mediated endocytotic delivery of the triplex-associated plasmid *(5)*. Such hybrids may, in principle, be designed to increase nuclear uptake *(6)*, facilitate endosomal escape, target gene delivery to the cell surface or to intracellular receptors, or provide artificial transcription activators.

The physical properties of DNA are not significantly modified by triplex hybridization with a PNA clamp. Intracellular delivery of the PNA-clamped DNA has been accomplished by the same nonviral methods used for DNA transfection including microinjection *(7)*, cell permeabilization *(7)*, cationic lipofection *(8–12)*, and cationic polymer association *(5,13,14)*. Other nonviral delivery methods (naked DNA and ballistic impingement) are presumably as effective with DNA–PNA complexes as with the corresponding DNA without a PNA clamp (although no literature evidence for DNA–PNA delivery using these other methods exists).

The procedure described in this chapter will allow an investigator to intracellularly deliver PNA-clamped plasmid DNA. This transfection can be monitored for both plasmid delivery and transgene expression by the fluorescence intensity of rhodamine (attached to the PNA clamp) and green fluorescent protein (GFP; the expressible reporter gene), respectively.

2. Material

2.1. PNA Clamps and PNA Clamp/Plasmid DNA Triplex Conjugates

1. Rhodamine-labeled PNA clamp triplex conjugated to plasmid DNA expressing GFP (Gene Therapy Systems, San Diego, CA; Cat. no. G101040) (*see* **Note 1**).

2.2. Nonviral Delivery System

1. DOTAP (1,2-dioleoyloxy-3-trimethylammonium-propane) (Avanti Polar Lipids, Alabaster, AL; Cat. no. 890890).
2. DOPE (dioleoyl phosphatidylethanolamine) (Avanti Polar Lipids; Cat. no. 850725) (*see* **Notes 2** and **3**).

2.3. Solutions and Culture Medium

1. Dulbecco's Modified Eagle's Medium (DMEM) Auto-Mod (Sigma, St. Louis, MO; Cat. no. D5280), supplemented with 10% (v/v) calf serum (Invitrogen, Carlsbad, CA; Cat. no. 16170-078), 1% L-glutamine 200 m*M* (Invitrogen; Cat. no. 25030-081), and 1% antibiotic-antimycotic (penicillin-streptomycin-amphotericin) (Invitrogen; Cat. no. 15240-062).
2. Serum-free Opti-Mem I (Invitrogen; Cat. no. 18292-011).
3. Sterile, distilled water.
4. Calcium- and magnesium-free phosphate-buffered saline (CMF-PBS) with ethylenediamine tetra-acetic acid (EDTA).
   a. Dissolve 1.14 g of sodium phosphate dibasic, anhydrous in 300 mL of sterile, distilled water.
   b. Adjust the pH to 7.4 with either 1 *M* HCl or 1 *M* NaOH.
   c. Dissolve 8.00 g of sodium chloride and 0.20 g of potassium chloride in 600 mL of sterile, distilled water.
   d. Dissolve 0.20 g of disodium EDTA in 50 mL of sterile, distilled water.
   e. Combine the three solutions.
   f. Adjust the total volume to 1000 mL with sterile, distilled water.
   g. Adjust the pH to 7.4 with 1 *M* NaOH.
5. Cell Scrub (Gene Therapy Systems; Cat. no. B100001).
6. Trypsin-EDTA, 1X (Invitrogen; Cat. no. 25200-056).

2.4. Cells

1. HeLa cells (human epithelioid, ATCC, Manassas, VA; Cat. no. CCL-2).

2.5. Equipment

1. Lipex liposome extruder (Northern Lipids, Inc., Vancouver, BC, Canada).
2. Rotary vacuum evaporator (Buchler Instruments / Labconco; Kansas City, MO; VV-micro).
3. Flow cytometer (Becton-Dickinson, San Jose, CA; FACScan).

2.6. Laboratory Supplies

1. Sterile 35-mm cell-culture dishes (BD Falcon, San Jose, CA; Cat. no. 108).
2. Chloroform (Fisher Scientific, Atlanta, GA; Cat. no. C574-1).

3. Polycarbonate microporous membrane (Poretics, Livermore, CA; Cat. no. 11050 [0.8-μm pore diameter], Cat. no. 11054 [1.0 μm pore diameter]).
4. Polyester support drain disc (Poretics; Cat. no. 87483).
5. Eppendorf centrifuge tubes (Fisher Scientific; Cat. no. 05-402 [1.5 mL]).
6. Cytometry tubes and caps (Elkay UK, Basingstoke, Hampshire, UK; Cat. no. 000-2052-BX1).
7. Formaldehyde, EM-grade, 16% solution (Electron Microscopy Sciences, Fort Washington, PA; Cat. no. 15710).
8. Microporous 70-μm filter mesh (Spectrum Laboratories, Rancho Dominguez, CA; Cat. no. 146490).

3. Methods

3.1. Preparation of Cells

1. Subculture HeLa cells (*see* **Note 4**) onto sterile 35-mm cell-culture dishes (*see* **Note 5**) at 2×10^4 cells/cm^2.
2. Incubate at 37°C in an environment containing 5% CO_2 and 95% relative humidity until 80% confluent (*see* **Note 6**).

3.2. Preparation of Liposomes

1. Dissolve 20 mg of an equimolar mixture of DOTAP and DOPE in a 100-mL round-bottom flask containing 5 mL of chloroform.
2. Attach the flask to a rotary vacuum evaporator and evaporate under vacuum (500 mm Hg) at 40°C at 10 rpm until no trace of the chloroform remains. This process typically requires less than 60 min.
3. Add 10 mL of CMF-PBS to the flask and resume rotation at 10 rpm and 40°C until the thin lipid film is completely hydrated. This process typically requires 60–120 min.
4. Place two stacked microporous filters with pore diameter of 1.0 μm in the Lipex liposome extruder above a polyester support drain disk and rinse with 5 mL of CMF-PBS extruded with nitrogen gas. Perfuse the extrusion barrel with 30°C water and adjust the nitrogen gas pressure to provide a CMF-PBS flow rate of approx 1 mL/min.
5. Introduce the multilamellar vesicle suspension from the rotary vacuum evaporator flask into the extrusion apparatus and extrude under the same conditions as for CMF-PBS extrusion. Recycle the product into the extruder, completing a total of five extrusion cycles.
6. Replace the 1.0-μm pore diameter filters with two 0.8-μm pore diameter filters stacked above a new polyester support drain disk.
7. Rinse the extrusion apparatus with 5 mL of CMF-PBS as before.
8. Introduce the vesicle suspension into the extrusion apparatus and extrude under the same conditions as previously. Recycle the product into the extruder, completing a total of five extrusion cycles.

9. Store the resulting unilamellar vesicle suspension, 10 mL total volume at an approximate concentration of 2 mg/mL, at 4°C in preparation for lipoplex formation.

3.3. Preparation of Lipoplex

1. Add 1.25 µg of PNA-clamped plasmid to 100 µL of serum-free Opti-Mem I at 37°C in a 1.5-mL conical Eppendorf centrifuge tube.
2. Add 6.25 µL of the unilamellar vesicle suspension (equivalent to 12.5 µg of lipid) to the Opti-Mem I containing the PNA-clamped plasmid, invert several times to mix and incubate at 37°C for 15–30 min (*see* **Notes 7–9**). This is the lipoplex suspension.

3.4. Transfection

1. Wash each 35-mm dish of cells with 1 mL of serum-free DMEM at 37°C.
2. Add 0.5 mL of fresh Opti-Mem I to each 35-mm dish of cells at 37°C.
3. Add 400 µL of fresh Opti-Mem I at 37°C to the lipoplex suspension and invert twice to mix.
4. Apply the diluted lipoplex suspension drop-wise to the cells (*see* **Note 10**).
5. Maintain the cells under standard culture conditions for 12 h in the presence of the lipoplex.
6. After 12 h, aspirate the lipoplex medium and replace with complete DMEM cell medium without lipoplex.
7. Maintain the cells under standard culture conditions for an additional 12 h in the absence of the lipoplex.

3.5. Recovery of Cells

1. Aspirate the cell culture medium and wash cells with 1 mL of CMF-PBS.
2. Aspirate the CMF-PBS and replace with 1 mL of Cell Scrub and incubate for 15 min at room temperature (*see* **Note 11**).
3. Aspirate the Cell Scrub and wash cells twice with 1 mL of CMF-PBS.
4. Aspirate CMF-PBS and replace with 1 mL of trypsin-EDTA.
5. Monitor cell detachment from the culture surface by phase microscope observation (*see* **Note 12**).
6. Immediately transfer the detached cell suspension into labeled cytometer tubes containing 1 mL of DMEM + 10% calf serum (*see* **Note 13**).
7. Add 125 µL of 16 % formaldehyde to the cell suspension and incubate for 5 min at room temperature (*see* **Note 14**).
8. Pellet cells by centrifugation at 850*g* for 5 min.
9. Aspirate and discard the cell-free supernatant and replace with CMF-PBS containing 1% formaldehyde.

10. Resuspend the cells by dragging the capped cytometry tube across the top of a wire centrifuge tube rack (20 cm long) 10 times with light to moderate force.
11. Pellet cells by centrifugation at 850*g* for 5 min.
12. Aspirate cell-free supernate and replace with CMF-PBS containing 1% formaldehyde.
13. Resuspend cells by dragging the capped cytometry tube across the top of a wire centrifuge tube rack (20 cm long) 10 times with light to moderate force.
14. Filter cells through 70 μm mesh (*see* **Note 15**).
15. Maintain washed and filtered cell suspension at 4°C until analysis (*see* **Note 16**).

3.6. Flow Cytometric Measurement of Cells Transfected With PNA-Clamped Plasmid

1. Analyze cell samples using a flow cytometer that has an argon laser exciting at a wavelength of 488 nm.
2. Identify the intact cell population in the forward light scatter (FLS) intensity vs side light scatter (SLS) intensity measurement space.
3. Collect FLS intensity, SLS intensity, and fluorescence emission intensities centered at 530 nm (FL1), 580 nm (FL2), and 610 nm (FL3) from each of at least 5000 intact cell events.

3.7. Analysis of Flow Cytometry Data: Estimate the Fraction of Transgene-Containing or Transgene-Expressing Cells

1. Identify the fluorescence emission wavelength associated with the desired parameter.
   a. GFP emission (transgene expression) is measured in the FL1 channel.
   b. Rhodamine emission (transgene delivery) is measured in the FL2 channel.
2. Identify the flow cytometry channel number that contains the 99th percentile of control (untreated) intact cell events (*see* **Note 17**).
3. Identify the negative percentile in the same channel number for the treated event population.
4. Calculate the percent positive treated events as: 100−negative percentile (*see* **Note 18**).

3.8. Analysis of Flow Cytometry Data: Simultaneous Measurement of Transgene Delivery and Transgene Expression

1. Identify the fluorescence-emission wavelength associated with the desired parameter.

    a. GFP emission (transgene expression) is measured in the FL1 channel.

    b. Rhodamine emission (transgene delivery) is measured in the FL2 channel.

2. Plot transgene-expression fluorescence intensity vs transgene-delivery fluorescence intensity (*see* **Notes 19–21**).

4. Notes

1. To form an effective clamp, the PNA sequence must be composed of two complementary sequences separated by a flexible linker and must also be complementary to the target DNA. Custom PNA sequences can be synthesized to conform to a specific DNA application (Applied Biosystems, Foster City, CA) but a flexible and modular PNA clamp system is available (Gene Therapy Systems, San Diego, CA) without the need to design and synthesize PNA fragments or targets. One member of that modular PNA-clamped plasmid DNA family is used here. This plasmid has an expressible GFP region and rhodamine labels associated with the PNA clamp.

2. Naked PNA or PNA clamp is essentially cell impermeant, but the PNA/plasmid DNA triplex conjugate can be delivered to cells using standard nonviral methods. The PNA–DNA hybrid described here is predominately composed of DNA (5757 base pairs) and forms lipoplexes with cationic lipids in essentially the same way and under the same conditions as the pure DNA plasmid (*see* Chapter 6). The PNA-clamped plasmid DNA used here contains 10 PNA clamp sites supporting 5 to 8 PNA clamps per plasmid. PNA clamps, including the attached rhodamine labels, increase the molecular weight of the plasmid DNA by approx 1%, insufficient to significantly modulate lipoplex formation. PNA oligomers alone are not polyanionic and cannot be formulated into effective transfection complexes with cationic delivery vehicles. Other methods have been demonstrated to provide cell-penetrating capability to PNA oligomers, including PNA coupled to peptides *(15,16)*, electroporation *(17)*, and microinjection *(7)*.

3. Substitution of a commercially prepared DOTAP/DOPE liposome suspension (Lipofectin® Reagent [Invitrogen, Carlsbad, CA; 18292-011]) yields equivalent results when the final lipid mass per unit volume in the transfection suspension is retained.

4. HeLa cells (human epitheliod, ATCC, Manassas, VA; CCL-2) and CV1 cells (monkey fibroblast, ATCC, Manassas, VA; CCL-70) have been used with this protocol. Cell type strongly modulates the extent of transgene expression, but most transformed, actively mitotic mammalian cell types transfected by this method can yield detectable transgene expression.

5. Any sterile substrate that promotes cell attachment and growth can be used. The minimum cell-culture surface area for robust and convenient analysis

of gene delivery and expression is approx 10 cm², yielding $1.0–2.5 \times 10^6$ cells at 80% confluence. One sample per well of a standard 6-well cell culture plate is appropriate for many applications.

6. Cells approaching 100% confluence typically yield reduced levels of transgene expression following lipoplex transfection. HeLa cells achieve confluence after approx 48–72 h in culture following standard subculture. The potential for cell toxicity is increased in samples with low confluence.

7. Mixing intensity applied during lipoplex formation can modulate transgene delivery and expression. Mix thoroughly and reproducibly by gentle inversion 10 times. Equivalent performance is obtained by dragging the capped tube once across the top of a wire centrifuge tube rack (20 cm long).

8. Transgene delivery and expression are sensitive functions of cell type and lipoplex composition. A comprehensive optimization protocol should be conducted to obtain the most effective transfection parameters for any particular system. The key variables include lipid/DNA ratio and lipoplex concentration. Optimized values of these parameters depend on lipid composition, plasmid characteristics, cell type, and incubation duration. Methods to optimize cationic liposome delivery to mammalian cells in vitro are detailed in Chapter 6.

9. Opti-Mem I containing the PNA-clamped plasmid is a light pink, water-clear solution. The unilamellar vesicle suspension is a milky white suspension. Addition of the small volume unilamellar vesicle suspension yields an immediate and obvious visual change in the large volume Opti-Mem I/PNA-clamped DNA solution from water-clear to a milky opaque suspension. This transition represents the successful interaction of lipid and DNA to form lipoplex. Solutions that fail to demonstrate this optical transition imply a failure in the preparation of one or more of the constituent components. The final lipoplex suspension will separate over the course of 15–60 min into multiple regions of a more condensed, milky white phase surrounded by a water-clear phase. This event presumably represents a secondary lipoplex transition, but the absence or presence of this event does not modulate transfection.

10. Gentle mixing of the lipoplex suspension by inversion just before addition to the cell culture ensures uniform delivery of lipoplex to cells. A gentle swirl (5 s) of the lipoplex-containing Opti-Mem I after addition to 35 mm dish before incubation is also recommended to ensure uniform lipoplex distribution.

11. Cell Scrub washing removes extracellular, cell-associated plasmid that might compromise subsequent measurements of intracellular plasmid concentration.

12. Continuous swirling motion of the 35 mm dish promotes cell detachment. If flat-sided cell culture vessels are used (such as T-flasks) a sharp impact of the T-flask against a solid object can also facilitate cell detachment.

13. Prolonged contact with the trypsin-EDTA will damage cells. Proteins in the calf serum quench the trypsin activity.

14. Subsequent analysis of transgene delivery and expression in this system does not require live cells, allowing the use of formaldehyde as a fixative. Detection of some transgene expression products (such as luciferase) depends on protein activity potentially compromised by the addition of any fixative, including formaldehyde. In these cases, the cell suspension is typically washed to remove calf serum proteins and trypsin and lysed by mechanical or osmotic methods. Cell lysis provides a homogeneous solution with unhindered association between transgenic protein and substrate. Lysed samples, however, do not allow measurement of transgene delivery or expression in individual cells.

15. Filtration removes cell aggregates incompatible with flow cytometric evaluation. This step may be omitted if subsequent analysis methods are not influenced by cell aggregates.

16. Sample quality deteriorates with time in storage. Analysis of transgene delivery and expression in intact cells are best performed within 24 h of preparation.

17. The 99th percentile criteria can be interpreted as allowing 1% of the untreated control event to be characterized as "positive." Other criteria can be selected to distinguish between "negative" and "positive" events, but the 99th percentile method is reproducible, comparable among all measured flow cytometry parameters, and reasonably sensitive.

18. HeLa cells transfected with this protocol were measured after 24 h (12 h transfection followed by 12 h of incubation in the absence of lipoplex) for intracellular plasmid delivery and transgene expression. The treated cells were exposed to lipoplex prepared from rhodamine-labeled PNA clamp triplex conjugated to plasmid DNA that includes an expressible GFP region. The control cells were exposed to lipoplex prepared from unlabeled plasmid lacking GFP expression, but having the same length as the labeled plasmid. The 99th percentile of 5000 control events includes 4950 events. The integrated control population in FL1 (GFP intensity as an index of transgene expression) yielded the 4950th event in flow cytometer channel number 60 (of 256 total channels). The integrated treated population through channel number 60 of FL1 included 3101 events (of 4965 total events). Thus, there are 37.4% positive transgene expression events in the treated sample: $(100 - [3101/4950 \times 100])$. Note that there are also 1% positive transgene expression events in the control sample using the 99th per-

centile convention. The same analysis in FL2 (rhodamine fluorescence intensity as an index of transgene delivery) yielded the 4950th event in channel 65. The integrated treated population through channel number 65 of FL2 included 3418 events (of 4965 total events). Thus, there are 30.9% positive transgene delivery events in the treated sample: $100 - [(3,418/4,950) \times 100]$. Note that there are also 1% positive transgene delivery events in the control sample using the 99th percentile convention. Note that these uncalibrated results suggest that there are more cells expressing transgene (37.4%) than cells possessing intracellular PNA-

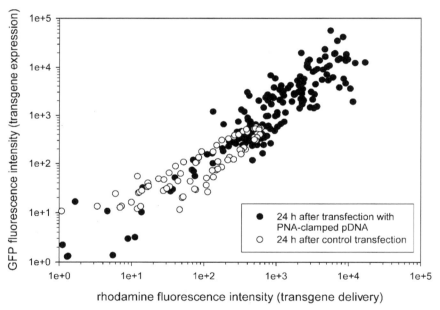

rhodamine fluorescence intensity (transgene delivery)

Fig. 1. Analysis of transgene expression as a function of transgene delivery in transfected HeLa cell populations. The treated cells (•) were exposed to lipoplex prepared from rhodamine-labeled PNA clamp triplex conjugated to plasmid DNA expressing GFP. The control cells (○) were exposed to lipoplex prepared from unlabeled plasmid lacking GFP expression, but having the same length as the labeled plasmid. The control cell population exhibits autofluorescence in both the rhodamine and GFP wavelengths consistent with unlabeled cells. The treated cell population contains significant intracellular plasmid and also demonstrates significant transgene expression as interpreted from the elevated levels of both rhodamine and GFP fluorescence relative to untreated cells. This measurement takes advantage of the multifunctional capabilities of PNA-clamped plasmids.

clamped plasmid DNA (30.9%)! This apparent impossibility is corrected in the calibrated results as described in **Note 19**. The lower limit of fluorescence intensity detection also must be considered in the interpretation of results from multiple fluors having different fluorescence characteristics.

19. All flow cytometric fluorescence intensity channels should be quantitatively calibrated in terms of the fundamental quantity of corresponding fluor. For example, the FL1 channel should be calibrated using microspheres having known GFP concentrations. In addition, the calibrations should be compensated for fluorescence intensity contributions from other simultaneously measured fluors. For example, the contribution of GFP fluorescence measured in FL2 should be subtracted from the measured FL2 fluorescence intensity to obtain an accurate measure of rhodamine fluorescence intensity. The details of such calibrations and compensations are beyond the scope of this protocol *(18–20)*.

20. HeLa cells transfected as described in **Note 18** were analyzed by flow cytometry. The real power of PNA-clamped plasmid is evident in the simultaneous, single-cell measurement of both transgene delivery (from rhodamine fluorescence associated with the PNA clamp) and transgene expression (from GFP fluorescence associated with the plasmid) shown as Fig. 1.

21. Cell samples were analyzed using a FACScan (Becton-Dickinson, San Jose, CA) flow cytometer. Winlist 3D (Verity Software House, Topsham, ME; version 4.0) software was employed to analyze the list-mode data saved from the flow cytometer.

Acknowledgments

The authors recognize Molly James for collecting experimental data that appears as Fig. 1. The methods of Srinivasa Chakravarthy and Molly James also appear in this work.

References

1. Nielsen, P. E., Egholm, E., Berg, R. H., and Buchardt, O. (1991) Sequence selective recognition of DNA by strand displacement with a thymine-substituted polyamide. *Science* **254**, 1497–1500.
2. Scarfì, S., Giovine, M., Gasparini, A., Damonte, G., Millo, E., Pozzolini, M., and Benatti, U. (1999) Modified peptide nucleic acids are internalized in mouse macrophages RAW 264.7 and inhibit inducible nitric oxide synthase. *FEBS Lett.* **451**, 264–268.
3. Pardridge, W. M., Boado, R. J., and Kang, Y. S. (1995) Vector-mediated delivery of a polyamide "peptide" nucleic acid analogue through the blood-brain barrier in vivo. *Proc. Natl. Acad. Sci. USA* **92**, 5592–5596.

4. Nielsen, P. E. (2000) Peptide nucleic acids: on the road to new gene therapeutic drugs. *Pharmacol. Toxicol.* **86**, 3–7.

5. Liang, K. W., Hoffman, E. P., and Huang, L. (2000) Targeted delivery of plasmid DNA to myogenic cells via transferrin-conjugated peptide nucleic acid. *Mol. Ther.* **1**, 236–243.

6. Brandén, L. J., Christensson, B., and Smith, C. I. E. (2001) In vivo nuclear delivery of oligonucleotides via hybridizing bifunctional peptides. *Gene Ther.* **8**, 84–87.

7. Wilson, G. L., Dean, B. S., Wang, G., and Dean, D. A. (1999) Nuclear import of plasmid DNA in digitonin-permeabilized cells required both cytoplasmic factors and specific DNA sequences. *J. Biol. Chem.* **274**, 22025–22032.

8. Zelphati, O., Liang, X., Hobart, P., and Felgner, P. L. (1999) Gene chemistry: functionally and conformationally intact fluorescent plasmid DNA. *Human Gene Ther.* **10**, 15–24.

9. Nastruzzi, C., Cortesi, R., Esposito, E., Gambari, R., Borgatti, M., Bianchi, N., et al. (2000) Liposomes as carriers for DNA-PNA hybrids. *J. Control. Rel.* **68**, 237–249.

10. Doyle, D. F., Braasch, D. A., Simmons, C. G., Janowski, B. A., and Corey, D. R. (2001) Inhibition of gene expression inside cells by peptide nucleic acids: effects of mRNA target sequence, mismatched bases and PNA length. *Biochem.* **40**, 53–64.

11. Norton, J. C., Piatyszek, M. A., Wright, W. E., Shay, J. W., and Corey, D. R. (1996) Inhibition of human telomerase activity by peptide nucleic acids. *Nat. Biotech.* **14**, 615–619.

12. Herbert, B., Pitts, A. E., Baker, S. I., Hamilton, S. E., Wright, W. E., Shay, J. W., and Corey, D. R. (1999) Inhibition of human telomerase in immortal human cells leads to progressive telomere shortening and cell death. *Proc. Natl. Acad. Sci. USA* **96**, 14267–14281.

13. Godbey, W. T., Wu, K. K., and Mikos, A. G. (1999) Tracking the intracellular path of poly(ethylenimine)/DNA complexes for gene delivery. *Proc. Natl. Acad. Sci. USA* **96**, 5177–5181.

14. Wightman, L., Kircheis, R., Rössler, V., Carotta, S., Ruzicka, R., Kursa, M., and Wagner, E. (2001) Different behavior of branched and linear polyethylenimine for gene delivery in vitro and in vivo. *J. Gene. Med.* **3**, 362–372.

15. Pooga, M., Soomets, U., Hällbrink, M., Valkna, A., Saar, K., Rezaei, K., et al. (1998) Cell penetrating PNA constructs regulate galanin receptor levels and modify pain transmission in vivo. *Nat. Biotech.* **16**, 857–861.

16. Simmons, C. G., Pitts, A. E., Mayfield, L. D., Shay, J. W., and Corey, D. R. (1997) Synthesis and membrane permeability of PNA-peptide conjugates. *Bioorg. Med. Che. Lett.* **7**, 3001–3006.

17. Shammas, M. A., Simmons, C. G., Corey, D. R., and Shmookler-Reis, R. J. (1999) Telomerase inhibition by peptide nucleic acids reverses "immortality" of transformed human cells. *Oncogene* **18**, 6191–6200.

18. James, M. B. and Giorgio, T. D. (2000) Nuclear-associated plasmid, but not cell-associated plasmid, is correlated with transgene expression in cultured mammalian cells. *Mol. Ther.* **1**, 339–346.

19. Tseng, W. C., Haselton, F. R., and Giorgio, T. D. (1997) Transfection by cationic liposomes using simultaneous single cell measurements of plasmid delivery and transgene expression. *J. Biol. Chem.* **272**, 25641–25647.
20. Tseng, W., Purvis, N. B., Haselton, F. R., and Giorgio, T. D. (1996) Cationic liposomal delivery of plasmid to endothelial cells measured by quantitative flow cytometry. *Biotech. Bioeng.* **50**, 548–554.

5

Dendrimer-Mediated Cell Transfection In Vitro

James R. Baker, Jr., Anna U. Bielinska, and Jolanta F. Kukowska-Latallo

1. Introduction

In recent years, Starburst® PAMAM dendrimers, a class of polyamidoamine polymers, have become an interesting alternative vector for nonviral delivery of DNA in vitro and in vivo *(1–6)*. These nanoscopic polymers, characterized by regular dendritic branching, radial symmetry, and uniform size ranging from 4–11 nm, have excellent water solubility and biocompatibility in a broad range of concentrations. Nonmodified dendrimers have a high density of positively charged primary amino groups on the surface, which is essential for their interaction with counter-charged nucleic acids. Positively charged dendrimers bind DNA through electrostatic interactions with negatively charged phosphates on the DNA molecule *(1)*. The consistent and predictable formation of dendrimer-DNA complexes allows for the design of efficient DNA transfections into variety of eukaryotic cell lines and primary cells in vitro *(1,2,4,5,7)*. The positively charged dendrimer-DNA complexes facilitate transfer of DNA into a cell primarily through endocytosis. DNA in a complexed form is protected from nuclease activity while the majority of the DNA template remains transcriptionaly active *(8)*. The well-designed dendrimer-based transfection is characterized by the lack or minimal cytotoxicity, high transfection efficiency, and stability of complexed plasmid DNA and oligonucleotides *(8,9)*. PAMAM dendrimers can be used for gene delivery in vivo and ex vivo *(6,10–12)*.

The majority of chemical and structural differences in PAMAM dendrimers relate to the core molecule, either ammonia (NH_3) as a trivalent initiator core

From: *Methods in Molecular Biology, vol. 245:*
Gene Delivery to Mammalian Cells: Vol. 1: Nonviral Gene Transfer Techniques
Edited by: W. C. Heiser © Humana Press Inc., Totowa, NJ

or ethylenediamine (EDA) as a tetravalent initiator core. The core starts the stepwise polymerization process and dictates several structural characteristics of the molecule, including the overall shape, density, and surface charge. With each new layer or generation, the diameter of molecule increases approximately 1 nm, the molecular weight of the dendrimer more than doubles, and the number of surface amine groups exactly doubles. Such regularity of the dendrimer architecture provides for the convenient calculations of electrostatic charge ratios between dendrimers and nucleic acids for transfection. All types of nucleic acids, including plasmid DNA, short single-stranded and double-stranded oligonucleotides (ODNs), and RNA, can be used for transfection with dendrimers in vitro and in vivo *(9,13)*. The quality of DNA preparation (e.g., presence of low molecular weight nucleic acids and bacterial endotoxins) affects the efficiency of transfection and cytotoxicity, especially gene transfer in vivo *(6)*.

2. Materials

2.1. Dendrimers

A number of dendrimers are available commercially. Generations (G) 1–4 of Starburst (PAMAM) dendrimers are available from Sigma-Aldrich (Milwaukee, WI). G5, G7, and G9 of PAMAM dendrimers are available from Dendritech (Midland, MI). In the method presented here, we have used dendrimers with EDA as a tetravalent polymerization initiator core. G5, G7, and G9 EDA core dendrimers have molar masses of 28,826, 116,493, and 467,162 kDa and the number of surface charges (amine groups) of 128, 512, and 2048, respectively.

2.2. Plasmids

The reporter gene plasmids encoding luciferase, chloramphenicol acetyltransferase (CAT), β-galactosidase, or green fluorescent protein (GFP) can be purchased (Promega, Madison, WI; Invitrogen, Rockville, MD; Clontech, Palo Alto, CA) or constructed specifically for the project (*see* **Note 1**).

Prepare plasmid DNA by any standard method *(14)*. The preferable method for purification used in the dendrimer-based protocol is double CsCl-gradient centrifugation and dialysis into sterile Tris-EDTA buffer or water.

2.3. Cell Lines and Cells

Transfections of Rat2 (rat embryonal fibroblast), normal human foreskin fibroblast (NHFF1), COS-1 (monkey kidney SV40 transformed fibroblast-like), and Jurkat (human acute T-cell leukemia) cell lines are presented in this protocol (*see* **Note 2**).

For optimum results, transfections should be performed on actively dividing cells, preferably from cultures at the logarithmic phase of growth.

2.4. Cell Culture Medium and Solutions

The cell culture media should be used as recommended by the supplier of cells or ATCC. Cells in the presented method are maintained in complete growth medium, either Dulbecco's Modified Eagle's Medium (DMEM) or RPMI 1640 (Invitrogen) supplemented with 10% fetal bovine serum (FBS) (HyClone, Logan, UT) and 1% penicillin-streptomycin solution (Invitrogen) at 37°C in 5% CO_2.

Molecular biology-grade PBS buffer, pH 7.4 (Invitrogen) is used for cell washes. Luciferin substrate and luciferase reporter buffer to assay luciferase gene expression is available from Promega. CAT enzyme-linked immunosorbent assay (ELISA) kit to quantify CAT protein expression is available from Roche Diagnostics (Indianapolis, IN). The protein concentration of the sample is measured with BCA Protein Assay Reagent (Pierce, Rockford, IL).

2.5. Equipment

1. The chemiluminometer LB96P is used for measurement of light emission (EG & G/Berthold, Gaithersburg, MD).
2. For spectrophotometric measurements of protein and ELISA assays SPECTRA MAX 340 microtiter plate reader is used (Molecular Devices Corp., Sunnyvale, CA).
3. Flow cytometry analysis is conducted using a Becton-Dickinson FACScan and CellQuest software (Becton-Dickinson, San Jose, CA).
4. In situ cell analysis is performed using a Nikon fluorescent microscope (Eclipse TE 200).

3. Methods

3.1. Preparation of Plasmid DNA-Dendrimer Complex

Prepare dendrimer and DNA complex based on the calculated dendrimer to DNA charge ratio. Mix 0.65 µg of dendrimer with 1.0 µg of DNA to obtain charge ratio of 1. Form complexes at various charge ratios of 1, 5, 10, and 20 at room temperature for 15 min before adding into transfection medium (*see* **Note 3**).

3.2. Transfection Protocol

1. Plate adherent cells (e.g., Rat2, NHFF1, COS-1) 12–24 h prior to transfection so that they are 60–70% confluent at the time of DNA delivery. Generally, seeding at 2–4×10^4 cells/cm² of plate surface is optimal for transfection of the majority of cell lines, however, the optimum cell density depends on the cell type and size, the growth rate, and serum concentration in the growth medium. Inoculate suspension cell lines (e.g., Jurkat) with 5–10×10^5 cells/mL 18–24 h prior to transfection from a culture that is in mid-logarithmic phase of growth.

2. If adherent cells are grown in medium containing more than 10% serum, wash the cells once with phosphate-buffered saline (PBS) or serum-free medium. Aspirate the wash. Add an appropriate volume of serum-free medium to the plate (e.g., 2–3 mL/100 mm plate, 0.3–0.5 mL/well of a 6-well plate, 0.2–0.3 mL/well of a 24-well plate). For suspension cultures, pellet the cells and resuspend $1–5 \times 10^6$ cells/mL in serum-free medium (or medium containing 1% FBS) medium for transfection.

3. Prepare the dendrimer/DNA complex. For example, for 60 μL of dendrimer/DNA complex at a charge ratio of 10, mix 50 μL of plasmid DNA at 0.02 μg/μL (up to 0.08 μg/μL) with 10 μL of dendrimer at 0.65 μg/μL (up to 2.6 μg/μL) (*see* **Note 3**).

4. Add an aliquot of dendrimer/DNA complex to the cells. For example, use 10–25 μL of the complex for transfection of 24-well plates or 10–50 μL for transfection in 6-well plates. Generally, complex made in water should not constitute more than 10% of the total volume of serum-free transfection medium.

    a. Incubate cells with transfection mixture for 3–6 h at 37°C, 5% CO_2 (*see* **Note 4**).

    b. Remove transfection medium and wash once gently with serum-free medium (however, wash is not necessary when a low concentration of dendrimer is used and no augmenting agents are added).

    c. Add fresh, complete growth medium.

    d. Incubate for the required time (24, 36, 48, 72 h, etc.) before harvesting cells (*see* **Note 5**). Prepare cells for assay of reporter gene expression (*see* **Subheading 3.3.**).

3.3 Measurement of Expression

Assay transfected cells for expression of the introduced transgene.

1. Measure luciferase activity in a chemiluminescence assay. Remove the medium, wash the cells twice with PBS, and prepare the cells for assay. For example, add 50 μL or 100 μL of luciferase reporter buffer in a well of a 24-well or 6-well plate, respectively, to lyse the cells for 5 min, or follow the supplier's protocol. Incubate the cell extract (typically 10 μL) with 2.35×10^{-2} μmol of luciferin substrate injected automatically in a volume of 50 or 100 μL. Measure light emission in the luminometer for 10 s and adjust for the protein concentration of the samples.

2. Assay CAT in an ELISA assay. Prepare a cell extract following the manufacturer's protocol for the CAT ELISA assay kit. Centrifuge the cell lysate at 12,000*g* for 10–15 min at 4°C. Assay the amount of CAT in 10 to 50 μL of the supernatant in an ELISA assay. Measure the absorbance of the samples at 405 nm using a microtiter plate reader and adjust for the protein concentration of the samples.

3. Analyze cells expressing GFP using flow cytometry. Wash harvested cells twice with PBS then fix in 2% paraformaldehyde for 15 min. Determine the fluorescence of GFP on FACScan from at least 10,000 cells per sample and analyze using CellQuest software.
4. Assess the expression of GFP in situ by fluorescence microscopy. Observe the cells using an inverted fluorescent microscope at an excitation wavelength of 450–480 nm and an emission wavelength of 515 nm. Take photographs at 20× magnification.
5. Measure the protein concentration of the cell extract (typically 10 μL) using the BCA Protein Assay Reagent. Assay sample absorbance at 562 nm using a microtiter plate reader.

4. Notes

1. Dendrimer-mediated DNA transfection is a straightforward technique that can be applied to gene delivery in order to modify cellular genetic makeup (i.e., to generate a stably transfected cell line), in studies on regulation of gene expression, and owing to lack of immunogenicity, for gene transfer in vivo. Reporter plasmids are very useful for optimizing transfection conditions. They can also be used in co-transfection with the DNA of interest for tracing the efficiency and kinetics of gene expression.
2. There is no cell origin or cell type limitation on the use of dendrimers for in vitro transfections. In our extensive studies on the efficiency and mechanism of dendrimer-mediated transfection, we have transfected a broad variety of cells *(1,7)*.
   a. Established cell lines that have been efficiently transfected include CCD-37Lu (human normal lung fibroblast), A549 (human lung carcinoma epithelial-like), COS-7 (monkey kidney SV40 transformed fibroblast-like), Clone9 (rat normal liver epithelium), BHK-21 (hamster kidney fibroblast-like), MC7 (human breast cancer line), B/6 (mouse melanoma), RAW264 (mouse monocyte-macrophage), NIH 3T3 (mouse embryonal), HeLa (human cervical adenocarcinoma epithelium), U937 (human histiocytic lymphoma), MDCK (dog normal kidney epithelium), and P815 (mouse mastocytoma).
   b. Primary cell cultures, efficiently transfected with dendrimer/DNA complex, include normal human bronchial epithelium (NHBE) and small airway epithelium (SAEC) grown in serum-free SABM/SAGM (collection and medium from Clonetics Co., San Diego, CA). Also, successful and efficient transfection was achieved in primary cells from human and mouse thyrocytes, normal human skin keratinocytes (NHSK) and human mucosal keratinocytes (NHMK), either isolated from patients or from Clonetics Co. collection *(7,12)*. Generally, the optimal conditions of cell culture growth are similar for adherent and suspension cell lines.

3. Preparation of the dendrimer/DNA complex for in vitro transfection is
   based on the charge ratio of both components. Calculations use maximal
   theoretical electrostatic charge present on each component *(1,15)*. For example, the number of phosphate groups (i.e., negative charges) in DNA
   equals 1.71×10^{15} per μg and the number of positive charges in a dendrimer equals 2.65×10^{15} per μg. This calculation is independent of the
   molecular properties of both components, including the size and form of
   nucleic acid or generation of dendrimer. Therefore, to obtain a charge ratio
   of 1, mix 0.65 μg of dendrimer with 1.0 μg of DNA. It was determined that
   for practical reasons, dendrimer/DNA complexes should be formed in
   water at appropriate DNA and dendrimer concentrations (*see* **Subheading
   3.2.**). For consistency, complexes are formed at room temperature before

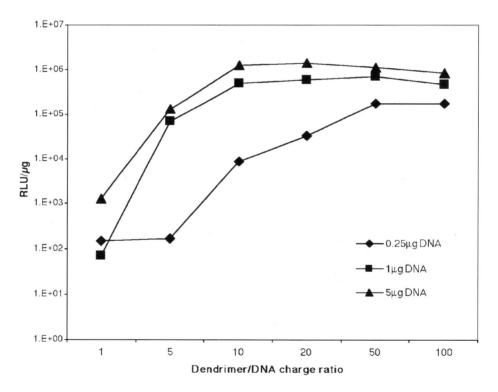

Fig. 1. The effect of DNA amount on dendrimer-mediated transfection in vitro. An
excess of positive charge is required for efficient transfection. The G9 EDA/DNA complexes for transfection of Rat2 embryonic fibroblasts were prepared with an increasing
amount of CMV-luciferase plasmid DNA (pCF1-Luc). The efficiency of transfection
was analyzed at a broad range (1–100) of dendrimer/DNA charge ratio. Luciferase activity (RLU) was measured 36 h after transfection and normalized per microgram of
protein in the cell lysate.

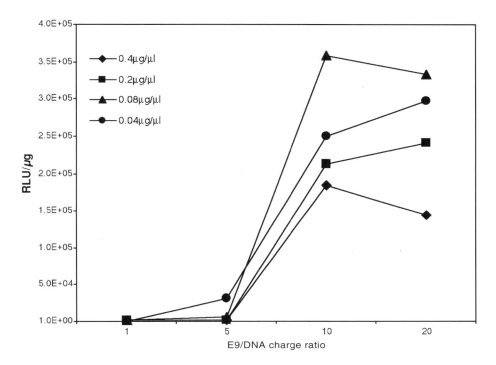

Fig. 2. The effect of DNA concentration during complex formation and dendrimer/DNA charge ratio on the efficiency of transfection. Rat2 cells were transfected with 1 μg of pCF1-Luc complexed with G9 EDA dendrimer. Complexes of equal amounts of DNA and the same dendrimer/DNA charge ratio differ in DNA concentration during complex formation. Data indicate that the optimal DNA concentration range from 0.04–0.08 μg/μL results in most efficient transfection.

adding to transfection medium. However, the dendrimer/DNA complex is stable for a prolonged time (up to few days) and can be used after remixing when precipitation occurs.

a. **Figure 1** illustrates the effect of increasing the dendrimer/DNA charge ratio on the efficiency of transfection. Most impressive improvement is obtained when a low amount (e.g., 0.25 μg per transfection) of DNA is used. With higher amounts of DNA (1–5 μg per transfection), the maximum efficiency is achieved at the relatively low charge ratio of 10. Maximal gene expression can be optimized for the amount of DNA available for delivery, the amount of dendrimer, and consequently dendrimer to DNA charge ratio. Generally, the lower the amount of DNA, the higher the dendrimer/DNA charge ratio (and *vice versa*) is required for optimum transfection.

b. In addition to charge ratio, the DNA concentration during complex formation significantly affects the efficiency of transfection (**Fig. 2**). Empir-

ical data indicate that complexes formed with DNA concentrations ranging from 20 to 80 µg/mL are most efficient. In such conditions, dendrimer/
DNA complexes formed at a broad range of charge ratios do not tend to ag-
gregate or precipitate and remain in suspension. This low-density complex
mediates the majority of transfections in vitro *(15)*. Soluble dendrimer formulations, obtained either in excess of dendrimer or DNA, also appear to
be more effective for in vivo transfection of solid tissues *(6,12)*.

  c. Another essential step in optimizing in vitro transfection is the choice of
generation (type) of dendrimer. As shown in **Fig. 3A**, the preferred dendrimer for transfection of Rat2 cells is G9 EDA core dendrimer. G5
EDA core dendrimer and to a lesser degree G7 EDA dendrimer are better for NHFF-1 cells (**Fig. 3B**). However, transfection obtained with the
intermediate G7 EDA dendrimer indicates that size preference may not
be an absolute property of either cells or dendrimer and can be affected
by other parameters of the complex formation (**Fig. 1** and **Fig. 2**).

4. The transfection can be carried out for 3–6 h. Additional time does not result in appreciable increase in transfection efficiency. After removal of
transfection solution, cells are washed once in serum-free medium. An appropriate amount of fresh complete medium (i.e., a 5- to 10-fold volume of
the transfection medium) is added to obtain cultures of approx $1–5 \times 10^5$
cells/mL. The single wash after removal of transfection medium is not necessary if no toxicity is observed during transfection or if cytotoxic augmentation reagents such as chloroquine or DEAE-dextran are not used. The
incubation time has to be assessed based on the results of cell expression
measured 24, 36, and 48 h after transfection.

  a. As with other polymer-based systems, dendrimer-mediated transfection
can be augmented with chloroquine and DEAE-dextran. A typical transfection experiment can be performed in serum-free medium augmented
with chloroquine at a final concentration of 50 µ*M* (the effective concentration varies from 10 to 100 µ*M* depending on cell type) or with
DEAE-dextran at a final concentration of 0.5 µ*M*. A combination of
chloroquine and DEAE-dextran at 50 µ*M* and 0.5 µ*M*, respectively, can
be used in the transfection medium. Chloroquine, a weak acidotropic
base, neutralizes the endosomal compartment and possibly enables endosomal escape of DNA. We have found that chloroquine alone does not
greatly improve dendrimer-mediated transfection, but in conjunction
with DEAE-dextran, the complex dispersing agent, the efficiency can be
enhanced up to two orders of magnitude (**Fig. 4A**). Unfortunately, both
of these agents are cytotoxic, requiring careful experimental adjustments, and cannot be applied in vivo.

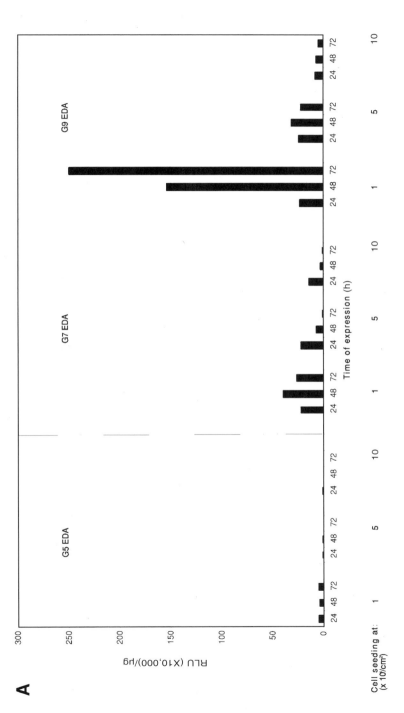

Fig. 3. Optimization of the transfection conditions for Rat2 (**A**) and NHFF1 (**B**) cells in vitro. Note that transfection conditions and the resulting kinetics of the transgene expression are cell-specific, including dendrimer requirements, cell density, and growth rate. Transfections were performed with 1 µg of pCF1-Luc DNA complexed with G5, G7, or G9 EDA dendrimers at a dendrimer/DNA charge ratio 10. Cells were seeded at 10^4, 5×10^4 and $10^5/cm^2$ 12 h before transfection, then incubated for 24, 48, and 72 h after transfection. For both cell lines, the highest luciferase expression was obtained in cells seeded at the lowest density. Optimal transfection in Rat2 cells (**A**) was achieved with G9 EDA dendrimer and in NHFF1 cells (**B**) with G5 EDA dendrimer. Kinetics of expression indicates that the maximum level of reporter protein production occurs 24–48 h after transfection. Relative light unit (RLU) is a measured light emission and is adjusted per µg of cell protein.

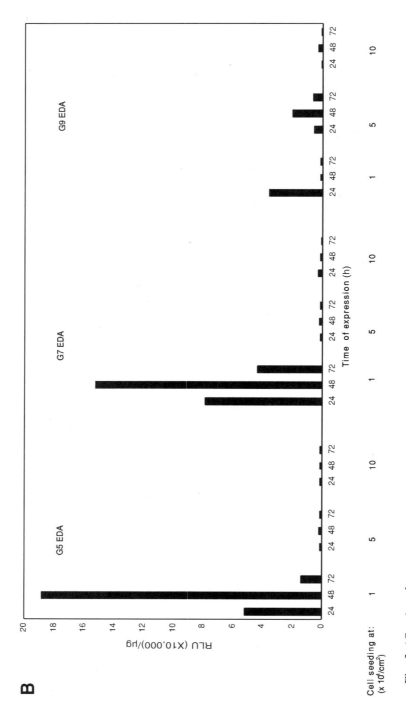

Fig. 3. (*Continued*)

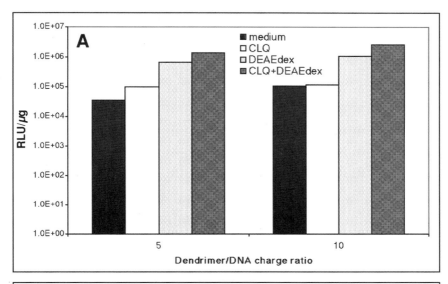

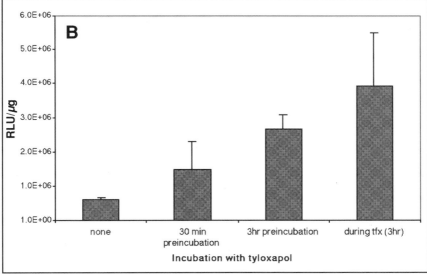

Fig. 4. Augmentation of the dendrimer-mediated transfection by the addition of chloroquine (CLQ) and DEAE-dextran (**A**) and tyloxapol (**B**). Rat 2 (A) and Jurkat (B) cells were transfected with 0.5 μg of pCF1-Luc DNA complexed with G9 EDA at charge ratios of 5 and 10 (A) and 10 (B). Luciferase activity was measured 36 h after transfection.

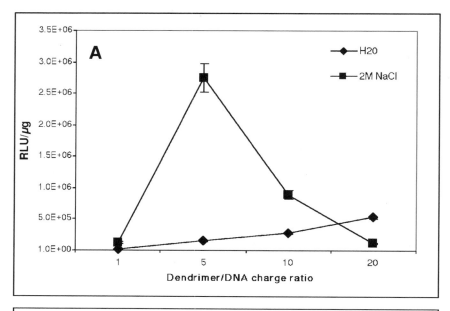

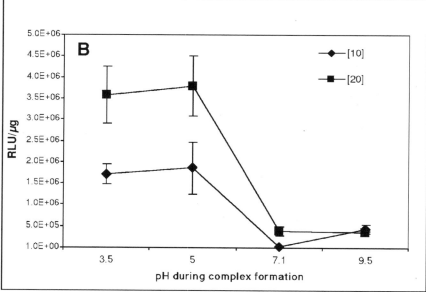

Fig. 5. Salt concentration (**A**) and pH (**B**) during dendrimer/DNA complex formation and their effect on transfection efficiency (*see* **Note 6**). COS-1 cells were transfected with 1 μg of pCF1-Luc DNA (0.05 μg/μL) complexed with G7 EDA at specified
dendrimer/DNA charge ratios 1, 5, 10, and 20 (**A**). Complexes (20 μl) formed either in
water or in the presence of 2 *M* NaCl were added to 400 μL of medium (1:20 dilution)

b. We have identified that incubation with the nonionic surfactant tyloxapol results in an increase in transfection efficiency in vitro, possibly through interaction with cellular membranes (**Fig. 4B**). Dendrimer/DNA complex can be added to medium containing tyloxapol at a final concentration of 0.1–0.5 mg/mL or dendrimer /DNA complex can be mixed with tyloxapol.

5. The design of the transfection-based experiments would not be complete without elementary consideration of cellular biology. Most model systems employ cell lines dividing and metabolizing rapidly.

a. The experiment presented in **Fig. 3** involves active cell cultures. However, growth, metabolic rate, and transcriptional milieu were manipulated by varying the initial density of cell seeding. Cultures of Rat2 and NHFF1 cells seeded at both 1×10^4 and 5×10^4 cells/cm^2 cells have similar growth rates but very different levels of gene expression. At 24 h after transfection, Rat2 cells at both densities express similar levels of luciferase, but no enzyme activity is detected in NHFF1 cells solely at the lowest cell density (10^4 cells/cm^2). Practically no expression is found in either of these cell types when they are transfected at near-confluent densities (initial seeding at 1×10^5 cells/cm^2).

b. Cellular differences can also be pronounced in the kinetics of transgene expression, with a maximum level for NHFF1 at approx 48 h and for Rat2 at 72 h after transfection. The time of harvest and analysis should be chosen depending on the goals of the experiment because the prolonged presence of the expressed protein not only may be a result of efficient transfection and transcription, but also may reflect a difference in the specific rate of degradation in a particular cell type.

6. The dendrimer/DNA complex composition and architecture seems to be critical for its transfection activity. The interaction of both components is affected by pH and/or salt concentration. Data in **Fig. 5** clearly shows that the initial condition of complex formation has consequences on transfection. A high concentration of NaCl present during complex formation at

---

to avoid a hypertonic effect of salt or hypotonic effect of water on cell physiology. Transfection efficiency predictably increased with the increasing charge ratios for the complexes formed in water. To achieve enhancement with monovalent salt, the condition would have to be carefully adjusted for the specific concentrations of dendrimer, DNA, as well as charge ratio because of the narrow functional optimum. In panel B, COS-1 cells were transfected with DNA complexed with G5 EDA in aqueous solution adjusted to below neutral, neutral, or alkaline pH. Complexes initially formed at low pH are the most efficient, despite the fact that transfections are performed routinely in standard growth media, buffered to neutral pH.

charge ratios of 5–10 initially resulted in enhanced DNA expression, but this effect disappeared with an increase in dendrimer concentration (**Fig. 5A**). For the formation of active complexes, we recommend using either water of lower than neutral pH or very dilute buffer (**Fig. 5B**). Our experience with various laboratory sources of distilled and deionized water indicate that all of them produce water at pH 5.0–5.5 which is suitable for preparation of dendrimer/DNA complexes.

References

1. Kukowska-Latallo, J. F., Bielinska, A. U., Johnson, J., Spindler, R., Tomalia, D. A., and Baker, J. R., Jr. (1996) Efficient transfer of genetic material into mammalian cells using Starburst polyamidoamine dendrimers. *Proc. Natl. Acad. Sci. USA* **93**, 4897–4902.
2. Baker, J. R., Jr., Bielinska, A., Johnson, J., Yin, R., and Kukowska-Latallo, J. F. (1996) Efficient transfer of genetic material into mammalian cells using polyamidoamine dendrimers as synthetic vectors: dendrimer-mediated transfection, in *Conference Proceedings Series: Artificial Self-Assembling Systems for GeneDelivery* (Felgner, P. L., Heller, M. J., Lehn, P., Behr, J. P., Szoka, F. C., Jr., eds.), American Chemical Society, Washington, DC, pp. 129–145.
3. Kukowska-Latallo, J. F., Bielinska, A. U., Chen, C., Rymaszewski, M., Tomalia, D. A., and Baker, J. R., Jr. (1998) Gene transfer using starburst dendrimers, in *Self-Assembling Complexes for Gene Delivery: From Chemistry to Clinical Trial* (Kabanov, A. V., Felgner, P. L., Seymour, L. W., eds.), John Wiley & Sons, Ltd. Sussex, UK, pp. 241–253.
4. Eichman, J. D., Bielinska, A. U., Kukowska-Latallo, J. F., and Baker, J. R., Jr. (2000) The use of PAMAM dendrimers for the efficient transfer of genetic material into cells. *Pharm. Sci. Technol. Today* **7**, 232–245.
5. Eichman, J. D., Bielinska, A. U., Kukowska-Latallo, J. F., Donovan, B. W., and Baker, J. R., Jr. (2001) Bioapplications of PAMAM dendrimers, in *Dendrimers and Other Dendritic Polymers* (Frechet, J. M. J. and Tomalia, D. A., eds.), John Wiley & Sons, Ltd. Sussex, UK, pp. 441–461.
6. Kukowska-Latallo, J. F, Raczka, E., Quintana A., Chen, C., Rymaszewski, M., and Baker, J. R., Jr. (2000) Intravascular and endobronchial DNA delivery to murine lung tissue using a novel, nonviral vector. *Hum. Gene Ther.* **11**, 1385–1395.
7. Kukowska-Latallo, J. F., Chen, C., Eichman, J., Bielinska, A. U., and Baker, J. R., Jr. (1999) Enhancement of dendrimer-mediated transfection using synthetic lung surfactant Exosurf Neonatal in vitro. *Biochem. Biophys. Res. Commun.* **264**, 253–261.
8. Bielinska, A. U., Kukowska-Latallo, J. F., and Baker, J. R., Jr. (1997) The interaction of plasmid DNA with polyamidoamine dendrimers: mechanism of complex formation and analysis of alterations induced in nuclease sensitivity and transcriptional activity of the complexed DNA. *Biochim. Biophys. Acta* **1353**, 180–190.
9. Bielinska, A., Kukowska-Latallo, J. F., Johnson, J., Tomalia, D. A., and Baker, J. R., Jr. (1996) Regulation of in vitro gene expression using antisense oligonu-

cleotides or antisense expression plasmids transfected using starburst PAMAM dendrimers. *Nucl. Acids Res.* **24**, 2176–2182.

10. Qin, L., Pahud, D. R., Ding, Y., Bielinska, A. U., Kukowska-latallo, J. F., Baker, J. R., Jr., and Bromberg, J. S. (1998) Efficient transfer of genes into murine cardiac grafts by starburst polyamidoamine dendrimers. *Hum. Gene Ther.* **9**, 553–560.

11. Wang, Y., Boros, P., Liu, J., Qin, L., Bai, Y., Bielinska, A. U., et al. (2000) DNA/dendrimer complexes mediate gene transfer into murine cardiac transplants ex vivo. *Mol. Ther.* **2**, 602–608.

12. Bielinska, A. U., Yen, A., Wu, H. L., Zahos, K. M., Sun, R., Weiner, N. D., et al. (2000) Application of membrane-based dendrimer/DNA complexes for solid phase transfection in vitro and in vivo. *Biomaterials* **21**, 877–887.

13. Raczka, E., Kukowska-Latallo, J. F., Rymaszewski, M., Chen, C., and Baker, J. R., Jr. (1998) The effect of synthetic surfactant Exosurf on gene transfer in mouse lung in vivo. *Gene Ther.* **5**, 1333–1339.

14. Sambrook, J., Fritsch, E. F., and Maniatis, T. (1989) in *Molecular Cloning: A Laboratory Manual*, 2nd ed. Cold Spring Harbor Laboratory Press, Cold Spring Harbor, NY, pp. 142–143.

15. Bielinska, A. U., Chen, C., Johnson, J., and Baker, J. R., Jr. (1999) DNA complexing with polyamidoamine dendrimers: implications for transfections. *Bioconjug. Chem.* **10**, 843–850.

16. Roessler, B. J., Bielinska, A. U., Janczak, K., Lee, I., and Baker, J. R., Jr. (2001) Substituted β-cyclodextrins interact with PAMAM dendrimer-DNA complexes and modify transfection efficiency. *Biochem. Biophys. Res. Comm.* **283**, 124–129.

6

DNA Delivery to Cells in Culture Using Cationic Liposomes

Shelby K. Wyatt and Todd D. Giorgio

1. Introduction

Following Fraley and colleagues' initial discovery of liposomes *(1)*, coupled with Behr's discovery of the ability of cationic lipids to interact with and condense negatively charged DNA *(2)*, Felgner et al. described the use of synthetic cationic lipids as a DNA delivery tool in 1987 *(3)*. The authors reported that a positively charged lipid, specifically N-[1-(2,3-dioleyloxy)propyl]-N,N,N-trimethylammonium chloride (DOTMA), could form liposomes under biologically relevant conditions either alone or in combination with neutral phospholipids. More importantly, these cationic liposome vesicles were demonstrated to react spontaneously with anionic DNA to form lipid–DNA complexes, or "lipoplexes" *(4)*. These lipoplexes demonstrated effective binding to cells in vitro and facilitation of intracellular delivery of a transgene. Felgner provides a detailed history of the progress leading up to and including lipofection *(5)*.

Attractive features of lipofection have supported extensive development of novel lipids and methods for lipid-mediated transfection. These developments recently have been reviewed in detail *(6,7)*. Essentially, lipoplexes are chemically pure with simple and reproducible formulation methods for both transient and stable expression *(3)*, facilitating their use as human pharmacological agents. Lipoplexes also lack specific surface proteins that can complicate in vivo application. Finally, functionally effective lipoplex concentrations are generally nontoxic to cultured cells.

From: *Methods in Molecular Biology, vol. 245:*
Gene Delivery to Mammalian Cells: Vol. 1: Nonviral Gene Transfer Techniques
Edited by: W. C. Heiser © Humana Press Inc., Totowa, NJ

The primary obstacle to in vivo lipofection is mirrored in the in vitro environment: relatively low transgene expression. This difficulty is characterized primarily by robust transgene expression, but from an insufficient fraction of treated cells. The mechanism(s) responsible for this behavior are currently under active investigation. Nuclear access *(8)* and/or intracellular degradation of the transgene *(9)* remain the primary suspects in modulating the relatively inefficient cellular response.

New, commercially available cationic liposome chemistries, designed to overcome these limitations, are widely available. Although some cationic liposome formulations are reported to significantly mediate increased transgene expression relative to other preparations, the underlying mechanisms for their improved performance are generally unknown *(7)*. The lack of correlation between lipoplex structure and function continues to hinder development of improved cationic liposomes for lipofection. Quantitative, single cell measurements of transgene delivery and expression represent one strategy for advancement of lipofection structure-function knowledge *(10)*.

Characterizing the lipoplex structure has also become a substantial field of interest since a better understanding should aid in establishing correlations between lipid-DNA complex assembly and biological activity *(6)*. Lipoplex visualization, for example, has been accomplished through various electron microscopy techniques *(11–13)*. Lipoplex binding to the cell membrane represents one obstacle affected by lipoplex structure and is essentially driven by electrostatic interactions. This binding is modulated by physical properties of the lipoplex, including size *(14)*, stability, and charge density, as well as lipoplex concentration *(15)* and incubation time *(15,16)*.

Evidence suggests that intracellular DNA transport involves progression through both endosomes and lysosomes *(17)*. Furthermore, cytoplasmic release of DNA occurs from an early endosomal compartment as determined via electron microscopy *(18,19)*. Facilitation of this endosomal release of plasmid DNA (pDNA) can significantly improve transgene expression. For example, the addition of neutral, "helper" lipids such as dioleoyl phosphatidylethanolamine (DOPE), cholesterol and dioleoylphosphatidylcholine (DOPC) *(20,21)* as well as pH-sensitive liposomes *(22)* both aid endosomal release.

Our laboratory has investigated a number of liposome chemistries. Our experience suggests that optimization of transfection parameters is at least as important as selection of the liposome chemistry in obtaining strong transgene expression in vitro. In this regard, the following protocol focuses on transgene expression optimization using Invitrogen's Lipofectin® reagent. This strategy is based on identification of optimum values of lipid:pDNA ratio, total pDNA, and incubation time. It is important to note, however, that other factors influence the efficiency of lipofection and should be taken into account when choosing the appropriate formulation. Particular considerations include cell type, culture environment, lipoplex construction, genetic material (plasmid, oligonucleotide, RNA, PNA, etc.), and reporter strategy.

The protocol described here in detail provides strong transgene expression of green fluorescent protein (GFP) from HeLa cells in vitro while simultaneously minimizing lipoplex toxicity. The outcome of this protocol is clearly evident through visualization or quantitative measurement *(10)* of GFP fluorescence by microscopy, fluorimetry, or flow cytometry.

2. Materials
2.1. Cells

1. HeLa cells (American Type Culture Collection [ATCC], Manassas, VA; Cat. no. CCL-2) (*see* **Note 1**).

2.2. Nonviral Delivery System

1. Cationic Lipid Formulation: Lipofectin® Reagent (Invitrogen, Carlsbad, CA; Cat. no. 18292-011), unmodified (*see* **Notes 2** and **3**).
2. Plasmid: Specifically a green fluorescent protein (GFP) expression plasmid, for example the pEGFP–N1 Vector (BD Biosciences Clontech, Palo Alto, CA; Cat. no. 6085-1), diluted in Tris-EDTA buffer (TE) (*see* **Notes 4** and **5** and **Subheading 2.3.5.**).

2.3. Solutions and Culture Medium

1. Tissue-culture medium: Dulbecco's Modified Eagle's Medium (DMEM) Auto-Mod™ (Sigma, St. Louis, MO; Cat. no. D5280), supplemented with 10% (v/v) calf serum (Invitrogen; Cat. no. 16170-078), 1% L-glutamine 200 mM (Invitrogen; Cat. no. 25030-081) and 1% antibiotic-antimycotic (penicillin-streptomycin-amphotericin) (Invitrogen; Cat. no. 15240-062). Both DMEM and calf serum should be stored between 2 and 8°C (*see* **Notes 1, 6–9**).
2. Sterile, distilled water.
3. Calcium- and Magnesium-Free Phosphate Buffered Saline (CMF-PBS) with EDTA. To prepare 1 L:
   a. Dissolve 1.141 g of sodium phosphate dibasic, anhydrous in 300 mL of distilled water.
   b. Adjust the pH to 7.4 with either HCl or NaOH.
   c. Dissolve 8 g of sodium chloride and 0.2 g of potassium chloride in 600 mL of distilled water.
   d. Dissolve 200 mg of disodium EDTA in 50 mL of distilled water.
   e. Combine the solutions.

    f. pH should drop to approx 6.8. Adjust the pH to 7.2–7.4 by slowly titrating NaOH.

    g. Adjust the total volume to 1 L.

4. Serum-Free Medium for Transfection: Opti-MEM® I Reduced Serum Medium (Invitrogen; Cat. no. 11058-021) (*see* **Note 9**).
5. TE buffer: 10 mM TrisCl, pH 7.4, 7.5, or 8.0 and 1 mM EDTA, pH 8.0.
6. Trypsin-EDTA, 1× (Invitrogen; Cat. no. 25200-056).
7. Formaldehyde EM-grade, 16% solution (Electron Microscopy Sciences, Fort Washington, PA; Cat. no. 15710). To fix the cells, it will also be necessary to prepare a 1% solution (v/v), diluted with CMF-PBS. Handle very carefully and with gloves because formaldehyde is an irritant, a carcinogen, and a mutagen.
8. Cell Scrub™ Buffer (Gene Therapy Systems, Inc., San Diego, CA; Cat. no. B100001).

2.4. Laboratory Supplies

1. 1.5-mL Eppendorf tubes (Fisher Scientific, Pittsburgh, PA; Cat. no. 05-402).
2. Falcon Multiwell Tissue Culture Plates (BD Falcon™, Bedford, MA; Cat. no. 353046) or 35-mm Petri dishes (BD Falcon™; Cat. no. 351008).
3. T-75 Tissue Culture Flasks (BD Falcon™; Cat. no. 353810).
4. Centrifuge tubes (Fisher Scientific; Cat. no. 05-539-9).
5. Disposable Polyethylene Transfer Pipets (Fisher Scientific, Pittsburgh, PA; Cat. no. 13-711-7).

3. METHODS

3.1. Tissue Culture

1. Thaw an ampoule of HeLa cells and plate in a T-75 tissue culture flask in 12–15 mL of DMEM supplemented with 10% (v/v) calf serum. Culture HeLa cells under standard incubation conditions: 37°C, 5% CO_2, and 95% humidity (*see* **Note 1**).
2. After 24 h, aspirate the medium and replace with 12–15 mL of fresh DMEM containing 10% calf serum.
3. Replace the medium two to three times per week; upon reaching confluency, split the cells 1:6.

3.2. General Transfection Technique

A number of variables influence the transfection efficiency and should first be optimized. These include the cell density, lipid to pDNA ratio, total pDNA, and incubation time of the lipoplexes with the cells. An optimization technique follows in **Subheading 3.4.**

Table 1
Surface Areas for Various Tissue Culture Plates *(25)*

Plate type	Surface area (cm²)	Volume (mL)
Multiwell–6	9.62	2.5
Multiwell–12	4.50	2.0
Multiwell–24	2.00	1.0

It is also important to note that the transfection procedure described in this section corresponds to delivery of genetic material to cells in 35-mm (6-well) plates. However, this protocol can easily be scaled according to the relative surface areas (**Table 1**).

1. Plate HeLa cells at 20,000 cells/cm² (or 1.9×10^5 cells/35-mm plate) and incubate overnight in normal culture medium at standard incubation conditions: 37°C, 5% CO_2, and 95% humidity to obtain 60–80% confluency (*see* **Notes 10** and **11**).
2. For each well to be transfected, dilute 3 µL of Lipofectin® Reagent in 100 µl of Opti-MEM into a 1.5-mL Eppendorf tube. Incubate for 30 min at room temperature (*see* **Notes 12** and **13**).
3. After the 30-min incubation, add 1.5 µg of pDNA (= 6µL of 0.25 µg/µL pDNA stock) per well to be transfected to the hydrated Lipofectin® solution. Mix by gently flicking the Eppendorf tube and incubate at room temperature for 15 min to allow complete formation of the lipoplexes (*see* **Notes 5** and **12**).
4. Immediately prior to transfection, wash each cell monolayer with 1.5 mL of Opti-MEM to eliminate any serum-containing medium remaining in the well.
5. Add 0.9 mL of Opti-MEM to each well.
6. Add 109 µL of the pDNA:liposome complexes (= 100µL Opti-Mem + 3µL Lipofectin® + 6µL pDNA) dropwise to each well (*see* **Note 14**).
7. Incubate at 37°C, 5% CO_2, and 95% humidity for 5 h.
8. After the 5-h incubation, aspirate the lipoplex solution and add 1.5 mL of tissue-culture medium. Incubate under standard incubation conditions for 43 h (for a total of 48 h transfection) (*see* **Note 15**).
9. Harvest the cells after 48 h of transfection (*see* **Subheading 3.3.**).

3.3. Cell Harvesting Technique

1. Aspirate the culture medium.
2. Wash the transfected cell monolayer once with 1 mL of CMF-PBS.

3. Aspirate the CMF-PBS and replace with 1 mL of Cell Scrub and incubate for 15 min at room temperature (*see* **Note 16**).
4. Aspirate the Cell Scrub and wash cells twice with 1 mL of CMF-PBS.
5. Aspirate the CMF-PBS and add 1 mL of trypsin. Incubate at room temperature for approx 5 min or until the cells completely detach from the well surface (*see* **Note 17**).
6. Add 1 mL of tissue culture medium to inactivate the trypsin.
7. Transfer the contents of each well to a separately labeled flow cytometer tube using transfer pipets.
8. Add 125 µL of 16% formaldehyde and mix well (*see* **Note 18**).
9. Pellet cells by centrifugation at 850g for 5 min.
10. Carefully pour out the supernatant (*see* **Note 19**).
11. Resuspend the pellet in 2 mL of 1% formaldehyde; this will act as a washing step (*see* **Note 20**).
12. Repeat **steps 9–11** twice.
13. Analyze fluorescence in the transfected cells (*see* **Notes 21** and **22**).

3.4. Optimization of Transfection Conditions

1. Following the transfection protocol as described in **Subheading 3.3.**, hold the total amount of pDNA constant at 1 µg per 35-mm well and the lipoplex-cells incubation time at 6 h. Vary the lipid:pDNA ratio from 0.5–12 µL lipid:1 µg pDNA. Harvest the cells at the same time, preferably 48-h posttransfection for maximum expression. At this time, an optimization curve can be made of expression vs lipid:pDNA ratio and a maximum determined. This represents the optimum lipid:pDNA ratio. An example is seen in **Fig. 1** (*see* **Notes 10** and **22**).
2. After determining the optimum lipid:pDNA ratio, hold the ratio constant at the optimum as well as the lipoplex-cells incubation time constant at 6 h. Vary the total amount of pDNA from 0.5 to 3 µg. Again, harvest the cells at the same time, preferably 48 h posttransfection for maximum expression. At this time, an optimization curve can be made of expression vs total pDNA and a maximum determined. This represents the optimum total pDNA (*see* **Fig. 2**).
3. Finally, hold the lipid:pDNA ratio and total pDNA constant at the determined optimum values. Vary the lipoplex incubation time with the cells between 2 and 7 h. Again, harvest the cells at the same time, preferably 48-h posttransfection for maximum expression. At this time, an optimization curve can be made of expression vs incubation time and a maximum determined. This represents your optimum incubation time.

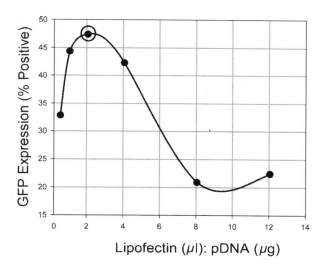

Fig. 1. Example optimization of lipid:pDNA ratio. While maintaining a constant cell density through a standard seeding protocol as well as the total DNA at 1 μg/35-mm well and incubation time at 6 h, the ratio of Lipofectin®:pDNA ratio was varied from 0.5 μL:1 μg to 12 μL:1 μg. An obvious maximum is present and is therefore regarded as the optimal ratio. This ratio is then utilized for subsequent optimization of total pDNA and incubation time.

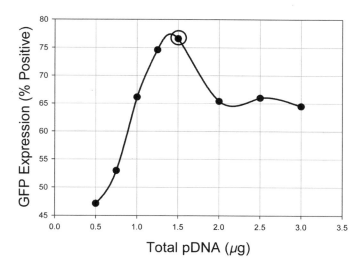

Fig. 2. Example optimization of total pDNA. While maintaining a constant cell density through a standard seeding protocol as well as the Lipofectin®:pDNA ratio at 2 μL:1 μg and incubation time at 6 h, the total pDNA was varied from 0.5 μg to 3.0 μg. An obvious maximum is present and is therefore regarded as the optimal total pDNA. This value is then utilized for subsequent optimization of incubation time.

4. Notes

1. The methods presented here are specific for HeLa cells. Cell type strongly modulates the extent of transgene expression, but most transformed, actively mitotic mammalian cell types transfected by this method can yield detectable transgene expression. It is important to note, however, that modifications may be necessary for other cell types, culture types (i.e., monolayer, suspension, clonal) and degree of chemical definition necessary. Cell-culture methods for other cell types can generally be found in the literature or directly from the cell source.

2. The cationic lipid reagent should be selected relative to your specific cell line. Invitrogen and Life Technologies provides a general guideline for their transfection reagents *(23)*. For reference, Lipofectin® Reagent is essentially a 1:1 (w/w) liposome formulation of the cationic lipid DOTMA, and DOPE in membrane filtered water.

3. Cationic lipid should not be frozen or stored in a section of the refrigerator where the temperature is less than 4°C.

4. Plasmid DNA should be diluted into and stored in TE buffer rather than distilled water for two main reasons. The Tris effectively acts as a buffer while the EDTA acts as a chelator, effectively minimizing plasmid degradation.

5. Diluting the pDNA into appropriately sized aliquots (based on experimentation frequency) beneficially allows for less repeated freezing and thawing of the plasmid. Also, diluting the pDNA to a concentration of 0.25 µg/µL provides for more accurate pipetting. Total amounts of pDNA per 35-mm well are generally on the order of 1–2 µg. Therefore, a pDNA concentration of 0.25 µg/µL allows for a more accurate measurement of 4–8 µL per well rather than a smaller, less accurate volume.

6. This specific tissue-culture medium is in powder form and must be prepared, as described in the accompanying "Product Information" sheet, before use. An equivalent, liquid form is also commercially available.

7. Monitor the following aspects to determine deterioration of the powder and liquid medium: (Powder) color change, granulation or clumping, insolubility and (Liquid) pH change, precipitate or particulate matter throughout the solution, cloudy appearance, color change.

8. Add calf serum only when immediately ready to use, since the addition decreases the shelf life of the tissue-culture medium.

9. Medium containing serum and/or antibiotics should not be used with this transfection protocol. The cationic lipid reagents cause cells' sensitivity to increase. Some protocols exist for transfection in the presence of serum, however, both serum and antibiotics do have a great impact on the transfection efficiency.

10. Reproducible results require a standard seeding protocol between experimental cohorts. In other words, consistently plate cells at a constant density and incubate for the same period of time overnight to obtain a constant cell density for transfection. The cell density can affect the transfection efficiency and could therefore also be optimized.

11. Cells approaching 100% confluency typically yield reduced levels of transgene expression following lipoplex transfection. The potential for cell toxicity is increased in samples with low confluence.

12. When mixing, avoid vortexing or agitating the lipid excessively as this may form cationic lipid reagent peroxides.

13. Allowing the lipid to rehydrate for 30 min prior to the addition of the pDNA has shown significant improvements in transfection efficiency *(24)*.

14. This is rather important in distributing the lipoplexes evenly throughout the monolayer.

15. Maximum expression levels are normally seen at 48 h posttransfection, however the cells can be analyzed prior to this and should demonstrate increasing efficiency up to approx 48 h.

16. Cell Scrub washing removes extracellular, cell-associated plasmid that might compromise subsequent measurements of intracellular plasmid concentration.

17. Monitor cell detachment from the well surface by phase microscope observation. If you do not trypsinize them long enough, the cells will not detach from the well surface or from each other. However, prolonged contact with trypsin can produce cellular injury. A sharp impact of the well against a solid object can also facilitate cell detachment.

18. It is normal for the culture medium to change from a red to a yellowish color. This results from the pH change caused by the formaldehyde. Please also note that this material should be handled with much caution because it is an irritant, a carcinogen, and a mutagen.

19. Aspirating the supernatant is too difficult and too risky at this point. The "pellet" can actually be seen as a cloudy film on the surface of the cytometer tube. Pour the liquid out very carefully; do not agitate the cytometer tubes, and the pellet will remain undisturbed.

20. The most effective way to resuspend the transfected cells involves dragging the capped tube across the top of a wire centrifuge rack (20 cm long).

21. Sample quality deteriorates with time in storage. Although the cells are "fixed," the formaldehyde will begin to degrade the cells. Analysis of transgene expression in intact cells is best performed within 24 h of cell harvesting.

22. The data presented in **Figs. 1** and **2** was obtained via flow cytometry. Other methods include fluorimetry and fluorescence microscopy.

References

1. Fraley, R., Subramani, S., Berg, P., and Papahadjopoulos, D. (1980) Introduction of liposome-encapsulated SV40 DNA into cells. *J. Biol. Chem.* **255**, 10431–10435.
2. Behr, J. P. (1986) DNA strongly binds to micelles and vesicles containing lipopolyamines or lipointercalants. *Tetrahedron Lett.* **27**, 5861–5864.
3. Felgner, P. L., Gadek, T. R., Holm, M., Roman, R., Chan, H. W., Wenz, M., et al. (1987) Lipofection: a highly efficient, lipid-mediated DNA-transfection procedure. *Proc. Natl. Acad. Sci. USA* **84**, 7413–7417.
4. Felgner, P. L., Barenholz, Y., Behr, J. P., Cheng, S. H., Cullis, P., Huang, L., et al. (1997) Nomenclature for synthetic gene delivery systems. *Hum. Gene Ther.* **8**, 511–512.
5. Felgner, P. L. (1999) Progress in gene delivery research and development, in *Nonviral Vectors for Gene Therapy* (Huang, L., Hung, M.-C., and Wagner, E., eds.), Academic Press, San Diego, CA., pp. 25–38.
6. Chesnoy, S. and Huang, L. (2000) Structure and function of lipid-DNA complexes for gene delivery. *Ann. Rev. Biophys. Biomol. Struct.* **29**, 27–47.
7. Pedroso de Lima, M. C., Simoes, S., Pires, P., Faneca, H., and Duzgunes, N. (2001) Cationic lipid-DNA complexes in gene delivery: from biophysics to biological applications. *Adv. Drug Deliv. Rev.* **47**, 277–294.
8. James, M. B. and Giorgio, T. D. (2000) Nuclear-associated plasmid, but not cell-associated plasmid, is correlated with transgene expression in cultured mammalian cells. *Mol. Ther.* **1**, 339–346.
9. Lechardeur, D., Sohn, K. J., Haardt, M., Joshi, P. B., Monck, M., Graham, R. W., et al. (1999) Metabolic instability of plasmid DNA in the cytosol: a potential barrier to gene transfer. *Gene Ther.* **6**, 482–497.
10. Tseng, W. C., Haselton, F. R., and Giorgio, T. D. (1997) Transfection by cationic liposomes using simultaneous single cell measurements of plasmid delivery and transgene expression. *J. Biol. Chem.* **272**, 25641–25647.
11. Gustafsson, J., Arvidson, G., Karlsson, G., and Almgren, M. (1995) Complexes between cationic liposomes and DNA visualized by cryo-TEM. *Biochim. Biophys. Acta* **1235**, 305–312.
12. Sternberg, B., Sorgi, F. L., and Huang, L. (1994) New structures in complex formation between DNA and cationic liposomes visualized by freeze-fracture electron microscopy. *FEBS Lett.* **356**, 361–366.
13. Templeton, N. S., Lasic, D. D., Frederik, P. M., Strey, H. H., Roberts, D. D., and Pavlakis, G. N. (1997) Improved DNA: liposome complexes for increased systemic delivery and gene expression. *Nat. Biotechnol.* **15**, 647–652.
14. Ross, P. C. and Hui, S. W. (1999) Lipoplex size is a major determinant of in vitro lipofection efficiency. *Gene Ther.* **6**, 651–659.
15. van der Woude, I., Visser, H. W., ter Beest, M. B., Wagenaar, A., Ruiters, M. H., Engberts, J. B., and Hoekstra, D. (1995) Parameters influencing the introduction of plasmid DNA into cells by the use of synthetic amphiphiles as a carrier system. *Biochim. Biophys. Acta* **1240**, 34–40.
16. Zabner, J., Fasbender, A. J., Moninger, T., Poellinger, K. A., and Welsh, M. J.

(1995) Cellular and molecular barriers to gene transfer by a cationic lipid. *J. Biol. Chem.* **270**, 18997–19007.

17. Coonrod, A., Li, F. Q., and Horwitz, M. (1997) On the mechanism of DNA transfection: efficient gene transfer without viruses. *Gene Ther.* **4**, 1313–1321.

18. Zhou, X. and Huang, L. (1994) DNA transfection mediated by cationic liposomes containing lipopolylysine: characterization and mechanism of action. *Biochim. Biophys. Acta* **1189**, 195–203.

19. Friend, D. S., Papahadjopoulos, D., and Debs, R. J. (1996) Endocytosis and intracellular processing accompanying transfection mediated by cationic liposomes. *Biochim. Biophys. Acta* **1278**, 41–50.

20. Hui, S. W., Langner, M., Zhao, Y. L., Ross, P., Hurley, E., and Chan, K. (1996) The role of helper lipids in cationic liposome-mediated gene transfer. *Biophys. J.* **71**, 590–599.

21. Farhood, H., Serbina, N., and Huang, L. (1995) The role of dioleoyl phosphatidylethanolamine in cationic liposome mediated gene transfer. *Biochim. Biophys. Acta* **1235**, 289–295.

22. Simoes, S., Slepushkin, V., Duzgunes, N., and Pedroso de Lima, M. C. (2001) On the mechanisms of internalization and intracellular delivery mediated by pH-sensitive liposomes. *Biochim. Biophys. Acta* **1515**, 23–37.

23. Invitrogen and Life Technologies (1999) *Guide to Eukaryotic Transfections with Cationic Lipid Reagents.* Rockville, MD, pp. 1–33.

24. Ciccarone, V. and Hawley-Nelson, P. (1995) Lipofectin transfection activity increased by protocol improvement. *Focus* **17**(3), 103.

25. Freshney, R. I. (1993) *Culture of Animal Cells.* Alan R. Liss, Inc., New York.

7

Formulation of Synthetic Gene Delivery Vectors for Transduction of the Airway Epithelium

John Marshall, Nelson S. Yew, and Seng H. Cheng

1. Introduction

The ability to mediate gene transfer to the lumen of the lung offers opportunities to treat diseases that affect the airways. Currently, several different gene-transfer vectors are being evaluated for delivery of a variety of therapeutic genes to the airways. These include recombinant viral (adenovirus, adeno-associated virus, retrovirus) and synthetic, self-assembling (cationic lipids, polymers) vectors. Although much of the early focus of these vectors has been directed toward a therapy for cystic fibrosis (1–5), other lung diseases, including asthma, chronic obstructive pulmonary disease, and cancer are being explored as viable disease targets.

Cationic lipid-based synthetic vectors, the focus of this chapter, are generated by the condensation of negatively charged plasmid DNA (pDNA) encoding the gene of interest, by the positively charged cationic moieties of the lipid. Depending on the cationic lipid species and formulation used, the resultant macromolecular complexes can range between 10 and several hundred nanometers (mostly multilamellar vesicles) in diameter and may also exhibit different net surface charges (6). These cationic lipid:pDNA complexes have been shown capable of transfecting not only cells in culture but also organs in vivo, albeit at lower efficiencies than with adenoviral vectors. Nevertheless, there is continued interest in the development of these synthetic gene-delivery vectors, borne primarily from the observation that unlike viral vectors, synthetic gene-

From: *Methods in Molecular Biology, vol. 245:*
Gene Delivery to Mammalian Cells: Vol. 1: Nonviral Gene Transfer Techniques
Edited by: W. C. Heiser © Humana Press Inc., Totowa, NJ

delivery systems do not elicit a host immune response that precludes their use in subsequent readministrations. As such, they are more conducive for use in the treatment of chronic diseases that require lifelong therapy. Manufacturing the components of the cationic lipid:pDNA complexes in large quantities is also relatively facile compared to the production of viral vectors.

To be considered for use in gene therapy of chronic lung diseases, the synthetic vectors ideally should have the following properties. For most applications, these formulations should be (1) capable of facilitating efficient gene transduction and persistent expression of the transgene product in the appropriate target cells, (2) relatively nontoxic, (3) conducive for use in repeated administrations, and (4) compatible with nebulization to the airways *(7)*. The identification of vector formulations that meet these criteria is not trivial and is compounded by our lack of understanding on how these synthetic gene-transfer complexes mediate cellular transfection. It is now possible to chart the several processes that are necessary in order for productive cellular transfection to occur. These include, for example, the uptake of the cationic lipid:pDNA complexes by the target cells, escape from the lysosomal compartment, and subsequent translocation of the complexes to the nucleus. However, it remains unclear which of these are the key limiting events, compounded by them perhaps being different for each cationic lipid species and formulation. Consequently, the development of these synthetic gene-delivery vectors has been largely reliant on empirical testing of many different polycations and formulations *(8–10)*. Despite these impediments, much progress has been made, particularly with cationic lipid-based vectors, to the point where several were deemed sufficiently efficacious to warrant testing in human clinical studies for cystic fibrosis *(1–5)*. The results of these early studies showed that although cationic lipid-based gene transfer to the airway epithelium is possible, the efficiency of gene transduction was low and was associated with mild toxicity *(5,11)*. This toxicity was caused partially by the cationic lipids, which typically have a detergent-like structure and properties, but was mostly attributed to the immunostimulatory CpG motifs found in the bacterially derived DNA *(12)*.

In this chapter we address many of the factors associated with the development and use of cationic lipid-based vectors, with efficient and safe gene transduction of the lung being the goals. These include methods for formulating and storing the components of this vector system (to reduce undesirable chemical reactions), methods to reduce the toxicity associated with pDNA, formulation modifications that support high concentrations of the complexes (to facilitate aerosolization), and methods for delivery of these synthetic vectors to the lung.

2. Materials

2.1. Formulation and Storage of Cationic Lipids

1. 1,2-dioleoyl-*sn*-glycero-3-phosphatidylethanolamine (DOPE) (Avanti Polar Lipids, Inc., Alabaster, AL).
2. 1,2-dimyristyl-*sn*-glycero-3-phosphtidylethanolamine-*N*-[poly(ethyleneglycol)5000] (DMPE-PEG$_{5000}$) (Avanti Polar Lipids).
3. 1,2-dioleoyl-3-trimethylammonium-propane (DOTAP) (Avanti Polar Lipids).
4. Cholesterol (Avanti Polar Lipids).
5. The GL-series of cationic lipids were synthesized at Genzyme Corporation, Cambridge, MA *(8,9)*.
6. RBS-35 (Pierce Chemical Co., Rockford, IL).
7. Chloroform, *t*-butanol, ethanol.
8. Argon.
9. Labline TransSonic bath sonicator model 820/H (Barnstead/Thermolyne, Dubuque, IA)

2.1.1. Formulation of Dry Lipid Films

1. 20-mm neck, 10-mL serum tubing vials, aluminum crimp (Wheaton, Millville, NJ).
2. Omniflex butyl-L stoppers (Helvoet Pharma, Pennsauken, NJ).
3. Dispensing pump model 505Di/RL (Watson-Marlow, Wilmington, MA)
4. DuraStop/DuraDry Lyophilizer (FTS Systems, Stone Ridge, NY).
5. 0.2-μm filter.
6. −70°C freezer.

2.1.2. Formulation of Aqueous Lipid Suspensions

1. Büchi rotavapor model R-114 (Brinkman, Westbury, NY).
2. Büchi water bath model B-480 (Brinkman).
3. Vacuum aspirator model B-169 (Brinkman).
4. 5% dextrose in water (D5W).
5. Filters (1.0 and 0.45 μm), Puradisc 25AS (Whatman, Clifton, NJ).
6. Filters (0.2 and 0.1 μm), Anotop 25 (Whatman).

2.3. Preparation of Cationic Lipid:pDNA Complexes

1. Water for irrigation (WFI) (VWR, West Chester, PA).
2. Polystyrene tubes (Falcon no. 2054) (Becton Dickinson, Franklin Lakes, NJ).
3. Multi-tube vortexer (VWR, West Chester, PA).

4. 30°C water bath.
5. *Limulus* amebocyte lysate (LAL) assay (BioWhittaker, Walkersville, MD).

2.4. Physical Characterization of Cationic Lipid:pDNA Complexes

1. Orange G.
2. Tris-borate-EDTA (TBE) buffer: 45 mM Tris-borate, 1 mM ethylenedi-amine tetraacetic acid (EDTA), pH 8.0.
3. Ethidium bromide (EB).
4. Sodium dodecyl sulfate (SDS).
5. Nicomp model 380ZLS zeta potential/particle sizer (Nicomp, Santa Barbara, CA).
6. Electrophoresis tank, power source, gel cassette, and comb.
7. Foto/Prep transilluminator (Fotodyne, Hartland, WI).

2.5. Aerosol Delivery of Cationic Lipid:pDNA Complexes to the Mouse Lung

1. Rhodamine-phosphatidylethanolamine (Avanti Polar Lipids).
2. TOTO-1 (Molecular Probes, Eugene, OR).
3. Octylglucoside.
4. Sodium butyrate.
5. Triton X-100.
6. Pari LC Jet Plus nebulizer (Pari Respiratory Equipment, Inc., Richmond, VA).
7. Pall BB-50T Breathing Circuit Filter (Pall Biomedical, Inc., Fajardo PR 00648).
8. Andersen IACFM cascade impactor (ThermoAndersen, Smyrna, GA).
9. All Glass Impinger (AGI) (Ace Glass Inc., Vineland, NJ).
10. Spex Fluoromax fluorometer (Spex, Metuchen, NJ).
11. Female BALB/c mice (Taconic, Germantown, NY); animal care in accordance with the AAALAC guidelines.

2.6. Insufflation of Complexes into Mouse Lung

1. Isoflurane (J.A. Webster, Sterling, MA)
2. Anesthesia chamber (Colonial Medical, Franconia, NH).

3. Methods

3.1. Formulation and Storage of Cationic Lipids

Several cationic lipids of different structure types have been synthesized and evaluated for use in transduction of the lung. Examples of cationic lipids that have been shown to be particularly active at mediating gene transfer to the lung include DC-Chol *(1,2)*, DOTAP *(3)*, and the GL-series of lipids *(4,5,7,9)*, (*see* **Note 1**). These cationic lipids are invariably formulated with a neutral co-lipid such as dioleoylphosphatidylethanolamine (DOPE) or cholesterol. To highlight examples of how such formulations are generated, we will describe two processes, one for generating GL-67:DOPE lipids as a dry film and another for formulating DOTAP:cholesterol as a stable aqueous suspension (*see* **Note 2**).

3.1.1. Formulation of Dry Lipid Films

1. Weigh 615 mg of the cationic lipid GL-67 and dissolve it in 50 mL of *t*-butanol:water (9:1, v/v) in a 100-mL volumetric flask. Sonicate the lipid in an ultrasonic waterbath until the lipid is completely in solution. Add *t*-butanol:water to bring the volume up to 100 mL and the lipid concentration to 10 m*M*.
2. Weigh 1488 mg of the neutral lipid DOPE and dissolve it in 50 mL of *t*-butanol:water (9:1, v/v) in a 100-mL volumetric flask. Complexes to be nebulized (*see* **Subheading 3.5.**) require the incorporation of DMPE-PEG$_{5000}$ into the formulation (*see* **Note 3**). Weigh 285 mg of DMPE-PEG$_{5000}$ and dissolve it in 50 mL of *t*-butanol:water along with the neutral lipid. Sonicate the lipid in an ultrasonic waterbath until the lipid is completely in solution. Add *t*-butanol:water to bring the volume up to 100 mL and the neutral lipid concentration to 20 m*M*.
3. Combine the neutral and cationic lipids and aliquot into glass vials (20 mm neck, 10 mL serum tubing vials; Wheaton) using a dispensing pump (typically 4 μmoles of cationic lipid is aliquotted into each vial). Loosely place unsiliconized inert butyl Omniflex stoppers in the tops of the vials to allow unobstructed airflow for subsequent lyophilization.
4. Place the vials on a shelf in a lyophilizer and cool to −30°C for 25 min to freeze the contents. Apply a vacuum until the pressure is below 100 mT, then raise the shelf temperature to −20°C and maintain the lyophilizer at this temperature with the vacuum at 0 mT for an additional 5 h. Raise the shelf temperature to −5°C and keep the vials under vacuum at 0 mT an additional 48–60 h to remove any residual solvent.
5. Following lyophilization, release the vacuum and back-fill the lyophilizer with argon gas that has been sterilized through a 0.2-μm sterile filter at 1 atmosphere pressure. Filling the vials with an inert gas such as argon is an important consideration especially for cationic lipids like GL-67 that contain spermine as a headgroup because this is reactive with carbon dioxide. Before opening the lyophilizer, raise the shelves inside to push the stoppers into place so that they become fully seated in the vials.
6. Open the lyophilizer and seal the stoppers in place with an aluminum crimp (Wheaton). Store the vials at −70°C until ready for use.

3.1.2. Formulation of Aqueous Lipid Suspensions

Although most cationic lipid formulations are best stored as dried films, some formulations have been described that are also relatively stable as liquid suspensions. An example of one such formulation involves the cationic lipid DOTAP which, when formulated with cholesterol and subjected to extrusion *(13)*, results in an aqueous suspension that is stable at 4°C for at least 1 mo. Aqueous suspensions of liposomes generated using this procedure are relatively homogenous in size (~200 nm diameter).

1. Weigh 140 mg of the cationic lipid 1,2-dioleoyl-3-trimethylammonium-propane (DOTAP) and 77 mg of cholesterol. Combine them in a 1-L round-bottom flask and dissolve in 10 mL of chloroform.
2. Attach the flask to a Büchi rotary evaporator at a 45° angle and maximally immerse in a 30°C water bath. Rotate the flask in the 30°C water bath for 30 min and place the resulting dried thin film under vacuum for 15 min (*see* **Note 4**).
3. Hydrate the dried lipid film by rapidly adding 10 mL of D5W, taking care to ensure resuspension of the lipids particularly from the top of the flask. Rotate the flask in a 50°C water bath for 45 min, gradually changing the angle from 45° to vertical over this period. Complete the hydration process by further rotating the flask at 35°C for 10 min and then allowing it to stand for 16 h at room temperature.
4. Sonicate the hydrated liposomes at 50°C in a bath sonicator at 35 kHz for 5 min to generate a nonviscous lipid suspension.
5. Extrude the lipid suspension sequentially through 1.0, 0.45, 0.2, and 0.1 μm filters using a 10-mL syringe. Perform the extrusions as rapidly as possible to minimize cooling of the lipids that can render them viscous and difficult to extrude.
6. Store the resulting liposome suspension under argon at 4°C.

Both of these formulation protocols are amenable to cationic lipid and/or neutral co-lipid substitutions. However additional care is necessary for some cationic lipids such as those of the GL-series that contain unprotonated primary amines on their cationic headgroups. These are particularly susceptible to a transacylation reaction *(9)* when they are co-formulated with a neutral lipid containing an ester-linked acyl chain, such as those in DOPE (*see* **Note 5**).

3.2. Plasmid Mutagenesis

A major contributor to the acute toxicity observed following administration of synthetic vectors to the lung is the bacterial-derived pDNA *(12)*. This toxicity has been shown to be owing primarily to the presence of immunostimulatory CpG sequences on the pDNA. This issue can be circumvented by directed mutagenesis of the plasmids' DNA sequence to remove these CpG sequences. Simultaneously, amino acid codons can be optimized through alteration of degenerate or "wobble" bases in order to utilize the most abundant mammalian transfer RNAs while maintaining the correct polypeptide sequence. Although these changes maybe affected through site-directed mutagenesis of the pDNA, because the number of CpG motifs that are invariably present on pDNA is quite high, a more expedient approach is to synthesize the gene that incorporates these changes. Below, we describe a strategy for designing these synthetic genes.

1. The amino acid sequence of the given protein to be expressed is sufficient to design a synthetic gene. The synthetic gene can, but does not need to, resemble the nucleotide sequence of the original cDNA. A table of the codon usage in highly expressed human genes is then used to assemble the nucleotide sequence; an example of such a table can be found in Kim et al. *(14)*. Assembly can most easily be achieved with a word processing program, using the find and replace feature to substitute each amino acid symbol with a corresponding nucleotide triplet. The basic criterion for selecting which codon to insert is to avoid rarely used codons. One could choose only the most commonly used codon, but this would introduce excessive codon bias. This may introduce small internal repeats that could be problematic when synthesizing the gene. In addition, the use of only one codon for a given amino acid may in theory lead to a transient shortage of certain isoacceptor tRNAs during translation, but whether this actually occurs is not known.

2. Scan the optimized sequence for CpG dinucleotides. For example, four of the possible codons for arginine are CGT, CGC, CGA, and CGG, and these must be substituted with either of the two remaining alternative codons for arginine, AGA or AGG. Other CpG motifs will occur at the junction between two codons when the first codon ends in a C and the second begins with a G. Frequently these can be removed by changing a "wobble" base C to a G, T, or A where possible.

3. A sequence optimized for expression in human cells will be GC rich, because the prevalent human codons usually have a C or G in the third degenerate position. Because a sequence with a high GC content will be difficult to amplify by polymerase chain reaction (PCR), a proportion of the codons should be substituted with codons containing fewer G and C nucleotides, to reduce the overall GC content. High GC content may also create secondary structure in the translated mRNA that may inhibit translation.

4. The sequence context of the translation initiation and termination codons can also be optimized. The consensus sequence for initiation in vertebrates

is (A/G)CCAUGG *(15)*. The most important residue is at position −3 relative to the A residue of the AUG codon, followed by the residue at position +4. The G at the +4 position may alter the coding for the second amino acid of the protein, and if so a choice must be made to determine the relative importance of preserving the amino acid sequence vs the benefit of the optimal residue. This may have to be determined empirically. The preferred translation termination codon in mammals is UGA *(16)*. The next base can also influence termination efficiency (A, G ≫ C, U). If termination is not optimized, there may be significant translational read-through and reduced protein expression.

An example of a synthetic CpG-reduced sequence that has been codon optimized for expression in human cells is shown (**Fig. 1**). This sequence codes for the *Escherichia coli* chloramphenicol acetyltransferase gene. The nucleotide sequence does resemble the wild-type sequence, but contains only one CpG motif vs 33 in the original sequence (counting one strand). In vitro transfections of plasmid vectors containing the wild-type vs the synthetic CAT gene demonstrate an approximate two- to threefold increase in CAT expression from the plasmid vector containing the synthetic gene (**Fig. 2**). This increased expression is also realized following intranasal or systemic administration of the synthetic pDNA vector into mice *(17)*. Importantly, the CpG-reduced pCFA-sCAT is also less inflammatory than its wild-type counterpart *(17)*.

3.3. Preparation of Cationic Lipid:pDNA Complexes

Cationic lipid and pDNA complexes are normally prepared just prior to use. The cationic lipid film is hydrated with sterile water at 4°C to a concentration that is double that required for the final concentration in the complex (*see* **Note 6**). Aqueous liposome suspensions are also diluted to be twice the final concentration in the complex, but the dilutions are made using D5W at room temperature. The pDNA is similarly diluted to a concentration that is twice that required in the final concentration using either sterile water or D5W, as appropriate. For the GL-series of cationic lipids all of the dilutions and complexes are prepared in sterile polystyrene tubes. For the aqueous liposome suspensions, complexes are prepared in 1.5-mL polypropylene microfuge tubes.

3.3.1. Preparation of Cationic Lipid:pDNA Complexes Using Cationic Lipid Films

1. Add water (WFI) to the dried lipid film (e.g., add 2 mL to a 4 µmole cationic lipid vial to generate a 2 m*M* suspension) and hydrate for 60 min at 4°C on a multi-tube vortexer set at approx 1500 cycles/min. Carefully evaluate the lipid in the vial visually to ensure that complete hydration has occurred and that the lipid vesicles are in a homogenous suspension.

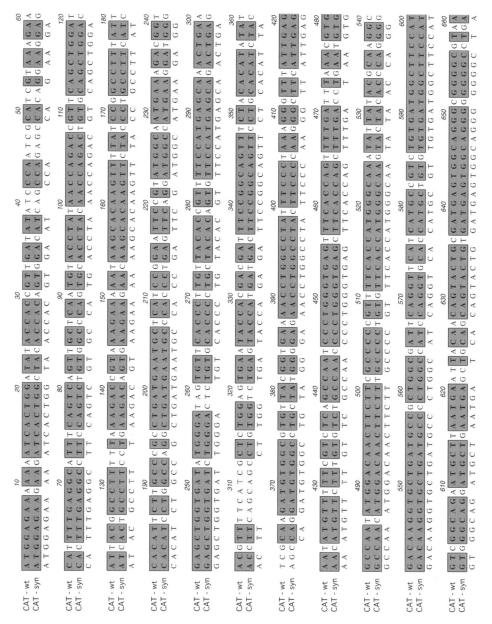

Fig. 1. Comparison of the wild-type *vs* the synthetic gene encoding *E. coli* chloramphenicol acetyltransferase.

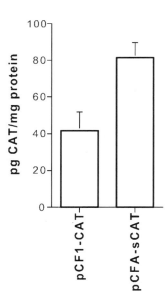

Fig. 2. Increased CAT expression from a plasmid DNA vector containing a codon-optimized and CpG reduced CAT cDNA (sCAT). Cells were transfected with either pCF1-CAT or pCFA-sCAT (GL-67:pDNA, 10.5:60 mM). Cell lysates were collected 2 d after transfection and CAT assays were performed. $n = 6$ wells per plasmid. Data shown are mean ± standard deviation.

2. Aliquot an appropriate volume of the liposome suspension into polystyrene tubes (e.g., for 100 μL mouse instillations, aliquot at least 55 μL for each mouse receiving that formulation).

3. Dilute the pDNA with water (WFI) at 4°C to the appropriate concentration. This has to be determined empirically but typically is in the 0.1–10 mM range (*see* **Note 7**). Aliquot a volume of the pDNA equal to the liposome suspension into polystyrene tubes.

4. Five minutes prior to mixing, warm the complex components to 30°C in a water bath and then gently pipet an equal volume of the cationic lipid into the pDNA solution, avoiding subsequent agitation (*see* **Note 8**). Allow the complexes to form at 30°C for a further 15 min. Once formed, the complexes are stable for up to 4 h at room temperature and can be used for aerosolization or for direct instillation into mouse lung.

3.3.2. Preparation of Cationic Lipid:pDNA Complexes Using Liposome Suspensions

1. Dilute the prepared liposomes with room temperature D5W to be double the final required concentration and dispense an appropriate volume into a

polypropylene microfuge tube (e.g., for 100 µL mouse instillations, aliquot at least 55 µL for each mouse receiving that formulation).

2. Dilute the pDNA with room temperature D5W to the appropriate concentration. This has to be determined empirically but typically is in the 0.1–10 m*M* range (*see* **Note 7**).
3. Rapidly pipet a volume of the pDNA equal to that of the liposome suspension into the liposomes and mix vigorously by pipetting 3 to 4 times.
4. Once formed, the complexes are stable for up to 24 h at room temperature or can be stored for several days at 4°C.

3.4. Physical Characterization of the Cationic Lipid:pDNA Complexes

The extent and nature of the electrostatic interactions between the cationic moiety of the lipid and the anionic phosphates of the pDNA can be variable depending on various experimental factors such as the temperature and vigor of agitation during complex formation, the concentration and purity of the components, and the resultant charge ratio of the complexes. To determine if the complexes generated are reproducible and conform to those of previous experiments, a number of physical characterizations can be performed (*6,18,19*). Dynamic light scattering can be used to assess the overall size of the liposomes and lipid:pDNA preparations using, for example, a NICOMP model 380ZLS particle-sizing system. Zeta potential measurements, which provide an indication of the net charge on the surface of the resultant complexes, can also be measured with this apparatus (*6,10*). Agarose gel electrophoresis can be used to measure the extent of pDNA association with the lipids, and freeze-fracture electron microscopy can be used to visualize the size and morphology of the resultant entities formed (*20*).

3.4.1. Size Determination of Cationic Lipid:pDNA Complexes

The size of the complexes that are formed is an important consideration for gene transduction. Depending on the desired target for genetic modification, differently sized complexes may be more suitable. Most cationic lipid:pDNA complexes generated by the simple mixing of the two components exhibit mean diameters ranging between 300 and 600 nm. These complexes are unlikely to be effective at penetrating the sinusoidal fenestrations (<100 nm) and transducing liver hepatocytes. Indeed, it has been shown that intravenous delivery of such complexes invariably results in localization of a proportion of the complexes in the pulmonary capillaries. The resulting gene transfection associated with systemic delivery of this synthetic vector is primarily in the lung endothelium and not in the epithelium. However, we have determined that lung epithelium can be transduced with complexes over a broad size range when delivered to the lumen of the lung.

The size of the liposomes and cationic lipid:pDNA complexes can be determined by measuring the dynamic light scattering using a NICOMP model 380 or similar instrument.

1. Prepare cationic lipid:pDNA complexes at a cationic lipid concentration of between 100 and 500 μM in water (*see* **Subheading 3.3.**).
2. Measure the dynamic light scattering using a NICOMP model 380 following the manufacturer's instructions. Make these measurements at 25°C, setting the slit width at 250 nm, the viscosity at 1.00, and the index of refraction at 1.00.
3. Take mean diameter measurements from at least three separate readings of 10 cycles.
4. Present the data in terms of either intensity vs mean diameter (i.e., the relative volume of liposomes at a given size distribution) or particle count vs mean diameter (i.e., the number of liposomes detected at a given size distribution). Typically, the mean diameter of GL-67:DOPE:pDNA complexes is between 400 and 500 nm but other cationic lipid:pDNA complexes may range from 100 nm up to 1 µm, depending on the cationic lipid species and also the formulation of the complex. Complexes that are over 1 µm in diameter are likely aggregates resulting from a rapid and irreversible nucleation event. These large complexes, which are visible as a fine precipitate, are not recommended for in vivo studies. With the GL-series of cationic lipids, this precipitation can be minimized by the inclusion of a small amount of DMPE-PEG$_{5000}$ in the lipid formulation (*see* **Note 3**).

3.4.2. Zeta Potential Measurements

The zeta potential of complexes tends to become more positive as the charge ratio of the cationic lipid:pDNA is increased to greater than 1, reflecting that more of the pDNA is present as a complex with increasing amounts of cationic lipid. Zeta potential measurements can be taken using a NICOMP model 380ZLS particle sizing system.

1. Prepare cationic lipid:pDNA complexes at a cationic lipid concentration of between 100 and 500 μM in water (*see* **Subheading 3.3.**).
2. Place the sample into a standard disposable fluorescence cuvet.
3. Using a zeta electrode that accompanies the NICOMP, determine the zeta potential of the samples and evaluate the measurements using the default software parameters for aqueous samples. The zeta potential of the cationic lipid:pDNA complexes is dependent on the molar ratio of the cationic lipid:pDNA complexes. GL-67:DOPE liposomes typically have a zeta potential of approx +55–65 mV. When present as a complex with pDNA, charge neutrality (zeta potential of 0 mV) is attained when the molar ratio of GL-67:pDNA is 1:1.

3.4.3. Extent of pDNA Complexation

1. Prepare complexes by adding 15 μL of cationic lipid (at twice the required final concentration) to 15 μL of 400 μ*M* pDNA (~1.5 μg; *see* **Note 7**).
2. Following complexation, add 5 μL of loading buffer (25% sucrose, Orange G) to each 25 μL sample and then load into the wells of a 0.75% (w/v) agarose gel.
3. Electrophorese the samples in TBE buffer containing 0.5 μg/mL ethidium bromide at approx 100 V for 1–2 h.
4. Visualize the ethidium bromide-stained pDNA on an ultraviolet transilluminator. Because of its size, cationic lipid-complexed pDNA remains in the well. Typically, complexes that are at charge neutrality or that contain an excess of cationic lipid (a positive zeta potential) do not migrate from the well. Free or uncomplexed pDNA, as present in formulations containing an excess of pDNA (a negative zeta potential), will co-migrate with pDNA alone. Quantitation of this band using known standards provides a means to estimate the amount of free pDNA and the extent of pDNA that is present as a complex in the formulation.
5. Determine the amount and integrity of the pDNA in a complex by adding SDS to 0.4% to disrupt the cationic lipid:pDNA interactions prior to electrophoresis.

3.5. Aerosol Delivery of Cationic Lipid:pDNA Complexes to the Mouse Lung

Aerosol delivery of synthetic vector systems to the lung requires that several physical properties of the complexes be taken into consideration *(21)*. Formulation modifications need to be evaluated to identify those that (1) prevent the degradation of pDNA caused by the shear forces generated during nebulization (*see* **Note 9**), and (2) stabilize the complex against lipid-phase separation, which could lead to precipitation. Careful evaluation of the available commercial nebulizers also needs to be performed to identify those that (1) provide an optimal delivery rate of the aerosolized complexes, (2) ensure the generation of appropriately sized respirable particles, and (3) minimize any changes in the physical characteristics of the complexes (e.g., degradation, aggregation, and precipitation).

3.5.1. Analysis of Nebulized Cationic Lipid:pDNA Complexes

As the lung is the target for the aerosolized synthetic vectors, nebulizers need to be evaluated for their ability to deliver the therapeutic within several defined specifications (e.g., particle size, integrity of complexes, rate of delivery). The particle size is determined using an Andersen cascade impactor. The integrity of the nebulized particles can be elucidated by collecting aerosolized complexes on an all glass impinger (AGI). The rate of delivery is a parameter defined by the nebulizer. We have determined that the Pari LC Jet Plus nebulizer *(22)* fulfills the criteria necessary for delivery and maintenance of integrity of cationic lipid:pDNA complexes (*see* **Note 10**). Measurement of the aerosol droplet size is normally performed by nebulizing rhodamine-labeled phosphatidylethanolamine-containing complexes (*see* **Note 10**) into an Andersen cascade impactor. The complexes that collect on the stages of the impactor are recovered by addition of 1% Triton X-100 and the amount of the fluorescent lipid on the different stages quantitated using a fluorometer in essentially the same manner as described for the AGI.

Nebulization of GL-67:DOPE:DMPE-PEG$_{5000}$:pDNA (1:2:0.05:1.3 molar ratio) using the Pari nebulizer under the conditions described does not result in measurable degradation of the pDNA nor does it alter the ratio of the components of the complexes. There is however a slight time-dependent increase in the concentration of the complexes remaining in the nebulizer reservoir, as predicted *(23,24)*. The ratio of cationic lipid and pDNA present in the aerosol can be determined by use of fluorescently labeled lipids (e.g., rhodamine-phosphatidylethanolamine) and pDNA (e.g., TOTO-1) (*see* **Note 10**). The ratio of lipid to pDNA in the nebulized complex is determined by collecting the aerosolized complexes on an AGI *(19,22)*.

1. Weigh the AGI and the nebulizer prior to commencement of aerosolization.
2. Prepare 4 mL of fluorescently labeled complexes (*see* **Subheading 3.1.1.** and **Note 10**). Pipet the complexes into the nebulization chamber of a Pari LC Jet Plus nebulizer. Operate the nebulizer at a flow rate of approx 7.8 L/min, provided by a standard air compressor (*see* **Note 11**).
3. Reweigh the nebulizer to determine the amount of complexes that have been aerosolized.
4. Recover the complexes that are deposited on the AGI by rinsing the impinger sequentially with 4 mL of octylglucoside buffer (100 m*M* octylglucoside in distilled water), then with 4 ml of phosphate-buffered saline (PBS), and again with 4 mL of octylglucoside buffer.
5. Reweigh the AGI to determine the total volume it contains.
6. Quantitate the fluorescence of both the Rh-PE and TOTO-1 present in the AGI washes using a fluorometer. Use the constant wavelength analysis program and standard curves generated using the starting materials diluted in a 1:1 mixture of PBS and octylglucoside buffer to quantitate the amounts of lipid and pDNA transferred to the AGI.
7. Assess the transfection activity of the complexes following nebulization by placing a plate of tissue-culture cells in an acrylic box connected to the nebulizer and vented through a Pall BB-50T Breathing Circuit Filter *(18)*. Following exposure of the cells to the aerosolized complexes, grow the cells

for a further 48 h, then lyse them and determine the extent of transduction and transgene expression.

3.5.2. Nebulization of Cationic Lipid:pDNA Complexes to the Mouse Lung

The ability of aerosolized cationic lipid:pDNA complexes to effect transfection in vivo can be tested in the lungs of mice. Because it is difficult to analyze statistically relevant numbers of mice using adapted masks to provide direct aerosol delivery, compounded by the need for them to be anesthetized, thus altering respiratory function, use of a whole body exposure chamber to deliver the complexes to the mice is preferred (*see* **Note 12**).

1. Place 10 female BALB/c mice (16–18 g) in an acrylic box (approx 20 × 20 × 15 cm) that is connected to the nebulizer using anesthesia tubing and vented through a Pall BB-50T Breathing Circuit Filter.
2. Aerosolize 10 mL of complex (GL-67:DOPE:DMPE-PEG$_{5000}$:pDNA) using 2-s bursts followed by 4-s rests (total time of 20–45 min). The airborne aerosol particles can be seen as a dissipating cloud entering the acrylic chamber, with the mice becoming moist as the droplets coalesce.
3. Following completion of the nebulization process, house the mice for a further 48 h, then sacrifice and analyze their lungs for expression of the transgene (*see* **Note 13**).

3.6. Administration of Cationic Lipid:pDNA Complexes to Mouse Airway Epithelia

We routinely assess the in vivo transfection activity of the synthetic gene-delivery vectors in the lungs of female BALB/c mice (4–6 wk old) by administering a bolus of 100 µL of the complexes via insufflation through the nose *(25)* or by direct injection through a trans-tracheal incision. The instillation technique is the preferred method for complex delivery because it (1) allows a larger cohort of animals to be treated in a given time, (2) results in less animal-to-animal variability in groups of mice receiving the same test article (smaller standard deviations), and (3) generally results in higher levels of transgene expression in the lung (unpublished observations).

3.6.1. Intranasal Instillation of Cationic Lipid:pDNA Complexes

1. Anesthetize mice by inhalation of isoflurane. Place the mice (≤10 at a time) in a rodent anesthesia machine and deliver oxygen at 800 mL/min passed over an isoflurane reservoir set to level 5 (results in approx 5% isoflurane content in the air) for 1 min. Keep the animals under anesthesia by maintaining isoflurane at 2.5% in the chamber.

2. Remove the mice individually and instill immediately by holding the animals upright with their noses up and applying pressure to the lower mandible to immobilize the tongue and prevent the reflex to swallow the complex. Administer the cationic lipid:pDNA complexes ($\leq$100 µL) dropwise (20–25 µL/drop over 15 s) into the nares of mice using a standard Pipetteman (0–200 µL) with an appropriate tip. Using this procedure, we estimate that approx 70% of the total complexes are delivered into the lung. The mice recover from the anesthesia within 30 s of removal from the isoflurane atmosphere.

3. House the mice for a further 48 h, then sacrifice and analyze their lungs for expression of the transgene (*see* **Note 13**).

4. Notes

1. For transfection of the lung, we have observed that significantly higher levels of transgene expression could be realized using the nonprotonated, or free-base forms of the GL-series of cationic lipids than with the salt forms of the protonated amines *(9)*.

2. Preparation of cationic lipid formulations should be performed sterilely to minimize contamination by potential pathogens or pyrogens. Glassware should be depyrogenated by heating at 250°C for 6 h. Where appropriate, materials should also be autoclaved. As an alternative, glassware can be cleaned with a nonresidue detergent (e.g., RBS-35), followed by thorough rinsing with ethanol.

3. The concentration of the complexes that can be aerosolized can be limiting (containing less than 0.5 mg/mL of pDNA) owing to nebulization-induced aggregation *(18)*. However, we have shown that with GL-67:DOPE:pDNA complexes, the inclusion of a small amount of DMPE-PEG$_{5000}$ can act to stabilize the complexes. By incorporating DMPE-PEG5$_{000}$ at as low as 0.05 molar ratio, formulations of complexes that contain as high as 6 mg/mL pDNA (18.2 m*M*) can be generated without aggregation. Addition of DMPE-PEG$_{5000}$ does not alter the size range or the in vivo activity of the nebulized particles. The development of formulations that support high concentrations of pDNA is important as it facilitates delivery of more complex per unit time.

4. Upon application of the vacuum, the residual solvent immediately comes off as a cloud. Following this step the result is a clear lipid film.

5. Cationic lipids, containing unprotonated primary amines, when formulated with a neutral lipid containing an ester-linked acyl chain (e.g., DOPE) can undergo a transacylation reaction. This occurs by nucleophilic attack of the lone pair of electrons of the primary amine on the δ^+ carbonyl group of the ester moiety. This results in the transfer of an acyl chain to the cationic lipid and the generation of an isomeric pair of lysophosphatidylethanolamines.

We have determined that this chemical reaction is time-, pH-, and temperature-dependent, and occurs irrespective of whether the lipids are in an aqueous suspension or a dry film. Aqueous suspensions of these lipid formulations at room temperature transacylate at a rate of approx 1% per hour. However this reaction can be reduced by storing the dry films at $-20°C$ (conversion at a rate of 2% per month over 6 mo) and can be effectively negated by storage at $-70°C$ *(9)*. Hence, for long-term storage of such unprotonated cationic lipids, it is recommended that they be kept as a dry film at $-70°C$. Following suspension in aqueous solution, the rate of transacylation can be reduced by maintaining the pH at least one unit below the pK_a of the primary amine. The primary amines of a cationic lipid possessing a spermine headgroup (such as GL-67) typically have a pK_a of approx 10 *(26)*, thus suspension of the cationic lipids in a relatively neutral (pH 7.0–7.4) buffer should be protective.

6. For the GL-series of cationic lipids, formulation of the lipid:pDNA complexes in excipients, even with the solutes as low as 10 mM, results in significantly reduced transgene expression when compared with complexes formulated in water *(9)*.

7. The molarity of the pDNA solution refers to the molarity of the nucleotides and is calculated based on an average nucleotide molecular weight of 330 Daltons. For example, a plasmid of 5000 base pairs at a concentration of 1 mg/mL is at the same molarity (3.03 mM) as a plasmid of 10,000 base pairs that is at a concentration of 1 mg/mL. Polynucleic acids of many forms and chain lengths have been successfully transfected. In addition to pDNA these include cosmids, mRNA, ribozymes, sheared genomic DNA, mammalian artificial chromosomes, and oligonucleotides (including S-oligos and RNA-DNA oligos). All of the nucleic acids should contain less than 5 μg endotoxin U/mg to prevent an adverse inflammatory response in the treated animals. Endotoxin levels can be determined using the *Limulus* amebocyte lysate (LAL) assay.

8. Agitation of the cationic lipid:pDNA mixture during complex formation can lead to a nucleation event that results in aggregation and precipitation of the complexes. Also, complexes are prone to aggregation and precipitation when the formulation has a net neutral charge. This problem can be ameliorated by the inclusion of DMPE-PEG$_{5000}$ *(18)* (*see* also **Note 3**).

9. The nebulization process rapidly degrades uncomplexed pDNA. To minimize this loss, the pDNA can be protected by completely complexing it with the cationic lipid *(18,19)*. Formulations of cationic lipid:pDNA complexes that are close to charge neutrality or that have a slightly positive charge ratio should conform to this requirement. The extent of pDNA complexation can be readily assessed by gel electrophoresis (*see* **Subheading 3.4.3.**).

10. The cationic liposomes can be formulated at a 2.5 mol% with rhodamine-phosphatidylethanolamine (Rh-PE) (32 mg in the example in **Subheading 3.1.1.**) and pDNA can be labeled with TOTO-1 (one TOTO-1 per thousand nucleotides).
11. A flow rate of 7.8 L/min results in delivery of 0.62 ± 0.06 mL of GL-67:DOPE:DMPE-PEG$_{5000}$:pDNA complexes/min. The mass median aerodynamic diameter (mmad) of the aerosol droplets generated under these conditions is approx 2.2 μm with a geometric standard deviation (gsd) of 3.0, which is within the respirable range (1–5 μm) *(22)*.
12. Aerosol delivery of cationic lipid:pDNA complexes to mouse lung is very inefficient with approx 2 μL being deposited onto the airway epithelium for every 10 mL nebulized.
13. Transcriptional activity of the commonly used CMV promoter can be enhanced by instilling 100 μL of 20 m*M* sodium butyrate to the mouse (*see* **Subheading 3.6.1.**). If this is performed 24 h prior to tissue harvest, a two- to fivefold elevation in transgene expression can typically be attained that can be beneficial when assessing the efficiency of transduction.

References

1. Caplen, N. J., Alton, E. W. F. W., Middleton, P. G., Dorin, J. R., Stevenson, B. J., Gao, X., et al. (1995) Liposome-mediated CFTR gene transfer to the nasal epithelium of patients with cystic fibrosis. *Nature Med.* **1,** 39–46.
2. Gill, D. R., Southern, K. W., Mofford, K. A., Seddon, T., Huang, L., Sorgi, F., et al. (1997) A placebo-controlled study of liposome-mediated gene transfer to the nasal epithelium of patients with cystic fibrosis. *Gene Ther.* **4,** 199–209.
3. Porteus, D. J., Dorin, J. R., McLachlan, G., Davidson-Smith, H., Stevenson, B. J., Carothers, A. D., et al. (1997) Evidence for the safety and efficacy of DOTAP cationic liposome mediated CFTR gene transfer to the nasal epithelium of patients with cystic fibrosis. *Gene Ther.* **4,** 210–218.
4. Zabner, J., Cheng, S. H., Meeker, D., Launspach, J., Balfour, R., Perricone, M. A., et al. (1997) Comparison of DNA-lipid complexes and DNA alone for gene transfer to cystic fibrosis airway epithelia in vivo. *J. Clin. Invest.* **100,** 1529–1537.
5. Alton, E. W. F. W., Stern, F., Farley, R., Jaffe, A., Chadwick, S. L., Phillips, J., et al. (1999) Cationic lipid-mediated CFTR gene transfer to the lungs and nose of patients with cystic fibrosis: a double-blind placebo-controlled trial. *Lancet* **353,** 947–954.
6. Eastman, S. J., Siegel, C., Tousignant, J. D., Smith, A. E., Cheng, S. H., and Scheule, R. K. (1997) Biophysical characterization of cationic lipid:DNA complexes. *Biochim. Biophys. Acta* **1325,** 41–62.
7. Marshall, J., Yew, N. S., Eastman, S. J., Jiang, C., Scheule, R. K., and Cheng, S. H. (1999) Cationic lipid-mediated gene delivery to the airways, in *Non-viral Vectors*

for Gene Therapy (Huang, L, Hung, M.C., and Wagner, E., eds.), Academic Press, San Diego, CA., pp. 39–68.

8. Lee, E. R., Marshall, J., Siegel, C. S., Jiang, C., Yew, N. S., Nichols, M. R., et al. (1996) Detailed analysis of structures and formulations of cationic lipids for efficient gene transfer to the lung. *Hum. Gene Ther.* **7**, 1701–1717.

9. Marshall, J., Nietupski, J. B., Lee, E. R., Siegel, C.S., Rafter, P. W., Rudginsky, S. A., et al. (2000) Cationic lipid structure and formulation considerations for optimal gene transfection of the lung. *J. Drug Target.* **7**, 453–469.

10. McCluskie, M. J., Chu, Y., Xia, J-L., Jessee, J., Gebyehu, G., and Davis, H. L. (1998) Direct gene transfer to the respiratory tract of mice with pure plasmid and lipid-formulated DNA. *Antisense Nucleic Acid Drug Dev.* **8**, 401–414.

11. Scheule, R. K., St. George, J. A., Bagley, R. G., Marshall, J., Kaplan, J. M., Akita, G. Y., et al. (1997) Basis of pulmonary toxicity associated with cationic lipid-mediated gene transfer to the mammalian lung. *Hum. Gene Ther.* **8**, 689–707.

12. Yew, N. S., Wang, K. X., Przybylska, M., Bagley, R. G., Stedman, M., Marshall, J., et al. (1999) Contribution of plasmid DNA to inflammation in the lung after administration of cationic lipid:pDNA complexes. *Hum. Gene Ther.* **10**, 223–234.

13. Smyth Templeton, N., Lasic, D. D., Frederik, P. M., Strey, H. H., Roberts, D. D., and Pavlakis, G. N. (1997) Improved DNA:liposome complexes for increased systemic delivery and gene expression. *Nature Biotech.* **15**, 647–652.

14. Kim, C. H., Oh, Y., and Lee, T. H. (1997) Codon optimization for high-level expression of human erythropoietin (EPO) in mammalian cells. *Gene* **199**, 293–301.

15. Kozak, M. (1986) Point mutations define a sequence flanking the AUG initiator codon that modulates translation by eukaryotic ribosomes. *Cell* **44**, 283–292.

16. McCaughan, K. K., Brown, C. M., Dalphin, M. E., Berry, M. J., and Tate, W. P. (1995) Translational termination efficiency in mammals is influenced by the base following the stop codon. *Proc. Natl. Acad. Sci. USA* **92**, 5431–5435.

17. Yew, N. S., Zhao, H., Wu, I-H, Song, A., Tousignant, J. D., Przybylska, M., and Cheng, S. H. (2000) Reduced inflammatory response to plasmid DNA vectors by elimination and inhibition of immunostimulatory CpG motifs. *Mole. Ther.* **1**, 255–262.

18. Eastman, S. J., Tousignant, J. D., Lukason, M. J., Murray, H., Siegel, C. S., Constantino, P., et al. (1997) Optimization of formulations and conditions for the aerosol delivery of functional cationic lipid:DNA complexes. *Hum. Gene Ther.* **8**, 313–322.

19. Eastman, S. J., Lukason, M. J., Tousignant, J. D., Murray, H., Lane, M. D., St. George, J. A., et al. (1997) A concentrated and stable aerosol formulation of cationic lipid:DNA complexes giving high-level gene expression in mouse lung. *Hum. Gene Ther.* **8**, 765–773.

20. Fisher, K. and Branton, D. (1974) Application of the freeze-fracture technique to natural membranes. *Methods Enzymol.* **32**, 35–44.

21. Cheng, S. H. and Scheule, R. K. (1998) Airway delivery of cationic lipid:DNA complexes for cystic fibrosis. *Adv. Drug Delivery Rev.* **30**, 173–184.

22. Eastman, S. J., Tousignant, J. D., Lukason, M. J., Chu, Q., Cheng, S. H., and Scheule, R. K. (1998) Aerosolization of cationic lipid:pDNA complexes: in vitro optimization of nebulizer parameters for human clinical trials. *Hum. Gene Ther.* **9,** 43–52.

23. Mercer, T. T., Tillery, M. I., and Chow, H. Y. (1968) Operating characteristics of some compressed-air nebulizers. *Am. Ind. Hyg. Assoc.* **29,** 66–78.

24. Dennis, J. H., Stenton, S. C., Beach, J. R., Avery, A. J., Walters, E. H., and Hendrick, D. J. (1990) Jet and ultrasonic nebulizer output: use of a new method for direct measurement of aerosol output. *Thorax* **45,** 728–732.

25. Jiang, C., O'Connor, S. P., Fang, S. L., Wang, K. X., Marshall, J., Williams, J. L., et al. (1998) Efficiency of cationic lipid-mediated transfection of polarized and differentiated airway epithelial cells in vitro and in vivo. *Hum. Gene Ther.* **9,** 1531–1542.

26. Geall, A. J., Taylor, R. J., Earll, M. E., Eaton, M. A. W., and Blagbrough, I. S. (2000) Synthesis of cholesteryl polyamine carbamates: pK_a studies and condensation of calf thymus DNA. *Bioconjug. Chem.* **11,** 314–326.

8

Cationic Liposome-Mediated DNA Delivery to the Lung Endothelium

Young K. Song, Guisheng Zhang, and Dexi Liu

1. Introduction

The lung endothelium has been studied as one of the most important sites for systemic DNA delivery using nonviral vectors. This is because not only does the lung play a critical role in maintaining the O_2/CO_2 concentration in blood, but it also has the largest capillary bed and an extensive cell surface area to allow DNA to bind to and enter the endothelial cells. An additional advantage of the lung endothelium for DNA delivery is that the entire blood volume circulates through the lung and thus, the intravenously injected DNA molecules have full excess to the lung endothelium. Furthermore, blood flow through the pulmonary capillaries is slow and discontinuous. On inspiration, as the alveoli fill with air, blood flow slows and stops. On expiration, blood flow resumes. Slower blood flow in the lung favors DNA binding to the endothelial cells.

Numerous studies have been conducted in the past decade to optimize DNA delivery to the lung endothelium utilizing cationic liposomes as carriers *(1–14)*. Many physicochemical parameters including cationic lipid structure, liposome composition, liposome size, cationic lipid to DNA ratio in the lipoplexes, and the injected dose were found to be important in determining the transfection activity of cationic liposomes *(4–8)*. It is believed that these physicochemical parameters influence the ultimate transfection activity by affecting the interaction between DNA and the lung endothelium *(8,15,16)*. In contrast to liposome-

From: *Methods in Molecular Biology, vol. 245:*
Gene Delivery to Mammalian Cells: Vol. 1: Nonviral Gene Transfer Techniques
Edited by: W. C. Heiser © Humana Press Inc., Totowa, NJ

mediated DNA delivery in vitro (cell culture), mechanistic studies revealed that the lipoplex structure is not critical for the transfection of lung endothelial cells upon intravenous administration into mice *(16)*. Injection of DNA in lipoplexes or free liposomes followed by naked DNA results in identical levels of transgene expression in the lung *(11,12)*. The function of cationic liposomes in the intravenous transfection of the lung endothelium appears to be prolonging the retention time of DNA in the lung capillary *(11)*.

The objective of this chapter is to provide researchers with the information necessary to carry out their own endothelium transfection studies with cationic liposomes. It includes the conditions for liposome preparation, preparation of DNA/liposome complexes, transfection of animals as well as the techniques for measurement of luciferase reporter gene expression. The procedure does not need special skills and training.

2. Materials

1. DNA: any plasmid containing a CMV promoter-driven reporter gene (e.g., luciferase, β-galactosidase, green fluoresence protein, or human α1-antitrypsin) may be used. Prepare plasmid by any standard method and purify using Qiagen columns or CsCl-ethidium bromide gradient centrifugation *(17)*. Store purified plasmid in TE at −20°C (*see* **Note 1**).
2. Cationic lipids: 1,2-dipalmitoyl-3-trimethylammoniumpropane chloride (DOTAP) is available from Avanti Polar Lipids (Atlanta, GA). *N*-[1-(2,3-dioleyloxy)propyl]-*N,N,N*-trimethylammonium chloride (DOTMA) was synthesized according to procedure described in *(18)*. Many other cationic lipids have been synthesized *(13,14,18–40)*, some of which are commercially available (*see* **Note 2**).
3. Phosphate-buffered saline (PBS): 10 mM Na_2HPO_4, 1.8 mM KH_2PO_4, 137 mM NaCl, 2.7 mM KCl, pH 7.4.
4. Tris-EDTA (TE) buffer: 10 mM Tris-HCl, 1 mM ethylenediaminetetracetic acid (EDTA), pH 8.0.
5. Luciferase lysis buffer: 0.1 M Tris-HCl, 2 mM EDTA, 0.05% Triton X-100, pH 7.8.
6. Luciferase Assay System: luciferase assay kit is available from Promega (Madison, WI).
7. Saline: 0.9% NaCl in filter-sterilized distilled water.
8. Anesthetization solution: 20 mg/mL of 2,2,2-tribromoethanol in saline.
9. Coomassie blue protein assay reagent is available from Bio-Rad (Hercules, CA).
10. Tissue Tearor is available from Biospec Products, Inc. (Racine, WI).
11. CD-1 male mice (18–20 g) are available from Charles River (Wilmington, MA).

3. Methods

3.1. Preparation of Cationic Liposomes

1. To a glass tube (12 × 75 mm) add an appropriate volume of solution containing 20 µmol of cationic lipids in chloroform (*see* **Note 3**).
2. Evaporate the organic solvent under a stream of nitrogen gas to form a thin lipid film by rotating the glass tube at a 45° angle during evaporation.
3. Vacuum desiccate for 2 h at room temperature to ensure a complete removal of the organic solvent.
4. Add 1 mL of PBS to make a solution with a lipid concentration of 20 µmol/mL. Keep the lipid solution at room temperature for 30 min with occasional agitation on a vortex mixer (*see* **Note 4**).
5. Sonicate the lipid suspension in a water bath sonicator for 5 min, homogenize for 2–3 min using a Tissue Tearor at maximal speed, or by extruding 10–20 times through two layers of polycarbonate membrane with a pore size of 0.4 µm using an extrusion device (*see* **Note 5**).

3.2. Preparation of DNA/Cationic Liposome Complexes

1. Calculate the amount of plasmid DNA needed for each experiment on the basis of 25 µg/mouse. Prepare DNA solution at a concentration of 250 µg/mL in PBS.
2. Calculate the amount of cationic liposomes needed for each experiment on the basis of 900 nmole/mouse. Dilute the cationic liposome solution with PBS to a concentration of 9 µmole/mL.
3. To the calculated volume of liposome solution add drop-by-drop and with gentle vortexing an equal volume of DNA solution. (Final concentration: [DNA] = 125 µg/mL, [Lipid] = 4.5 µmol/mL, [DNA/lipid] = 1 µg/36 nmole, [Charge ratio, +/−] = 12:1) (*see* **Notes 6** and **7**).
4. Keep the DNA/liposome mixture on ice for 30 min.

3.3. Intravenous Administration of DNA/Cationic Liposome Complexes

1. Restrain male CD-1 mice (18–20 g) in a standard mouse restrainer, locate the tail vein (*see* **Note 8**) and inject 200 µL of the DNA/liposome mixture by means of a sterile 1-cc syringe and 27G1/2 needle (*see* **Note 9**).

3.4. Determination of Luciferase Expression

1. Anesthetize mice by intraperitoneal injection of 600 mg/kg of tribromoethanol.
2. Harvest the lungs (and other organs if needed) from the anesthetized mice. Place the lung from each mouse into a glass tube and add 1 mL of ice-cold luciferase lysis buffer. Keep the samples on ice.

3. Homogenize the lung using a Tissue Tearor at maximum speed for 20–30 s. Keep the sample on ice.
4. Transfer the tissue homogenate to a microcentrifuge tube and centrifuge at 10,000g for 10 min at 4°C.
5. Assay luciferase activity in 10 µL of supernatant using the Luciferase Assay System and a luminometer. Measure light output for 10 s following injection of luciferin.
6. Assay the protein content in 1 µL of supernatant by the Coomassie blue protein staining assay.
7. Calculate the level of luciferase gene expression as relative light units (RLU) per mg of protein in the lung extract. The amount of luciferase protein in the lung can be calculated based on an established standard curve developed using purified luciferase enzyme (*see* **Note 10** and **11**).

4. Notes

1. Verify the purity of plasmid DNA spectrophotometrically and by 1% agarose gel electrophoresis. The DNA solution should have an OD_{260}/OD_{280} ratio greater than 1.8 and be free of genomic DNA contamination as indicated by electrophoresis. Plasmids purified by Qiagen columns or CsCl-etheidium bromide density gradient centrifugation are equally active. Supercoiled or linear DNA exhibits similar activity.
2. Cationic lipids with two hydrocarbon chains (C18Δ9) as the hydrophobic anchor, a quaternary ammonium as the cationic head group and ether bonds to connect the head group and hydrophobic moieties appear to be superior to the cholesterol-based and polycationic lipids for transfection of the lung endothelium *(8,14)*. Although many cationic lipids with these preferred structural features have been synthesized *(13,14,18–40)*, very few are commercially available. For some of our experiments, we use DOTAP. We have found that the transfection activity of DOTAP is 10-fold lower than its counterpart, *N*-[1-(2,3-dioleyloxy) propyl]-*N*,*N*,*N*-trimethylammonium chloride (DOTMA) under optimal conditions *(8)*. The only difference between DOTMA and DOTAP is that DOTAP has two ester linkage bonds in contrast to two ether bonds in DOTMA. It is also worth noting that most commercially available transfection reagents that work well in cell culture do not necessarily work well for systemic DNA delivery to the lung endothelium.
3. Inclusion of additional lipids into the liposome composition can be accomplished by mixing the lipids before the removal of organic solvent. At a lower +/– ratio, inclusion of cholesterol into the liposomes at a molar ratio of 2:1 (cationic lipid:cholesterol) enhances the transfection activity of cationic liposomes *(8)*.

4. The temperature for lipid hydration and liposome preparation should be at least 10° above the phase-transition temperature (T_m) of the lipids.

5. As self-assembling structures in aqueous solution, liposomes can be prepared in many different ways. The most convenient methods include water bath sonication, membrane extrusion, or simple agitation *(41)*. Our experience with intravenous transfection of the lung endothelium is that liposomes with an average diameter of less than 400 nm performed better than those of smaller size (<400 nm). Although emphasized in a few studies, the size homogeneity of liposomes does not seem to be important with regard to the level of transgene expression in the lung. For this reason, we often use a Tissue Tearor to make cationic liposomes *(6)*. To obtain liposomes with a narrow size distribution, the easiest method is extrusion using a LiposoFast extrusion device (Avestin, Ottawa, Canada). Two layers of Nucleopore polycarbonate membrane (Costar, Cambridge, MA) of defined pore size of 0.4 μm are commonly used. Multiple extrusions (10–20 times) are needed to ensure size homogeneity.

6. Transfection activity of cationic liposomes is dependent on the ratio of cationic lipid to DNA *(6,8)*. While the charge ratio of 12:1 (+/−) described here is optimal for DOTMA and DOTAP liposomes *(8)*, different ratios may need to be examined when new lipids or liposome compositions are utilized. To prepare lipoplexes with various charge ratios, we usually prepare liposome solution of different lipid compositions and keep the DNA concentration and the mixing volumes the same. Owing to the polyionic nature of DNA and cationic liposomes, a common problem for preparation of lipoplexes is the formation of large aggregates when mixing DNA with liposome solutions, especially at lower charge ratio (e.g., +/− = 1). A higher charge ratio as suggested in this protocol solves the aggregation problem to a certain extent. If aggregation becomes a problem, we recommend gentle shaking instead of vortexing for mixing liposomes and DNA solutions.

7. Alternatively, similar levels of luciferase gene expression have been obtained by a sequential injection procedure *(11,12,42)* where liposomes (100 μL) are injected into an animal 5 min prior the injection of DNA solution (100 μl). Although one additional injection is performed on the animal, this procedure eliminates the need for the preparation of lipoplexes. Results from a recent study showed that the sequential injection protocol significantly reduces the acute immune response induced by lipoplexes *(42)*.

8. Although tail-vein injections are easy and routine in many laboratories, we found that warming the animals briefly (1–2 min) under an infrared heating lamp enhances visualization of the tail vein and allows for easy insertion of the needle.

9. The amount of DNA and cationic liposomes needed for transfection has been optimized and is a balance between efficiency of transfection and safety for animals. Exceeding these limits may result in significant toxicity to the animals.

10. Converting the enzyme activity of luciferase to the amount of luciferase protein as a measure of luciferase gene expression requires a standard curve. We have found that standard curves generated by using pure luciferase vary significantly depending on the buffer system used. We recommend that an identical buffer composition should be used for both sample and standard in order to avoid errors in the estimation of luciferase level in the sample. In particular, for studies aimed at estimating the level of lu-

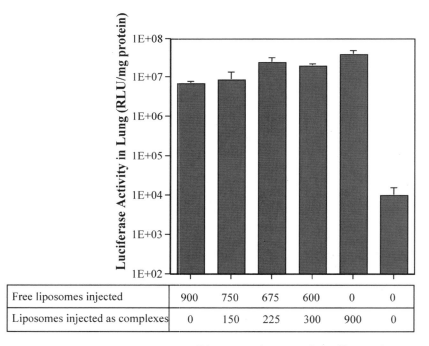

Free liposomes injected	900	750	675	600	0	0
Liposomes injected as complexes	0	150	225	300	900	0

Liposome Amount (nmol/mouse)

Fig. 1. Cationic liposome-mediated transfection of the lung. Mice were injected with either DNA/DOTMA complexes, or free DOTMA liposomes first followed by 25 µg of free pCMV-Luc plasmid or same amount of plasmid complexed with different amounts of DOTMA liposomes. In all cases, 900 nmol of DOTMA was injected into each mouse. Eight hours following plasmid injection the mice were sacrificed and luciferase activity was determined in the lungs. Data represent mean ±SEM of values from three to six animals.

ciferase gene expression in vivo where the lysis buffer has to be used to release the luciferase from the cells, it is crucial to use the tissue extract, not the lysis buffer alone, as the dilution buffer for establishment of the standard curve in order to avoid a two to three orders of magnitude overestimation of the luciferase level in tissues.

11. **Figure 1** shows the effect of total amount of cationic lipids on the level of luciferase gene expression in the lung after intravenous transfection. It is evident that it is the total amount of cationic liposomes delivered that determines the level of luciferase gene expression. DNA injected in free form following injection of free liposomes or in lipoplex form with different amount of liposomes gives almost identical levels of gene expression, supporting the conclusion that cationic liposome-mediated gene transfer into the lung endothelium does not depend on the structure of injected lipoplexes *(8,16,42)*.

Acknowledgment

We would like to thank Drs. Joseph E. Knapp and William C. Heiser for critical reading of the manuscript. The original work was supported by grant from NIH (CA72529) and research contract from Targeted Genetics Corporation.

References

1. Brigham, K. L., Meyrick, B., Christman, B., Magnuson, M., King, G., and Berry, L. C., Jr. (1989) In vivo transfection of murine lungs with a functioning prokaryotic gene using a liposome vehicle. *Am. J. Med. Sci.* **298**, 278–281.
2. Zhu, N., Liggitt, D., Liu, Y., and Debs, R. (1993) Systemic gene expression after intravenous DNA delivery into adult mice. *Science* **261**, 209–211.
3. Thierry, A. R., Lunardi-Iskandar, Y., Bryant, J. L., Rabinovich, P., Gallo, R. C., and Mahan, L. C. (1995) Systemic gene therapy: biodistribution and long-term expression of a transgene in mice. *Proc. Natl. Acad. Sci. USA* **92**, 9742–9746.
4. Hong, K., Zheng, W., Baker, A., and Papahadjopoulos, D. (1997) Stabilization of cationic liposome-plasmid DNA complexes by polyamines and poly(ethylene glycol)-phospholipid conjugates for efficient in vivo gene delivery. *FEBS Lett.* **400**, 233-237.
5. Li, S. and Huang, L. (1997) In vivo gene transfer via intravenous administration of cationic lipid-protamine-DNA (LPD) complexes. *Gene Ther.* **4**, 891–900.
6. Liu, F., Qi, H., Huang, L., and Liu, D. (1997) Factors controlling the efficiency of cationic lipid-mediated transfection in vivo via intravenous administration. *Gene Ther.* **4**, 517–523.
7. Liu, Y., Mounkes, L. C., Liggitt, H. D., Brown, C. S., Solodin, I., Heath, T. D., and Debs, R. J. (1997) Factors influencing the efficiency of cationic liposome-mediated intravenous gene delivery. *Nat. Biotechnol.* **15**, 167–173.

8. Song, Y. K., Liu, F., Chu, S., and Liu, D. (1997) Characterization of cationic liposome-mediated gene transfer in vivo by intravenous administration. *Hum. Gene Ther.* **8**, 1585–1594.

9. Templeton, N. S., Lasic, D. D., Frederik, P. M., Strey, H. H., Roberts, D. D., and Pavlakis, G. N. (1997) Improved DNA: liposome complexes for increased systemic delivery and gene expression. *Nat. Biotechnol.* **15**, 647–652.

10. Barron, L. G., Meyer, K. B., and Szoka, F. C., Jr. (1998) Effects of complement depletion on the pharmacokinetics and gene delivery mediated by cationic lipid-DNA complexes. *Hum. Gene Ther.* **9**, 315–323.

11. Song, Y. K., Liu, F., and Liu, D. (1998) Enhanced gene expression in mouse lung by prolonging the retention time of intravenously injected plasmid DNA. *Gene Ther.* **5**, 1531–1537.

12. Song, Y. K. and Liu, D. (1998) Free liposomes enhance the transfection activity of DNA/lipid complexes in vivo by intravenous administration. *Biochim. Biophys. Acta.* **1372**, 141–150.

13. Ren, T., Zhang, G., Song, Y. K., and Liu, D. (1999) Synthesis and characterization of aromatic ring-based cationic lipids for gene delivery in vitro and in vivo. *J. Drug Target* **7**, 285–292.

14. Ren, T., Song, Y. K., Zhang, G., and Liu, D. (2000) Structural basis of DOTMA for its high intravenous transfection activity in mouse. *Gene Ther.* **7**, 764-768.

15. McLean, J. W., Fox, E. A., Baluk, P., Bolton, P. B., Haskell, A., Pearlman, R., et al. (1997) Organ-specific endothelial cell uptake of cationic liposome-DNA complexes in mice. *Am. J. Physiol.* **273**, H387–404.

16. Liu, D., Knapp, E. J., and Song, Y. K. (1999) Mechanisms of cationic liposome-mediated transfection of the lung endothelium, in *Nonviral Vectors for Gene Therapy* (Huang, L., Wagner, E., and Hung, M. C., eds.), Academic Press, San Diego, CA, pp. 313–335.

17. Sambrook, J., Fritsch, E. F., and Maniatis, T. (1989) *Molecular Cloning: A Laboratory Manual*, 2nd ed. Cold Spring Harbor Laboratory Press, Cold Spring Harbor, New York.

18. Felgner, P. L., Gadek, T. R., Holm, M., Roman, R., Chan, H. W., Wenz, M., et al. (1987) Lipofection: a highly efficient, lipid-mediated DNA-transfection procedure. *Proc. Natl. Acad. Sci. USA* **84**, 7413–7417.

19. Behr, J. P., Demeneix, B., Loeffler, J. P., and Perez-Mutul, J. (1989) Efficient gene transfer into mammalian primary endocrine cells with lipopolyamine-coated DNA. *Proc. Natl. Acad. Sci. USA* **86**, 6982–6986.

20. Leventis, R. and Silvius, J. R. (1990) Interactions of mammalian cells with lipid dispersions containing novel metabolizable cationic amphiphiles. *Biochim. Biophys. Acta.* **1023**, 124–132.

21. Gao, X. and Huang, L. (1991) A novel cationic liposome reagent for efficient transfection of mammalian cells. *Biochem. Biophys. Res. Commun.* **179**, 280–285.

22. Hawley-Nelson, P., Ciccarone, V., Gebeyehu, G., and Jessee, J. (1993) Lipofectamine reagent: a new, higher efficiency polycationic liposome transfection reagent. *Focus* **15**, 73–79.

23. San, H., Yang, Z. Y., Pompili, V. J., Jaffe, M. L., Plautz, G. E., Xu, L., et al. (1993) Safety and short-term toxicity of a novel cationic lipid formulation for human gene therapy. *Hum. Gene Ther.* **4**, 781–788.
24. Le Bolc'H, G., Lebris, N., Yaouanc, J. J., Clement, J. C., and des Abbayes, H., (1995) Cationic phosphonolipids as non viral vectors for DNA transfection. *Tetrahedron Lett.* **36**, 6681–6684.
25. Solodin, I., Brown, C. S., Bruno, M. S., Chow, C. Y., Jang, E. H., Debs, R. J., and Heath, T. D. (1995) A novel series of amphiphilic imidazolinium compounds for in vitro and in vivo gene delivery. *Biochemistry* **34**, 13537–13544.
26. Lee, E. R., Marshall, J., Siegel, C. S., Jiang, C., Yew, N. S., Nichols, M. R., et al. (1996) Detailed analysis of structures and formulations of cationic lipids for efficient gene transfer to the lung. *Hum. Gene Ther.* **7**, 1701–1717.
27. Wheeler, C. J., Felgner, P. L., Tsai, Y. J., Marshall, J., Sukhu, L., Doh, S. G., et al. (1996) A novel cationic lipid greatly enhances plasmid DNA delivery and expression in mouse lung. *Proc. Natl. Acad. Sci. USA* **93**, 11454–11459.
28. Deshmukh, H. M. and Huang, L. (1997) Liposome and polylysine mediated gene transfer. *New J. Chem.* **21**, 113–124.
29. Byk, G. and Scherman, D. (1998) Novel cationic lipids for gene delivery and gene therapy. *Expert Opin. Ther. Patents* **8**, 1125–1141.
30. Byk, G., Dubertret, C., Escriou, V., Frederic, M., Jaslin, G., Rangara, R., et al. (1998) Synthesis, activity, and structure-activity relationship studies of novel cationic lipids for DNA transfer. *J. Med. Chem.* **41**, 229–235.
31. Floch, V., Legros, N., Loisel, S., Guillaume, C., Guilbot, J., Benvegnu, T., et al. (1998) New biocompatible cationic amphiphiles derivative from glycine betaine: a novel family of efficient nonviral gene transfer agents. *Biochem. Biophys. Res. Commun.* **251**, 360–365.
32. Tang, F. and Hughes, J. A. (1998) Introduction of a disulfide bond into a cationic lipid enhances transgene expression of plasmid DNA. *Biochem. Biophys. Res. Commun.* **242**, 141–145.
33. Bhattacharya, S. and Dileep P. V. (1999) Synthesis of novel cationic lipids with oxyethylene spacers at the linkages between hydrocarbon chains and pseudoglyceryl backbone. *Therahedron Lett.* **40**, 8167–8171.
34. Nazih, A., Cordier, Y., Bischoff, R., Kolbe, H. V. J., and Heissler, D. (1999) Synthesis and stability study of the new pentammonio lipid pcTG90, a gene transfer agent. *Tetrahedron Lett.* **40**, 8089–8091.
35. Ren, T. and Liu, D. (1999) Synthesis of cationic lipids from 1,2,4-butanetriol. *Therahedron Lett.* **40**, 209–212.
36. Ren, T. and Liu, D. (1999) Synthesis of targetable cationic amphiphiles. *Therahedron Lett.* **40**, 7621–7625.
37. Bessodes, M., Dubertret, C., Jaslin, G., and Scherman, D. (2000) Synthesis and biological properties of new glycosidic cationic lipids for DNA delivery. *Bioorg. Med. Chem. Lett.* **10**, 1393–1395.
38. Frederic, M., Scherman, D., and Byk, G. (2000) Introduction of cyclic guanidines into cationic lipids for non-viral gene delivery. *Tetrahedron Lett.* **41**, 675–679.

39. Ren, T., Zhang, G., Liu, F., and Liu, D. (2000) Synthesis and evaluation of vitamin D-based cationic lipids for gene delivery in vitro. *Bioorg. Med. Chem. Lett.* **10**, 891–894.

40. Choi, J. S., Lee, E. J., Jang, H. S., and Park, J. S. (2001) New cationic liposomes for gene transfer into mammalian cells with high efficiency and low toxicity. *Bioconjug. Chem.* **12**, 108–113.

41. Woodle, M. C. and Papahadjopoulos, D. (1989) Liposome preparation and size characterization. *Methods Enzymol.* **171**, 193–217.

42. Tan, Y., Liu, F., Li, Z., Li, S., and Huang, L. (2001) Sequential injection of cationic liposome and plasmid DNA effectively transfects the lung with minimal inflammatory toxicity. *Mol. Ther.* **3**, 673–682.

9

Delivery of DNA to Tumor Cells Using Cationic Liposomes

Duen-Hwa Yan, Bill Spohn, and Mien-Chie Hung

1. Introduction

Cancer is a genetic disease. A cancer cell usually contains DNA abnormalities that may activate oncogenes or inactivate tumor suppressors or both. The identification of these cancer-causing genes and the signal-transduction pathways involved has allowed the development of cancer gene-therapy based on the idea that, by introducing certain therapeutic genes, cancer cells can be eliminated. Despite the promise of this simple but powerful concept, the clinical efficacy remains to be demonstrated. Thus far, the majority of cancer gene-therapy clinical trials used viral vectors to deliver DNA because of their high efficiency of gene transfer. However, the biological problems associated with viral vectors, such as the powerful immune responses to viral vectors and the ability of many cell types to turn off viral promoters, have limited their clinical usefulness. To circumvent these barriers, nonviral delivery systems such as naked DNA injection, physical (gene gun or electroporation), and chemical (cationic liposome or polymer) approaches have been developed (1). In particular, since the discovery of its ability to complex and condense DNA and the subsequent demonstration of efficient DNA delivery in vitro (2,3), cationic liposomes have been widely used to introduce DNA into cells (4). The efficiency of cationic liposome-mediated DNA delivery is thought to be attributed to the following: (1) an efficient condensation of DNA by electrostatic interaction

From: *Methods in Molecular Biology, vol. 245:*
Gene Delivery to Mammalian Cells: Vol. 1: Nonviral Gene Transfer Techniques
Edited by: W. C. Heiser © Humana Press Inc., Totowa, NJ

between the positively charged liposomes and the negatively charged DNA; (2) an efficient interaction between a net positive charge of the cationic liposome/DNA complex with the negatively charged cell membrane; and (3) an efficient intracellular DNA release owing to the fusogenic properties of cationic liposomes that fuse and/or destabilize the plasma membrane. Thus, cationic liposomes are generally considered to be a safe and versatile DNA delivery system. Given that, cationic liposomes still suffer from low transfection efficiency in clinical trials as compared with viral vectors. Much effort has been given to optimize the efficiency of DNA transfer by developing better formulations for cationic liposomes. In this chapter, we describe the preparation of an efficient in vivo DNA delivery system, that is, Stabilized Non-viral (SN) cationic liposomes, developed and routinely used in our laboratory *(5–7)*. The purpose of this chapter is to provide for the scientific community a step-by-step protocol using common laboratory equipment to prepare a lipid formulation that works well in our laboratory. The SN cationic liposome formulation is composed of 1,2-Dipalmitoyl-sn-Glycero-3-Ethylphosphocholine (PC), 1,2-Dipalmitoyl-sn-Glycero-3-Phosphoethanolamine-*N*-Polyethylene glycol-5000 (PEG), and polyethyleneimine (PEI). SN liposomes are prepared by a thin-lipid film hydration method followed by extrusion through a filter with 0.2-µm diameter pores *(5)*. In addition, we provide laboratory protocols for using SN cationic liposomes as a gene-delivery system to tumor cells in vitro and in vivo. We also provide examples of preclinical cancer gene therapy models that demonstrate treatment efficacy using SN cationic liposome/therapeutic gene complexes.

2. Materials

2.1. Cell Lines

PC3, DU145, SK-BR-3, MDA-MB-231, MDA-MB-435, MDA-MB-468, PANC-1, and SK-OV-3 can be obtained from ATCC (Manassas, VA).

2.2. Preparation of SN Cationic Liposomes and Determination of Lipid Concentration

1. PC and PEG were purchased from Avanti Polar Lipids, Inc. (Alabaster, AL). PEI was purchased from Sigma-Aldrich (St. Louis, MO).
2. Rotary evaporator: Buchi Rotavapor R110 (Brinkmann Instruments, Inc., Westbury, NY).
3. DU-70 spectrophotometer (Beckman Instruments, Inc., Fullerton, CA).
4. 28 mM ammonium heptamolybdate (Sigma, St. Louis, MO, Cat. no. A7302): prepare in 2.1 M H$_2$SO$_4$.

5. 0.76 m*M* Malachite green solution (Sigma, Cat. no. M9636): prepare in 0.35% polyvinyl alcohol.

2.3. Preparation of Plasmid DNA

1. LB medium: 1% Bacto-tryptone (DIFCO Laboratories, Detroit, MI, Cat. no. 0123-07-5), 0.5% yeast extract (DIFCO Laboratories, Cat. no. 0127-17-9), and 1% NaCl.
2. Antibiotics: 50 mg/mL ampicillin (Invitrogen, Carlsbad, CA, Cat. no. Q100-17), 50 mg/mL kanamycin (Invitrogen, Cat. no. Q100-18), sterilize by filtration and store in aliquots at −20°C.
3. Resuspension Buffer: 50 m*M* Tris-HCl, pH 8.0, 10 m*M* ethylenediaminetetraacetic acid (EDTA), 100 µg/mL RNase A. Store at 4°C.
4. Lysis Buffer: 200 m*M* NaOH, 1% SDS.
5. Neutralization Buffer: 3 *M* potassium acetate, pH 5.5.
6. Equilibration Buffer: 750 m*M* NaCl, 50 m*M* MOPS, pH 7.0, 15% isopropanol, 0.15% Triton X-100.
7. Wash Buffer: 1 *M* NaCl, 50 m*M* MOPS, pH 7.0, 15% isopropanol.
8. Elution Buffer: 1.25 M NaCl, 50 mM MOPS, pH 8.5, 15% isopropanol.
9. QIAGEN-tip 500 (Qiagen, Inc., Valencia, CA).
10. Plasmids, e.g., CMV-p202, -luc, and -bik *(5,6)* described in this protocol possess genes driven by CMV immediate early promoter.

2.4. Solutions and Culture Medium

1. DMEM (high glucose)/F12 (1:1) medium (incomplete medium) (Hyclone Laboratories, Inc., Logan, UT, Cat. no. SH30003.03 and SH30010.03); store at 4°C.
2. Fetal bovine serum (FBS) (Invitrogen/GIBCO, Carlsbad, CA, Cat. no. 26140-079); store at −20°C.
3. 100X Antibiotics (Penicillin, Streptomycin and Fungizone, Invitrogen/GIBCO, Cat. no. 15240-062); store at −20°C.
4. Complete medium: 450 mL mix DMEM/F12, 50 mL FBS, and 5 mL 100X Antibiotics.
5. Trypsin/EDTA (0.2%) (Invitrogen/GIBCO, Cat. no. 25300-054); store at 4°C.
6. 1X PBS: 137 m*M* NaCl, 2.7 m*M* KCl, 10 m*M* Na_2HPO_4, 1.76 m*M* KH_2PO_4, pH 7.4.
7. 75% ethanol.
8. 20 m*M* HEPES buffer, pH 7.8; store at 4°C.
9. Luciferase Assay Kit (Promega, Madison, WI).
10. 5X Reporter Lysis Buffer (Promega).

3. Methods

3.1. Preparation of SN Cationic Liposomes

1. Weigh 5.9 mg of PC. Dissolve it in 2 mL of solvent (90% chloroform, 10% methanol) and pipet it into a 500-mL round-bottom flask. Wash the tube with an additional 2 mL of solvent and transfer it into the flask.
2. Weigh 4.6 mg of PEG. Dissolve it in 1 mL of methanol and add it to the flask with the PC. Rinse the tube with an additional 1 mL of methanol and add it to the flask.
3. Rotate the flask on a rotary-evaporator at 55°C under vacuum until the liquid is gone. Continue to dry under vacuum without heat for an additional 90 min. Each 30 min return the flask to the heat for 5 min. Keep overnight at 4°C wrapped in foil.
4. Put the flask under vacuum for 30 min at room temperature (RT). Add 6 mL of solvent (PBS containing 1 mg/mL PEI). Rotate slowly for four or five cycles of 10 min of heat followed by 1 h at RT.
5. Keep at RT overnight in the dark without rotating.
6. Measure the volume and adjust to 6 mL with water.
7. Freeze/thaw five times, transferring the flask from −20°C to 4°C
8. Pass the solution five times through a series of filters, 0.8 μ, 0.45 μ, and 0.22 μ, while keeping it at 60°C in a water bath. The resulting concentration of SN liposomes is approx 2.75 mg/mL (*see* **Note 1**). One microliter of SN liposomes contains 1.75 mg of lipid and 1 mg of PEI. Store at 4°C protected from light.

3.2. Plasmid Purification

1. Pick a single colony from a freshly streaked selective plate and inoculate a starter culture of 2–5 mL of LB medium containing appropriate antibiotics. Incubate for 8 h at 37°C with vigorous shaking.
2. Dilute the starter culture 1/500 to 1/1000 into 500 mL of LB medium with an appropriate antibiotic. Shake for 12–16 h at 37°C.
3. Centrifuge the cells at 6000*g* for 15 min at 4°C. Resuspend the bacterial pellet in 10 mL of Resuspension Buffer.
4. Add 10 mL of Lysis Buffer; mix gently and thoroughly. Incubate at room temperature for 5 min.
5. Add 10 mL of Neutralization Buffer, mix gently, and incubate on ice for 20 min.
6. Centrifuge at 20,000*g* at 4°C for 30 min. Transfer the *clear* supernatant to a clean tube. It may be necessary to repeat this step to remove any precipitate that may clog the column in the next step.

7. Equilibrate a QIAGEN-tip 500 by applying 10 mL of Equilibration Buffer onto the column and allowing the buffer to flow through by gravity. Transfer the supernatant from **step 6** to the QIAGEN-tip 500 and allow the supernatant to flow through the column by gravity.
8. Wash the QIAGEN-tip twice with 30 mL of Wash Buffer.
9. Add 15 mL of Elution Buffer to the column and collect the eluent (about 15 mL).
10. Add 10.5 mL of isopropanol to the eluent. Mix well and centrifuge at 15,000g at 4°C for 30 min. Carefully decant the supernatant.
11. Wash the DNA pellet with 5 mL of 70% ethanol and centrifuge at 15,000g for 10 min. Carefully decant the supernatant and air-dry the pellet for 10 min. Resuspend the DNA pellet in 0.5 mL of sterile distilled water and store at 4°C.
12. Use a spectrophotometer to determine DNA concentration by measuring the optical density of the DNA sample (e.g., at 1:1000 dilution) at 260 nm and 280 nm (1 OD$_{260nm}$ = 50 μg/mL). The ratio of OD$_{260nm}$/OD$_{280nm}$ should be greater than 1.8.

3.3. Preparation of SN Liposome/DNA Complexes

1. Dilute SN liposomes and DNA separately to equal volumes with PBS or serum-free medium. The optimal ratio depends on the cell line being tested. We commonly complex 1 μL of SN liposomes (2.75 mg/mL) to 1 μg of DNA after diluting each to 100 μL with either PBS or serum-free medium. This ratio works well for the majority of cell lines tested. More difficult cell lines may require 1.5 or 2 μL of SN liposomes to 1 μg of DNA. Conditions may be optimized quickly by changing the ratio of SN liposomes to DNA using a plasmid containing the gene encoding the green fluorescent protein (GFP) and observing the percentage of transfected cells 24 h after transfection (*see* **Note 2**).
2. Gently drop the DNA into the liposome suspension. Incubate 30 min before use.

3.4. In Vitro Transfection

This method is applicable to most of the cell lines such as PC3, DU145, SKBR3, MDA-MB-231, MDA-MB-435, PANC-1, SK-OV3-ip1, and 2774.

1. Seed cells (~ 2.5 × 10^5 cells per well) into 6-well plates so that they are semi-confluent (50–70%) on the day of transfection.
2. Aspirate the medium and wash the cells with 5 mL of PBS. Remove the PBS and add 500 μL of DMEM/F12 medium or complete medium.

3. For each well of cells to be transfected, prepare the Liposome/DNA complex in a total volume of 200 µL of either PBS or serum-free medium (*see* **Subheading 3.3.**). The optimum amount of DNA is generally 1–3 µg/well. Incubate the Liposome/DNA complex for 30 min at room temperature, and then add it to the cells. Incubate the plate at 37°C in a humidified incubator with 5% CO_2 for 2–6 h. The optimal time is cell line-dependent.

4. Monitor the cells for toxicity during transfection by checking the dish every 30 min by light microscopy and, at the first sign of toxicity, stop the incubation. Most cell lines will not show any signs of toxicity, but some lines are very sensitive (*see* **Note 3**).

5. Replace the transfection medium with complete DMEM/F12 medium (if serum free medium was used during transfection) or if the cells show signs of toxicity.

3.5. In Vivo Transfection

3.5.1. Intra-Tumoral (it) Treatment (Applicable to Breast and Pancreatic Tumor Models)

Example 1: it injection of SN cationic liposomes carrying an interferon-inducible gene, CMV-p202, in pancreatic cancer xenografts (*6*).

1. Tumor cell preparation: inoculate human pancreatic cancer cell line (PANC-1) cells into a 10-cm plate and incubate at 37°C in a humidified incubator with 5% CO_2 so that the cells are approx 70% confluent 24–48 h after seeding. Wash the cells with sterile PBS. Add 1 mL of Trypsin/EDTA to the cells and incubate the plate in a 37°C incubator for 2–5 min. Add 3 mL of complete medium to the cells. Gently shake until all cells have detached from the plate. Transfer the cell suspension into a 15-mL conical tube. Mix well. Remove 20 µL of the cell suspension and determine the cell number using a hemocytometer. Centrifuge the remaining cells at 400*g* for 2–5 min. Resuspend the cell pellet in cold (4°C), sterile PBS at a density of 5×10^6 cells/ml. Store the cell suspension on ice for tumor inoculation.

2. Tumor inoculation: Inject 15 nude mice (nu/nu) with 200 µL (1×10^6) PANC-1 cells at each site subcutaneously on both sides of the abdomen.

3. Treatment: Divide the mice into three groups (five mice per group). Treatments begin when tumors reach 5 mm in diameter. For each injection, prepare SN liposome/DNA complexes using 15 µg of CMV-p202 diluted in 100 µL of PBS and 15 µL of SN liposomes diluted in 100 µL of PBS (*see* **Subheading 3.3.**). As controls, use 15 µg of a control vector (e.g., a plasmid containing the luciferase gene driven by the CMV promoter [pCMV-luc]) in place of CMV-p202, or SN liposomes alone. Administer the CMV-p202, control vector, or no DNA samples twice a week for 7 wk (*see* **Note 4**).

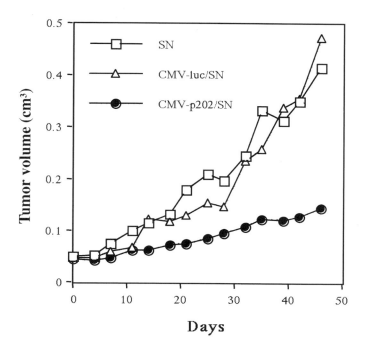

Fig. 1. Anti-tumor effect of p202/SN complex by intratumoral injection in a subcutaneous xenograft model. Tumors were produced by subcutaneously implanting PANC-1 cells into both flanks of each nude mouse. Tumor-bearing mice were divided into three treatment groups (five mice per group and two tumors per mouse): SN liposome alone (SN), CMV-luc/SN, and CMV-p202/SN. SN (15 µL) with or without DNA (15 µg) in 100 µL PBS was injected twice a week into each tumor. Tumors were measured twice a week after treatment began, and the average tumor volume per treatment group at the indicated time is presented *(6)*. Used with permission from American Association for Cancer Research.

4. Estimate the tumor size by measuring weekly using a caliper. Calculate the tumor volume using the formula: Tumor Volume = S × S × L/2, where S is the shortest length of the tumor in cm and L is the longest length of the tumor in cm (*see* **Fig. 1**) *(6)*.

3.5.2. Intravenous (iv) Treatment (Applicable to Most Solid Tumor and Metastatic Cancer Models Except Certain Brain Tumor Models)

Example 2: iv injection of SN cationic liposomes carrying the luciferase reporter gene (pCMV-luc) in orthotopic breast cancer xenografts *(5)*.

1. Tumor inoculation: grow the human breast cancer cell line, MDA-MB-231, to exponential phase and prepare at $5–25 \times 10^6$ cells/mL. Inoculate 200 µL

($1-5 \times 10^6$ cells) into mammary fat pads (m.f.p.) of 6–8-wk-old female nude mice using a 1-cc syringe with a 25 G needle.

2. Treatment: When the tumor becomes 5–8 mm in diameter (usually 5–7 wk after inoculation), prepare the SN liposome/pCMV-luc complex using 60 μL of SN liposome diluted to 100 μL in PBS and 100 μL (60 μg) of pCMV-luc DNA (*see* **Subheading 3.3.**) and deliver via tail-vein injection using a 1-cc syringe with a 27 G needle. Pre-heat the tail with a heat lamp for several minutes prior to injection. The injection volume is determined by the body weight of the animal being treated and should be 100 μL/10 g of body weight (*see* **Note 5**).

3. Preparation of tissue samples: One day after the injection, sacrifice the mice. Remove the tumors and the appropriate organs and immediately freeze on dry ice. Suspend the tissues in 1X Reporter Lysis Buffer in a volume equivalent to five times the tissue weight (mg) and then homogenize using a Dounce homogenizer. Freeze thaw the tissue suspensions by transferring them between an ethanol-dry ice bath and a 37°C water bath three times, then centrifuge at 2500*g* for 10 min.

4. Determination of the gene expression: Assay luciferase activity in the supernatant using a luminometer. Compare gene expression in different tissues by standardizing to the luciferase activity in 100 mg of wet tissue (*see* **Fig. 2 *[5]*** and **Note 6**).

Example 3: iv injection of SN cationic liposomes carrying the pro-apoptotic gene, bik, (pCMV-bik) in orthotopic breast cancer xenografts *(5)*.

1. Tumor inoculation: grow the human breast cancer cell line, MDA-MB-231, to exponential phase and prepare at 40×10^6 cells/mL. Inoculate 50 μL (2×10^6 cells) into the m.f.p. of thirty 6–8-wk-old female nude mice using a 0.5-cc syringe with a 25 G needle.

2. Treatment: Two weeks postinoculation, when the tumor size reaches 4 × 4 mm, divide the tumor-bearing mice into three groups with 10 mice in each. Prepare the SN liposome/DNA complexes with either pCMV-bik, pCMV-luc, or PBS alone using 30 μL of SN liposomes diluted to 100 μL in PBS and 30 μg of DNA diluted to 100 μL (*see* **Subheading 3.3.**). Deliver via tail-vein injection using a 1-cc syringe with a 27 G needle to inject 100 μL/10 g mouse body weight twice a week for 3 wk.

3. Determination of the tumor size: Estimate tumor volume weekly by measuring with a caliper (*see* **Fig. 3 *[5]***).

4. Notes

1. If desired, the exact concentration of SN liposomes can be determined by the Malachite green assay (8): Dilute 5–10 μL of liposome suspension

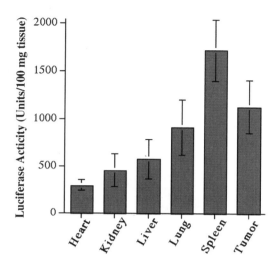

Fig. 2. SN liposomes are efficient for systemic gene delivery to breast tumors in vivo. MDA-MB-231 cells (2×10^6 cells/mouse) were inoculated into the mammary fat pads of nude mice. Six weeks later, each tumor-bearing mouse received a single iv injection of SN liposome/CMV-luc. Twenty-four hours after the injection, the luciferase activity in tumor and normal tissues were determined. The luciferase activity (luciferase units in 100 mg of wet tissue) is indicative of the strength as well as the distribution of gene expression in tissues. The data presented here are the mean ± standard deviation from three mice in each group (*5*). Used with permission from American Association for Cancer Research.

100–1000× with deionized water. Add 344 µL of 28 mM ammonium heptamolybdate and 256 µL of 0.76 mM Malachite green solution to 500 µL of the diluted liposome sample. (Blue-green and yellow colors indicate the presence and absence of phosphate, respectively.) Incubate at room temperature for 20 min. Measure the absorbance at 610 nm using a DU-70 spectrophotometer. Prepare a standard curve by reading the absorbance at 610 nm of a set of sodium phosphate solutions ranging from 1–20 µM. Calculate the liposome concentration of the SN liposome sample based on the standard curve.

2. When the DNA and SN liposomes are mixed, it is important to avoid vigorous shaking or vortexing to prevent the destruction of the SN liposome/DNA complexes.

3. Although serum is generally considered to be a negative factor for in vitro transfection, SN liposomes can tolerate serum during transfection. However, the SN liposome/DNA complexes should be formed in serum-free conditions.

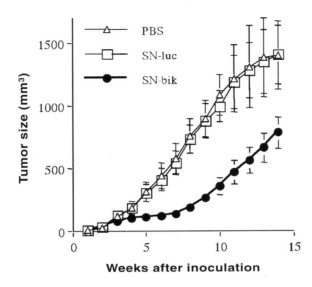

Fig. 3. The therapeutic effect of SN liposome/CMV-bik complex on an orthotopic breast cancer xenograft by iv injection. The human breast cancer cell line, MDA-MB-231, was inoculated into mammary fat pads of female nude mice. Two weeks postinoc-ulation, the tumor-bearing mice were injected with SN liposome/CMV-bik, SN lipo-some/CMV-luc, or phosphate-buffered saline (PBS). The treatment was twice a week for 3 wk at a dose of 15 µg DNA/mouse/injection. Tumor volume was measured weekly. The results shown are the average tumor size ± standard deviation (from five mice) vs time ($p < 0.001$, SN liposome/CMV-bik vs controls by two-side log rank test) (5). Used with permission from American Association for Cancer Research.

4. The maximum injection volume is limited, i.e., approx 50 µL per tumor, owing to the high internal pressure of solid tumors. Immediate leakage may take place at the site of injection when the injection volume exceeds the limit.
5. Injecting an excessive volume (e.g., >1 mL) will cause death to the animal.
6. Injecting a large volume may give a high luciferase activity in liver and tumor. However, it is not necessarily indicative of high transfection effi-ciency because it is more likely owing to the high pressure resulting from the injection than the transfection. Hence, injection of large volume should be avoided.

References

1. Nishikawa, M. and Huang, L. (2001) Nonviral vectors in the new millennium: de-livery barriers in gene transfer. *Hum. Gene Ther.* **12,** 861–870.

2. Behr, J. P. (1986) DNA strongly binds to micelles and vesicles containing lipopolyamines or lipointercalants. *Tetrahedron Lett.* **27**, 5861–5864.
3. Felgner, P. L., Gadek, T. R., Holm, M., Roman, R., Chan, H. W., Wenz, M., et al. (1987) Lipofection: a highly efficient, lipid-mediated DNA-transfection procedure. *Proc. Natl. Acad. Sci. USA* **84**, 7413–7417.
4. Pedroso de Lima, M. C., Simoes, S., Pires, P., Faneca, H., and Duzgunes, N. (2001) Cationic lipid-DNA complexes in gene delivery: from biophysics to biological applications. *Adv. Drug Deliv. Rev.* **47**, 277–294.
5. Zou, Y., Peng, H., Zhou, B., Wen, Y., Wang, S. C., Tsai, E. M., and Hung, M. C. (2002) Systemic tumor suppression by the proapoptotic gene bik. *Cancer Res.* **62**, 8–12.
6. Wen, Y., Yan, D.-H., Wang, B., Spohn, B., Ding, Y., Shao, R., et al. (2001) p202, an interferon-inducible protein, mediates multiple anti-tumor activities in human pancreatic cancer xenograft models. *Cancer Res.* **61**, 7142–7147.
7. Zou, Y., Peng, H., Zhou, B. H., Wen, Y., Tsai, Y.-M., Wang, S.-C., and Hung, M.-C. (2002) Systemic tumor suppression by the pro-apoptotic gene bik. (Correction). *Cancer Res.* **62**, 4167.
8. Van Veldhoven, P. P. and Mannaerts, G. P. (1987) Inorganic and organic phosphate measurements in the nanomolar range. *Anal. Biochem.* **161**, 45–48.

10

Delivery of Transposon DNA to Lungs of Mice Using Polyethyleneimine-DNA Complexes

Lalitha R. Belur and R. Scott McIvor

1. Introduction

The ability to introduce new gene sequences into the tissues of experimental animals allows investigations into the regulation and effect of gene expression in that tissue as well as providing model systems for human gene therapy. In this regard, there has been considerable interest in the expression of genes newly introduced into lung tissues as a means of understanding gene expression in the lung and to explore the possibility of developing gene transfer as a therapeutic approach for diseases of the lung *(1)*. Although much of this interest has focused on gene delivery through the airway by both viral and nonviral vectors *(2,3)*, gene transfer in the lung can also be accomplished by vascular delivery. DNA complexes with cationic lipids *(4)* and polycations *(5,6)* have all been reported as effective methods for intravascular gene delivery to the lung. Of the different nonviral approaches, DNA complexes with polyethyleneimine (PEI) have been reported to provide the most effective gene delivery to the lung.

Our primary interest in developing this technique has been to test the effectiveness of a new transposon system, *Sleeping Beauty (7)*, for achieving gene integration and long-term expression in a variety of tissues, including the lung. This transposon system consists of two components: first, a transposon comprising a gene of interest, including regulatory elements, flanked by transposon inverted repeat sequences that define the boundary of the transposon; and second, a transposase, or transposase-encoding polynucleotide (DNA or RNA). Once

From: *Methods in Molecular Biology, vol. 245:*
Gene Delivery to Mammalian Cells: Vol. 1: Nonviral Gene Transfer Techniques
Edited by: W. C. Heiser © Humana Press Inc., Totowa, NJ

expressed, the transposase component excises the transposon from the plasmid on which it has been delivered and inserts it into the chromosomal DNA of the host cell. In order to evaluate the effectiveness of the *Sleeping Beauty* transposon system in the lung, we needed a method for efficient delivery of plasmid DNA to the lung. In-depth description of the *Sleeping Beauty* system and its effect on prolonged gene expression in the lung is beyond the scope of this chapter, and thus here we will focus solely on PEI-mediated intravascular delivery of plasmid DNA complexes.

2. Material
2.1. Sources of Chemicals

1. A549 lung epithelial cells, from American Type Culture Collection (ATCC, Manassas, VA).
2. Dulbecco's Modified Eagle's Medium (DMEM), penicillin-streptomycin-fungizone (PSF); (Invitrogen/Gibco Carlsbad, CA).
3. Fetal calf serum (FCS); (Biowhittaker, Walkersville, MD).
4. Linear 22 kD PEI (ExGen 500, solution of 5.47 mM for in vitro transfection and solution of 100 mM for in vivo transfection); (MBI Fermentas, Hanover, MD).
5. Cell lysis buffer and luciferase substrate from Promega (Madison, WI).

2.2. Animals

1. C57Bl/6 mice (Jackson Laboratories, Bar Harbor, ME).
2. Ketamine and Acepromazine (Phoenix Pharmaceuticals, St. Joseph, MO).
3. Butorphanol (Fort Dodge Animal Health, Fort Dodge, IA).
4. Endotoxin-free saline (Baxter Chemicals, Deerfield, IL).
5. Animals were provided food and water ad libitum.

2.3. Plasmids

1. Luciferase transposons (pTL, pTCAL) contain the firefly luciferase gene from plasmid pGL3C, from Promega. pTL contains the luciferase transgene under transcriptional control of the Rous sarcoma virus (RSV) promoter *(8)*, while in pTCAL, the luciferase gene is under transcriptional regulation of the hybrid CAGGS (CMV enhancer, chicken beta actin promoter, beta globin intron, and SV40 polyA) promoter *(9)*.
2. The RSV long-terminal repeat promoter from plasmid pREP10 is available from Invitrogen.
3. DNA is prepared using the Endo-free maxiprep kit from Qiagen (Valencia, CA).

2.4. Solutions and Culture Medium

1. DMEM is supplemented with 10% FCS and 1X antibiotic antimycotic agent (10,000 U/mL penicillin, 10,000 U/mL streptomycin and 25 µg/mL Fungizone).
2. Endotoxin-free 5% dextrose is available from Baxter Chemicals (Deerfield, IL).
3. Endotoxin-free 0.1 *M* NaCl solution.
4. Luciferase assay reagents including cell lysis buffer and luciferin substrate and buffer are available from Promega. The lyophilized luciferin substrate is reconstituted using 10 mL of assay buffer supplied in the kit.

3. Methods
3.1. In Vitro Transfection of A549 Cells

1. Seed 2×10^4 A549 cells per well in a 24-well plate the day before transfection; the cells should be about 50–70% confluent the following day.
2. Preparation of PEI/DNA complexes: Mix 0.5, 2, and 5 µg of DNA with PEI at varying ratios ranging from 1–12 as follows:
   a. Mix appropriate amount of DNA in 100 µL of 0.1 *M* NaCl, vortex, and centrifuge briefly.
   b. Add PEI solution to the DNA to give desired N/P ratio.
   c. Vortex immediately for 10 s and centrifuge briefly.
   d. Incubate at room temperature for 10 min and overlay onto cells (*see* **Note 1**).
3. At 2 d posttransfection, assay for luciferase gene expression as follows:
   a. Aspirate cell-culture medium from cells.
   b. Wash cells with PBS.
   c. Add 0.1 mL of lysis buffer, scrape cells from dish, and collect cells and buffer in a 1.5-mL microfuge tube.
   d. Vortex cells for 30 s.
   e. Centrifuge cells for 1 min and collect supernatant.
   f. Assay supernatant for luciferase activity using a Berthold Lumat LB 9507 luminometer. Briefly, a borosilicate glass test tube containing 20 µL of cell supernatant is placed in the luminometer and 100 µL of the reconstituted luciferase substrate is injected into the supernatant. Light emission is counted for 10–15 s.
   g. Express luciferase activity as relative light units per mg protein (*see* **Note 2**).
   h. Assay protein content using the Bio-Rad DC assay reagent according to the manufacturer's instructions.

3.2. In Vivo *Transfection of Mouse Lung*

3.2.1. Preparation of DNA/PEI Complexes and Injection into Mice

Prepare PEI/DNA complexes using 50 µg of luciferase DNA at an N/P ratio of 7 (*see* **Note 1**). One µL of PEI used for the in vivo transfection is 100 mM in nitrogen residues.

1. Mix 50 µg of luciferase DNA in 0.25 mL of 5% dextrose, vortex, and centrifuge briefly.
2. Mix 10.5 µL of PEI in 0.25 mL of 5% dextrose, vortex, and centrifuge briefly. This will give an N/P ratio of 7.
3. Add PEI to DNA, vortex immediately, and centrifuge briefly.
4. Incubate at room temperature for 10 min.
5. Inject 0.5 mL of complexes into the tail vein of mice.

3.2.2. Perfusion of Mouse Lungs

1. Anesthetize mice using a combination of ketamine (80 mg/kg), acepromezine (1 mg/kg) and butorphanol (0.1 mg/kg) by i.p.injection.
2. Once mouse is under anesthesia, perfuse lungs as follows:

   a. Make an incision in the abdomen to locate the inferior vena cava.
   b. Using a small scissors, lacerate the vena cava and aorta distal to the renal vessels. Ensure that there is no clot in the vena cava to allow for good perfusion.
   c. Cut through the sternum and dissect through the pericardium.
   d. Inject 5 mL of saline into the left atrium using a 12 cc syringe attached to a 25-G needle.
   e. Then inject 5 mL of saline into the right atrium.

3.2.3. Luciferase Assay of Lung Homogenates

1. Remove lungs after perfusion and wash with ice cold 0.1 M NaCl.
2. Homogenize lungs in a 12-mL polypropylene tube using a PowerGen125 homogenizer for 20 s in 200 µL of lysis buffer on ice.
3. Transfer the homogenate to a pre-chilled 1.5-mL microfuge tube. Centrifuge lung homogenate in a microfuge tube at 4°C for 10 min.
4. Assay 20 µL of supernatant for luciferase activity using the Berthold Lumat luminometer.
5. Determine protein concentration using the Bio-Rad DC assay.
6. Express luciferase activity in relative light units per mg protein (*see* **Notes 3** and **4**).

4. Notes

1. Preparation of PEI/DNA complexes (*see* **Subheadings 3.1.** and **3.2.1.**): The amount of nitrogen in the PEI and phosphate in the DNA determines the overall charge of the transfection complex. N/P ratios are calculated as follows:

$$\text{No. of PEI equivalents} = \frac{\mu\text{l of PEI solution} \times 5.47}{\mu\text{g of DNA} \times 3}$$

(Note: 1 µL of PEI is 5.47 m*M* in nitrogen residues. One µg of DNA is 3 nmol of phosphate.)

Example: If an N/P ratio of 6 and 1 µg of DNA is required, then calculate volume of ExGen 500 (PEI) as follows:

$$6 \text{ PEI equivalents} = \frac{\mu\text{l of PEI solution} \times 5.47}{1 \text{ µg of DNA} \times 3}$$

Therefore, add 3.3 µL of PEI solution to 1 µg of DNA.

2. The in vitro transfection of A549 cells is carried out in the presence of serum (*see* **Subheading 3.1.**). As can be seen in **Fig. 1**, the highest luciferase activity was seen upon transfection of 2 µg of DNA at an N/P ratio of 3. Increasing N/P ratios to 9 and above increased cell death owing to both DNA and PEI mediated toxicity. When 5 µg of DNA was transfected, the highest level of expression was seen at an N/P ratio of 3. However, when 2 µg of DNA was

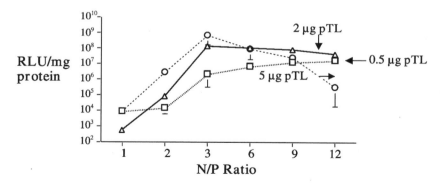

Fig. 1. Expression of luciferase in A549 lung epithelial cells. A549 cells were transfected with varying amounts of luciferase transposon DNA complexed with PEI at different N/P ratios. Forty-eight hours post-transfection, cell lysates were prepared and assayed for luciferase expression.

used the luciferase expression level observed was similar with lesser toxicity than observed with 5 µg of DNA. Transfection of 0.5 µg of DNA gave low levels of expression compared to the two other DNA concentrations used. These results show that a positive charge ratio is an important parameter for efficient PEI-mediated transfection of airway epithelial cells.

3. While preparing lung homogenates, it is critical to keep all reagents and reactions on ice to avoid protein degradation by proteases (*see* **Subheading 3.2.3.**). A protease inhibitor cocktail can also be included in the lysis buffer.

4. Expression of luciferase in the lung and relative promoter strength can be seen in **Fig. 2**. Background luciferase activity was observed when DNA was delivered without PEI. The highest level of luciferase expression was observed when transcription was regulated by the CAGGS promoter, with reduced levels observed for RSV and SV40, at 1 d post-injection (*see* **Subheading 3.2.3.**).

5. The pTL *(8)* and pTCAL plasmids (unpublished data) are not available commercially.

6. Among the various molecular weights and branched or linear forms that are available, the linear 22 kD PEI (ExGen 500) has been shown to be the most

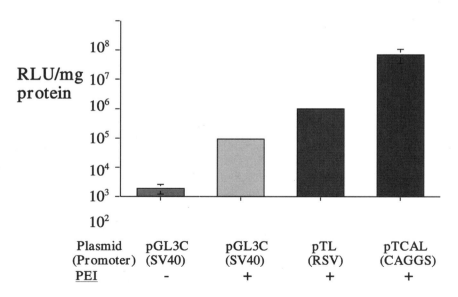

Fig. 2. Expression of luciferase in mouse lungs. Three plasmids expressing luciferase under transcriptional regulation of three different promoters, SV40 (simian virus 40), RSV (Rous sarcoma virus) and the hybrid CAGGS promoter were complexed with PEI at an N/P ratio of 7 and injected intravenously into mice. Twenty-four hours post-injection, mice were sacrificed and whole lung homogenates assayed for luciferase activ-

effective. It is an efficient delivery agent for gene transfer to a variety of different organs and cell types, but especially to lung cells both in vitro and in vivo. Following intravascular injection of PEI/DNA complexes, highest expression is seen in the lung, which is 2–3 orders of magnitude higher than the liver, the spleen, and the heart, and 3–4 orders of magnitude higher than the kidney and the brain *(10,11)*. Preparation of the transfection complexes is easy and transfection is very efficient even in the presence of serum. PEI is able to deliver DNA efficiently and reproducibly to lung epithelial cells, in particular to alveolar type 2 pneumocytes by a single noninvasive systemic tail vein injection.

References

1. Geddes, D. M. and Alton, E. W. (1999) The CF gene: 10 years on. *Thorax* **54**, 1052–1054.
2. Carter, P. J. and Samulski, R. J. (2000) Adeno-associated viral vectors as gene delivery vehicles. *Int. J. Mol. Med.* **6**, 17–27.
3. Uduehi, A. N., Stammberger, U., Kubisa, B., Gugger, M., Buehler, T. A., and Schmid, R. A. (2001) Effects of linear polyethylenimine and polyethylenimine/DNA on lung function after airway instillation to rat lungs. *Mol. Ther.* **4**, 52–57.
4. Li, S., Wu, S. P., Whitmore, M., Loeffert, E. J., Wang, L., Watkins, S. C., et al. (1999) Effect of immune response on gene transfer to the lung via systemic administration of cationic lipidic vectors. *Am. J. Physiol.* **276**, L796–804.
5. Rudolph, C., Lausier, J., Naundorf, S., Muller, R. H., and Rosenecker, J. (2000) In vivo gene delivery to the lung using polyethylenimine and fractured polyamidoamine dendrimers. *J. Gene Med.* **2**, 269–278.
6. Goula, D., Benoist, C., Mantero, S., Merlo, G., Levi, G., and Demeneix, B. A. (1998) Polyethylenimine-based intravenous delivery of transgenes to mouse lung. *Gene Ther.* **5**, 1291–1295.
7. Ivics, Z., Hackett, P. B., Plasterk, R. H., and Izsvak, Z. (1997) Molecular reconstruction of Sleeping Beauty, a Tc1-like transposon from fish, and its transposition in human cells. *Cell* **91**, 501–510.
8. Belur, L. R., Frandsen, F., Dupuy, A., Ingbar, D., Largaespada, L., Hackett, B., et al. (2002) Gene insertion and long-term expression in lung mediated by the Sleeping Beauty transposon system. *Mol. Ther.* In press.
9. Okabe, M., Ikawa, M., Kominami, K., Nakanishi, T., and Nishimune, Y. (1997) 'Green mice' as a source of ubiquitous green cells. *FEBS Lett.* **407**, 313–319.
10. Ferrari, S., Moro, E., Pettenazzo, A., Behr, J. P., Zacchello, F., and Scarpa, M. (1997) ExGen 500 is an efficient vector for gene delivery to lung epithelial cells in vitro and in vivo. *Gene Ther.* **4**, 1100–1106.
11. Goula, D., Becker, N., Lemkine, G. F., Normandie, P., Rodrigues, J., Mantero, S., et al. (2000) Rapid crossing of the pulmonary endothelial barrier by polyethylenimine/DNA complexes. *Gene Ther.* **7**, 499–504.

II

DELIVERY USING PHYSICAL METHODS

11

Gene Delivery Using Physical Methods

An Overview

Te-hui W. Chou, Subhabrata Biswas, and Shan Lu

1. Introduction

The study of gene transfer has always been a major component of modern molecular biology. Initially, transfer of exogenous genes into cells was considered a critical technical step toward understanding the function and regulation of newly discovered genes. With the rapid advancement in human genome research and improved success in gene therapy and nucleic acid vaccination, gene delivery is becoming part of an overall strategy to design and develop an effective prophylactic or therapeutic intervention toward many human diseases. Many biological and nonbiological techniques have been developed to achieve optimal expression of exogenous genes in targeted mammalian cells. Viral vectors have been used most extensively as a biological approach, but success with bacterial vectors has also been reported *(1)*. In following the classification of this book, the nonbiological techniques can be further divided into chemical and physical methods. The purpose of this chapter is to provide a simple overview on the latter type. However, it is important to understand that such a distinction may not be very clear, and different methods can often be combined to achieve the maximum expression of transferred genes.

Gene transfer occurs naturally, especially in prokaryote systems. The phenomenon of "transformation" in *Streptococcus pneumoniae* was described as early as in 1928 by Frederick Griffith. In 1951, Joshua Lederberg and Norton Zinder discovered a new type of gene transfer mediated by viruses. During the lytic infection, the virus picked up some host genes that were then inserted into

From: *Methods in Molecular Biology, vol. 245:*
Gene Delivery to Mammalian Cells: Vol. 1: Nonviral Gene Transfer Techniques
Edited by: W. C. Heiser © Humana Press Inc., Totowa, NJ

another host. They called this process "transduction." The study of gene transfer in mammalian cells was greatly aided with research on tumor viruses. Cultures of uninfected cells produced fully infectious viruses after being transferred with the genetic material of tumor viruses *(2,3)*. The gene-mediated transfer of infectious virus was termed "transfection" to distinguish from infection by a natural route of viral entry.

The potential of purified DNA to express in vivo was demonstrated by Ito *(4)*, who showed that tumors were induced after phenol-extracted DNA from papillomavirus was injected into cotton-tail rabbits. Soon after, it was demonstrated that subcutaneous injection of polyomavirus DNA from cultured cells could induce tumors and also generate anti-polyoma antibodies in hamsters *(5)*, truly the first evidence for nucleic acid vaccination. One of the earliest chemical methods to introduce DNA into cells consisted of incubating negatively charged DNA with an inert carbohydrate polymer (dextran) to which a positively charged group (DEAE) was coupled *(6)*. This method is still in use today but it is inadequate for many cell types. A breakthrough came with the study that DNA could be taken into cells in the form of a precipitate with calcium phosphate *(7)*. With this method, the efficiency of transfection was significantly improved over the DEAE-dextran method. Although the mechanism of calcium phosphate-mediated transfection has not been worked out in detail, it is presumed that after the calcium-DNA complex precipitates on the cell-surface it is internalized by nonspecific endocytosis. The efficacy varies, but in general 10–50% cells in culture can be transfected, and the transfected DNA usually persists for 3–5 d.

These early experiments, both in vitro and in vivo, clearly indicated that exogenous DNA could be taken up by mammalian cells and the expression of encoded genes can be detected. Therefore, gene transfer not only can help scientists understand the function of different genes, but also has tremendous clinical applications toward many human diseases. The classical concept of gene therapy, delivering the functioning gene to correct the aberrant one and to cure inheritable genetic disorders (e.g., adenosine deaminase deficiency and Duchenne muscular dystrophy), is already being tested in humans, although there are still major obstacles to overcome. At the same time, delivery of gene-based treatment for chronic diseases and cancer has gained more acceptance in recent years. The report by Wolff et al. *(8)* demonstrating the high-level expression of reporter genes in mouse skeletal muscle after being injected with purified "naked" DNA plasmids opened a new era for nucleic acid vaccination. Driven by the need for novel vaccines, especially for a worldwide effort to stop HIV epidemics and other major infectious diseases, a number of clinical trials with nucleic acid vaccination have been conducted in humans. The progress in the development of gene-based therapeutics and vaccines in the second half of the

20th century has been truly remarkable. Thus, the practical aspects of gene delivery are becoming increasingly more important.

Although many gene transfer methods can work well in vitro for cultured cells, gene delivery for in vivo applications is more challenging. A number of physical methods have been developed mainly to improve the efficacy of in vivo gene transfer. Usually, some type of physical force (pressure, sound, shock wave, or electric pulse) is used to overcome the physical barriers presented by mammalian tissues. The immediate objective is to achieve a higher percentage of transfection compared with nonphysical methods on the basis of an equal amount of DNA molecules used. In this way, even if the final total amount of proteins produced by one individual or the magnitude of a particular biological response is the same, the amount of DNA required to achieve such an outcome may be reduced. For example, if gene-gun inoculation of 5 µg DNA in mice can induce the same level of immune response as a 100 µg intramuscular (im) inoculation, the gene-gun approach is actually more efficient than the im delivery. This can lead to significant savings in cost and will also reduce any potential risk associated with the use of large amount of DNA. Most of the physical methods use some type of instrument, which can lead to more quantitative and standardized gene delivery. Such an "industrialized" approach is important for large-scale clinical applications and will minimize the common operator variation frequently discovered in the stage of laboratory research with nonphysical methods.

Table 1 compares physical methods with biological and chemical methods of gene delivery. The primary goal of gene delivery is to achieve high transfection efficiency. Chemical methods can reliably achieve this objective in cultured cells, but this goal becomes more challenging when it is used in vivo as a systemic delivery approach owing to the large volume of tissue needed to be transfected. Chemical methods may work for selected organs such as lung or airway with mucosal tissue as the target site, or in a fairly localized organ, such as intratumoral inoculation. The biological vector approaches use the natural cell receptor or other entry mechanism for the original biological agents (virus or bacteria). Therefore, it is more likely to achieve a systemic infection (when using a live vector) or one-round transfection (when using inactivated or defected vectors). Some bacterial vectors can be delivered through the gastrointestinal (GI) tract by infecting cells at mucosal surfaces. On the other hand, the physical methods employ completely different mechanisms to "force" cells or tissues to take up the genetic materials. The end result is higher transfection efficiency, but only in a local area subjected to the physical force.

Biological approaches bring the most concern about safety. First, some vectors are similar to the original pathogens and the threat of disease is clear and immediate. Second, the process of producing biological vectors, including the

Table 1
Comparison of DNA Delivery Systems

Gene delivery system	Advantages	Disadvantages
Biological vectors	• High transfection efficacy • Suitable for systemic delivery • Potential for targeting selected cell types	• Complicated manufacturing process • High quality-control requirement • High cost • Interference with pre-existing immunity against the biological vectors • Safety concerns • Require low temperature storage
Chemical methods	• Highly effective with cultured cells • Relatively simple manufacturing for gene-based products • Less limit on gene size • Easy storage and quality control	• Limited clinical applications so far • Challenge to prepare consistent formulations
Physical methods	• High local transfection efficiency • Not cell type-dependent • Easy to standardize the process • Less limit on gene size • Useful for ex vivo application	• Usually require specific instruments

cell lines used, add additional concerns, such as the potential for tumorigenicity. Third, live vectors in general are contraindicated in immunodeficient candidates. Finally, there is a possibility of allergic reaction to biological vectors, mainly through other proteins still remaining within these vectors, especially when the dose of the biological vectors is high. The manufacturing process for biological vector products tends to be more complicated than the process for producing naked DNA plasmids. A higher quality control standard and potentially higher cost may be required. Another challenge is the preexisting immune responses toward vectors. For example, a person with previous adenoviral infection may have anti-adeno humoral and cell-mediated immune responses, which can severely limit the efficacy of gene therapy using adenovectors with the same serotype. Biological gene transfer approaches usually require that the product be transported and stored cold in order to preserve its activity; this is difficult, if not impossible, for many developing countries.

Chemical and physical approaches were first used for in vitro gene transfer because they were effective methods in "carrying" DNA plasmids into cells. However, these approaches became important for in vivo applications after the discovery that these plasmids could be effectively delivered and expressed in vivo. The process of producing DNA plasmids is much more straightforward than the process of producing biological vector products. The challenge to the chemical approach is how to prepare a consistent formulation with DNA stably mixed in the delivery media. For physical approaches, some type of instrument is needed. In some cases, this adds additional steps to the preparation of the DNA so that it is in a form that can then be readily delivered by a particular physical force. Both chemical and physical approaches have the problem that they can only be used effectively in a fairly localized tissue. They have the advantage of transfecting different cell types, but this is also their disadvantage in that they are not suitable for specific cell targeting.

It is possible that the best results may be achieved by combining different methods of gene transfer. Physical or chemical methods can be used to deliver a biological vector-based gene transfer. For example, we have demonstrated that inoculation of a vector encoding for the full length of proviral SIV genome using a gene gun at the mucosal surface of nonhuman primate was able to produce a full SIV infection in the host *(9)*. By using the same technique, a gene gun is able to deliver a plasmid that can express a live-attenuated vaccine.

Even though DNA plasmids are frequently mentioned in this review as the form of genetic material being transferred, genes may also be delivered in the form of RNA. Additionally, RNA can be used to turn off a specific mRNA through complementary base-pair hybridization. Ribozymes provide another therapeutic design. After binding to the target mRNA sequence, the catalytic activity of the ribozyme cleaves the mRNA, thus preventing its translation into protein. Although using RNA omits the step of transcription from DNA and minimizes the sequence error, RNA is much less stable and more costly to make than DNA-based approaches. Thus, although for the discussion of delivery approaches there is no significant difference between DNA and RNA, the latter can be problematic in terms of manufacture and distribution.

This overview summarizes the major physical methods to be discussed in the following chapters: (1) microinjection, (2) conventional needle injection, (3) high-pressure needle injection (hydrodynamics), (4) particle-bombardment ("gene gun"), (5) electroporation, (6) ultrasound, and (7) encapsulated microspheres. **Table 2** compares different physical methods, their principles, instruments needed, DNA amount used, advantages, and limitations. To enter targeted cells, the negatively charged DNA macromolecule has to cross various biological barriers. Because of its size, DNA cannot effectively enter intact tissue. The

Table 2
Comparison of Different Physical Methods for Gene Delivery

Physical methods	Principles	Instruments	DNA amount used	Advantages	Limitations
Conventional needle injection	Physical force	Syringe	High	Low cost, simple, easily available in clinical practice	Low efficiency
Microinjection	Inject DNA intracellularly	Microscopic injection capillaries; Microscope	Low	Good for individual cell application No enzymatic degradation	Laborious procedure Not suitable for clinical use Operator experience is important
Particle bombardment	Micro-carriers Accelerated by high pressure gas	Various delivery instruments	Low	High efficiency at targeted tissue	Require additional equipment and steps Cost
Electroporation	Electric pulse	Electroporator	Low	Improved delivery efficiency	To be tested in humans Public acceptance
Hydrodynamics	Hydrodynamics force	Special syringes	Moderate	Simple and easy to use Improved delivery efficiency	Limited tissue access
Encapsulated microspheres	Phagosytosis of micro-encapsulated DNA		Moderate	Controlled delivery and release Less degradation	Formulation control
Ultrasound	Cavitations (irradiation with ultrasound)	Sonicator	Low to moderate	Safe Less tissue damage	Evolving technology; Limited experience

general principle underlying physical methods is similar in that a physical force is used to disrupt the tissue or cell membrane. Furthermore, DNA is likely to be opsonized and cleared from the body by reticuloendothelial cells after parenteral administration. Similar to conventional medications, DNA delivered in vivo should also have its own pharmacokinetics with a half-life yet to be determined. Quick transfer of DNA molecules into cells by physical methods should, in theory, should reduce the natural clearance or reduce the dose needed to achieve a desired biological response.

2. Microinjection

One widely used procedure to deliver genes directly into the cells is microinjection. Capillary microinjection into cultured somatic cells growing on a solid support has rapidly developed since the introduction of this technique (for an overview *see* **ref. *10***). This is now established as one of the most versatile methods of introducing genetic materials into living cells. Microinjection makes it possible to use single cells to study complicated cellular processes, structure, and function in vitro. Microinjection remains the most widely used method of generating transgenic animals. An important improvement in this respect is the introduction of automation in the micromanipulation and microinjection processes as well as the control and standardization of cell preparation or the production of injection capillaries *(11,12)*. The development of computer-assisted and microprocessor controlled injection systems makes high injection rates with optimal reproducibility feasible and allows for quantitative microinjection *(13)*.

In addition to the microinjection of nuclear DNA (nDNA), mitochondrial microinjection and cytoplast fusion have been used for gene therapy against non-Mendelian genetics diseases caused by mitochondrial DNA (mtDNA) mutations. In this fusion method, mtDNA in the cytoplast is transferred into mutant cells via the formation of cybrids *(14)*. However, mitochondrial microinjected plasmid DNA is rapidly degraded in the cytoplasm with an apparent half-life of 50–90 min. Thus, microinjection of naked DNA directly into the nucleus, bypassing cytoplasmic degradation, has a much higher level of gene expression than injection into the cytoplasm *(15,16)*.

Despite its straightforward approach, microinjection is a laborious procedure. Only one cell at a time can be injected, and many injections may be needed before there is a successful DNA delivery. This is a problem if a large number of cells need to be injected in a limited time. Therefore microinjection with its current technology has little application for in vivo gene therapy.

3. Direct DNA Inoculation by Conventional Needle Injection

Gene transfer with naked DNA, injecting locally with a standard hypodermic needle, eliminates the use of infectious agents and is likely to pose the lowest risk of toxicity or other unwanted reactions. Surprisingly, naked DNA gives virtually no transfection for cells ex vivo if not helped by any chemical or physical method. Hence, it was believed that naked DNA could not enter into cells unless assisted by a carrier. Countering this argument was the report by Wolff et al. *(8)*. They demonstrated that gene expression in skeletal muscle could be achieved by im injection of naked plasmid DNA. Injection of plasmid DNA intradermally (id) or into the skin subcutaneously (sc) transfects mainly skin fibroblasts and keratinocytes, whereas im injection largely transfects myocytes. However, direct gene transfer into normal mature muscle is not very efficient—at best 1–2% with plasmid DNA—although still far higher than for the adenoviral or retroviral vectors *(17)*. In addition to skeletal muscle, intra-organ injections including liver *(18)*, thyroid *(19)*, heart muscle *(20)*, brain *(21)*, and urological organ *(22)* have been demonstrated. Also, intra-tumor injection of naked DNA has induced transgene expression and elicited therapeutic effect *(23,24)*.

The mechanism of naked DNA uptake by cells in vivo is unknown. Various methods have been explored for enhancing the uptake of purified DNA by muscle. Using dry DNA pellets produces higher levels of expression than DNA in aqueous solutions *(25)*; preinjection of muscles with various agents, e.g., bupivicaine or hypertonic solutions, increases the efficiency of subsequent DNA injections *(17,26,27)*. The disadvantages of direct injection are the low transfection efficiency and the brief expression in most tissues. However, the initial success of DNA delivery in vivo by direct inoculation has opened a new field to develop better tools for gene delivery applications.

Hence a delivery system, jet injection, has been designed to deliver medications without the use of a needle. The jet gun can force DNA in solution through a tiny orifice that creates a very fine, high-pressure stream to penetrate the skin, depositing naked DNA in the tissue beneath. Jet injection of DNA into muscle leads to uptake and expression of the administered gene in muscle *(28)* as well as expression in skin *(29)*. DNA priming by id or im routes using a jet injection device, followed by recombinant modified vaccinia Ankara (rMVA) booster generated high frequencies of virus-specific T cells and controlled a high pathogenic immunodeficiency virus challenge in a rhesus macaque model *(30)*.

4. Hydrodynamics

Systemic administration of naked plasmid DNA by conventional needle injection is prone to degradation. A tail-vein injection of naked DNA into mice resulted in no gene expression in major organs *(31)* because of the rapid in vivo degradation of the DNA by nucleases, and clearance by the mononuclear phagocyte system (e.g., Kupffer cells in the liver) *(32)*. Hence, the "hydrodynamics gene delivery method" has been developed *(28,29,33)*. It involves a rapid injection of a large volume of naked DNA (e.g., 5 µg of DNA in 1.6 mL of saline solution for a

20 g mouse, which is almost equivalent to the total blood volume of the animal) into the tail vein in order to induce efficient gene transfer in internal organs including the lung, spleen, heart, kidney, and liver, with the highest level of expression being observed in the liver *(34,35)*. It is proposed that the injected DNA accumulates in the inferior vena cava and flows back to tissues that are directly linked to the vascular system, including the liver. Plasmid DNAs encoding the luciferase and the β-galactosidase reporter genes have been shown to be expressed in high amounts in the muscle tissue when injected intra-arterially. This gene-transfer strategy was not only successful in rodents *(36)* but also in nonhuman primates *(37)*. Although it is debatable whether this method can be used for human gene therapy, this procedure may be a valuable alternative for in vivo transfection of the liver in animals. With this technique, it is possible to screen many different genes for their therapeutic activity in animal models without involving any complicated surgical procedures for intra-organ gene delivery *(38,39)*.

5. Particle Bombardment

Sanford and colleagues *(40)* developed an alternative method for DNA delivery. The genetic shotgun, "biolistic" (for biological ballistic), consists of a cylinder with a cartridge containing a nylon projectile. It carries millions of microscopic tungsten spheres that are coated with DNA. When the cartridge is fired at a colony of cells, the projectile shoots down the cylinder, and the gene-carrying spheres enter the cytoplasm of the cells. DNA is gradually released, then expressed. This method has been used to transfect plants cells such as corn, wheat, rice, soybean *(41,42)*, and various cultured cells, including epithelial cells, endothelial cells, fibroblasts, lymphocytes, and monocytes *(43)*. Biolistics has also been used to transfect cells that are resistant to transfection by other means, such as multinucleated muscle fibers; mammalian neurons in primary culture; and neuronal cells, including PC12 cells *(44)*.

The Helios gene gun currently marketed by Bio-Rad Laboratories was developed by contributions from many scientists working in both industry and academic institutions. The gene gun uses a high-pressure, helium gas-powered, ballistic device to drive the DNA/gold mix into the targeted cells or tissues. The hand-held device provides rapid and direct gene transfer into a range of targets in vivo. It has been reported that delivery of less than microgram amounts of DNA to mouse skin using the gene gun could protect mice from influenza virus challenge *(45)*. Both antibody and cell-mediated immunity responses have been induced in animals following nucleic acid vaccination with this technology. Thus, gene- gun delivery to skin with plasmid DNA has become a promising alternative to nucleic acid vaccination by intramuscular or intradermal inoculation. A DNA vaccine against the hepatitis B virus, delivered by needle-free PowderJect™ system into skin cells, has demonstrated induction of both humoral and cell-mediated immune responses in humans *(46)*. At this point, experience with this type of gene delivery in humans is still limited, because the gene gun made by Bio-Rad is for research use only.

6. Electroporation

This physical approach to DNA transfection is based on the finding that a short-pulsed electric field can result in the cellular uptake of DNA. Two decades ago, researchers discovered that briefly applying an electric field to a living cell causes a transient permeability in the outer membrane of the cell. This permeation is manifested by the appearance of pores across the membrane. After the field is discontinued, the pores close in approx 1–30 min without significant damage to the exposed cells.

Electroporation is the most versatile method of DNA transfection because it has been shown to work for such a wide variety of cell types, including primary cell from tissue isolates, plant protoplasts, and bacterial cells. Electroporation of eukaryotic cells in vitro is performed by suspending cells in a buffered cell–DNA mixture, which is placed in a special cuvet connected to a power supply. By varying the electric field strength and the length of time the cells are exposed to the electric field, it is possible to optimize the parameters for essentially any cell type. Electroporation has been used successfully since the early 1980s, and has gone through a number of improvements *(47–50)*. With the optimum electric parameters, virtually all of the cells were permeablized and mostly survived the shocking conditions. Both transient expression and permanent expression of DNA were observed. The limitation that electroporation could only be applied to cells in suspension was overcome when a new procedure for in situ electroporation of cells grown on microporous membranes of polyethylene terephthalate or polyester was reported *(51)*. The use of dimethyl sulphoxide (DMSO) in mammalian cells to increase the electroporation efficiency has also been demonstrated *(52)*.

In the case of in vivo electroporation, the negatively charged DNA can enter the cells through a concentration gradient that is facilitated by electrophoretic and electroosmotic transport *(53)*. This technique has been applied to introduce plasmid DNA into tissues such as skin *(54)*, liver *(55)*, muscle *(56)*, and melanoma *(57)*. In these cases, naked plasmid DNA is injected into the interstitial spaces of the tissue and the required electric pulses are applied with needle or caliper-type electrodes. Applying electric pulses can increase gene expression up to 1000-fold compared with needle injection alone. Topical gene delivery into rat skin by electroporation following shaving and mild abrasions has also been demonstrated *(58)*. Rapid and localized gene expression was observed in the keratinocytes in the epidermis, and skin integrity was not compromised. It was reported that this method could effectively deliver exogenous genes into human hematopoetic precursor cells *(59)*. This offers a means to correct inborn genetic errors and to protect human stem cells from chemotherapy damage. The effect of interleukin-12 (IL-12) gene transduction on the growth of experimental murine tumors was studied using electroporation *(60)*. Recently, the development of an electroporation device (Medpulser) for human electroporation therapy was announced (www.genetronics.com).

Electroporation for in vivo use has yet to be fully explored. Because of the complexity of tissue architecture, parameters such as electrode configuration, pulse width, and field strength need to be optimized before electroporation can be used to augment the efficiency of DNA delivery. It is also important to understand the tissue differences so such devices can be used for gene delivery to various organs.

7. Ultrasound

A recently developed nonviral gene delivery method that makes biological membranes transiently permeable is ultrasound; this method facilitates the transfer of DNA into cells and across tissues. Applying ultrasound to a liquid leads to the formation of vapor-filled bubbles, or cavities, in the solution. The collapse of these bubbles can be violent enough to lead to interesting physical effects, including DNA delivery. The formation and collapse of the ultrasound-induced bubbles is called cavitation. It has been demonstrated that application of low frequency ultrasound effectively delivered macromolecules both in vitro as well as in vivo *(61)*. Application of 20 kHz ultrasound to a suspension of yeast cells by a sonicator was shown to facilitate the delivery of plasmid DNA *(62)*. The structural integrity of the plasmid DNA remained unaffected after sonication. Low-intensity ultrasound signals mediating differential gene transfer and transient expression of the green fluorescent protein (GFP) reporter in two human prostate cancer cell lines, LnCaP and PC-3, were among the first ultrasound deliveries to mammalian cells *(63)*. Ultrasound waves have no significant adverse effects when focused on different anatomic locations in the human body. Thus, ultrasound-mediated gene delivery, which raises few safety concerns, can be an alternative to the viral mode of gene delivery for the treatment of respiratory disorders. Ultrasound exposure in the presence of microbubble echo-contrast agents *(64)* led to enhanced acoustic cavitations and resulted in higher levels of transgene expression in vascular cells compared with naked DNA transfection alone *(65)*. Intratumoral injection of DNA followed by focused ultrasound resulted in a 10- to 15-fold increase in the reporter gene expression *(66)*.

8. Encapsulated Microspheres

Controlled drug-delivery technology, comprised of micro-encapsulation techniques, application of polymer science, and site-directed drug-delivery systems, have made great advancements in the last decade. There are many advantages to the polymer-encapsulated microsphere delivery systems: (1) DNA can be delivered to specific sites by implantation or injection; (2) DNA is protected from degradation by nucleases; and (3) sustained release of DNA can be continuous without repeated administration.

Microspheres—assembly of amphiphilic molcules with hydrophilic group outside and hydrophobic group inside—are generally formed by emulsion and internal gelation. High molecular-weight cationic polymers are more effective than cationic liposomes in condensing DNA. Lipid-coated micro-gels combining the characteristics of liposomes and polymeric beads were described in some of the first reports of successful DNA delivery using such encapsulated systems (67). In this case, DNA was immobilized within chitosan coated alginate microspheres with diameters ranging from 20 to 500 µm. In addition to natural polymers, various synthetic polymers have been developed for increasing transfection efficiency, biodegradability, thermo-sensitivity, and for decreasing cytotoxicity and immunogenicity (68). Polymers used in controlled release systems include poly-L-lysine (PLL), poly-L-ornithine, polyethyleneimine (PEI), chitosan, starburst dendrimers, and poly-(lactide-co-glycolic acid) (PLGA) microspheres.

Cationic polymers and liposomes (and combinations of the two) have shown high transfection efficiencies. They enhance the cellular uptake of plasmid DNA by nonspecific adsorptive endocytosis (69). When administered in vivo, encapsulated DNA is engulfed by mononuclear phagocytic cells and macrophage cells. These cells represent a population of professional antigen-presenting cells (APCs) that are required for the activation of specific immune responses (70). Within days of the phagocytosis, the DNA molecules released into the cytoplasm are expressed intracellularly and cytotoxic T-lymphocytes (CTLs) are activated (71,72). The encapsulated DNA was shown to be directed to macrophages that were phagocytosed, resulting in delivery of functional plasmid DNA at controlled rates (72,73). The kinetics of the release of the DNA from these microspheres once inside the cells have been studied and have been shown to be diffusion controlled. The larger the DNA, the slower was its rate of release (74).

In an attempt to target a specific population of cells, polymers such as PLL and PEI were covalently modified with ligands such as asialoglycoproteins (75), carbohydrates (76–79), transferrin (80,81), antibodies (82,83), or lung-surfactant proteins (84). Specific receptors for each of these ligands are present on various cell types. The results of all of these studies have demonstrated that this strategy could be used to improve gene transfection (75–84). In addition, when endososome-disrupting molecules were attached to cationic polymers, these compounds were able to increase the release of the entrapped plasmid DNA from endosomes/lysosomes (85,86). The design of ligand-conjugated polymeric DNA carriers helps to overcome the obstacles of delivery of genes to specific target cells by receptor-mediated endocytosis, and escape from endosomal degradation to obtain nuclear entry.

Plasmid DNA extracted from the PLGA microspheres retained both structural and functional integrity and can transfect cells in vitro *(72)*. This coating protected encapsulated DNA from degradation in the gastrointestinal tract. A recently developed formulation wherein an optimized water-in-oil-in-water double emulsion process was used for the micro-encapsulation of plasmid DNA in PLGA was shown to incorporate a higher amount of DNA, a greater retention of plasmid DNA integrity, and a rapid rate of DNA release *(87)*. Tissue distribution of PLGA-encapsulated plasmid DNA by an im or sc injection was studied in a murine model. The DNA could be detected for 100 d after injection, and was distributed primarily at the site of injection and in the lymphoid organs. Intravenous administration resulted in more widespread dissemination with long-term persistence in the lymphoid organs and cells of the reticuloendothelial system *(88)*. The distribution studies support the fact that DNA can be specifically directed to the APCs such as macrophages and dendritic cells (DCs), suggesting that they have a great utility for nucleic acid immunization purposes. These results, as well as those of other researchers, suggest that biodegradable PLGA microspheres may be the ideal carrier to target therapeutics into phagocytic cells such as macrophages and DCs *(89–91)*.

9. Summary

The history of using physical methods for gene transfer is short, but the experience learned from these practices, especially those used for in vivo purposes, have greatly expanded our understanding of the potential of what normal tissues can do in expressing exogenous genes. Future progress is expected to be made in two directions: further improvement and new inventions on physical-delivery technology to make this technology closer to eventual clinical applications; and expansion on the applications of physical methods to more internal organs or tissues that are difficult to reach. The boundary between biological and nonbiological approaches will be more obscured and both chemical and physical principles will be employed more seamlessly in serving the ultimate goal of developing simple, effective, and safe gene-based therapeutics and vaccines.

References

1. Nishikawa, M. and Huang, L. (2001) Nonviral vectors in the new millennium: delivery barriers in gene transfer. *Hum. Gene Ther.* **12**, 861–870.
2. Manker, R. A. and Groupe, V. (1956) Discrete foci of altered chicken embryo cells associated with Rous sarcoma virus in tissue culture. *Virology* **2**, 838–840.

3. Temin, H. M. and Rubin, H. (1958) Characteristics of an assay for Rous sarcoma virus and Rous sarcoma cells in tissue culture. *Virology* **6**, 669–688.
4. Ito, Y. (1960) A tumor reducing factor extracted by phenol from papillomatous tissues of cotton tail rabbits. *Virology* **12**, 596–601.
5. Goldner, H., Girardi, A. J., and Hilleman, M. R. (1965) Enhancement in hamsters of virus oncogenesis attending vaccination procedures. *Virology* **27**, 225–227.
6. Vaheri, A. and Pagano, J. S. (1965) Infectious poliovirus RNA: a sensitive method of assay. *Virology* **273**, 434–436.
7. Graham, F. L., and van der Eb, A. J. (1973) A new technique for the assay of infectivity of human adenovirus 5 DNA. *Virology* **52**, 456–467.
8. Wolff, J. A., Malone, R., Williams, W. P., Chong, W., Acsadi, G., Jani, A., and Felgner, P. L. (1990) Direct gene transfer into mouse muscle in vivo. *Science* **247**, 1465–1468.
9. Wang, S., Fuller, D., Manson, K., Wyand, M., and Lu, S. (2000) A new animal model to test the AIDS vaccine efficacy in non-human primates: gene gun-mediated mucosal infection with proviral DNA plasmids. *Antiviral Ther.* **5**, 29.
10. Graessmann, M. and Graessmann, A. (1983) Microinjection of tissue culture cells. *Methods Enzymol.* **101**, 482–492.
11. Ansorge, W. (1982) Improved system for capillary microinjection into living cells. *Exp. Cell Res.* **140**, 31–37.
12. Ansorge, W. and Pepperkok, R. (1988) Performance of an automated system for capillary microinjection into living cells. *J. Biochem. Biophys. Methods* **16**, 283–292.
13. Pepperkok, R., Zanetti, M., King, R., Delia, D., Ansorge, W., Philipson, L., and Schneider, C. (1988) Automatic microinjection system facilitates detection of growth inhibitory mRNA. *Proc. Natl. Acad. Sci. USA* **85**, 6748–6752.
14. Kagawa, Y., Inoki, Y., and Endo, H. (2001) Gene therapy by mitochondrial transfer. *Adv. Drug Del. Rev.* **49**, 107–119.
15. Capecchi, M. R. (1980) High efficiency transformation by direct microinjection of DNA into cultured mammalian cells. *Cell* **22**, 479–488.
16. Zabner, J., Fasbender, A. J., Moninger, T., Poellinger, K. A., and Welsh, M. J. (1995) Cellular and molecular barriers to gene transfer by a cationic lipid. *J. Biol. Chem.* **270**, 18997–19007.
17. Davis, H. L., Whalen, R. G., and Demeneix, B. A. (1993) Direct gene transfer into skeletal muscle in vivo: factors affecting efficiency of transfer and stability of expression. *Hum. Gene Ther.* **4**, 151–159.
18. Hickman, M. A., Malone, R. W., Lehmann-Bruinsma, K., Sih, T. R., Knoell, D., Szoka, F. C., et al. (1994) Gene expression following direct injection of DNA into liver. *Hum. Gene Ther.* **5**, 1477–1483.
19. Sikes, M. L., O'Malley, B. W., Jr., Finegold, M. J., and Ledley, F. D. (1994) In vivo gene transfer into rabbit thyroid follicular cells by direct DNA injection. *Hum. Gene Ther.* **5**, 837–844.
20. Ardehali, A., Fyfe, A., Laks, H., Drinkwater, D. C., Jr., Qiao, J. H., and Lusis, A. J. (1995) Direct gene transfer into donor hearts at the time of harvest. *J. Thorac. Cardiovasc. Surg.* **109**, 716–719; discussion 719–720.

21. Schwartz, B., Benoist, C., Abdallah, B., Rangara, R., Hassan, A. Scherman, D., and Demeneix, B.A. (1996) Gene transfer by naked DNA into adult mouse brain. *Gene Ther.* **3**, 405–411.
22. Yoo, J. J., Soker, S., Lin, L. F., Mehegan, K., Guthrie, P. D., and Atala, A. (1999) Direct in vivo gene transfer to urological organs. *J. Urol.* **162**, 1115–1118.
23. Vile, R. G. and Hart, I. R. (1993) Use of tissue-specific expression of the herpes simplex virus thymidine kinase gene to inhibit growth of established murine melanomas following direct intratumoral injection of DNA. *Cancer Res.* **53**, 3860–3864.
24. Nomura, T., K., Yasuda, T., Yamada, S. Okamoto, R. I. Mahato, Y., et al. (1999) Gene expression and antitumor effects following direct interferon (IFN)-gamma gene transfer with naked plasmid DNA and DC-chol liposome complexes in mice. *Gene Ther.* **6**, 121–129.
25. Wolff, J. A., Williams, P., Acsadi, G., Jiao, S., Jani, A., and Chong, W. (1991) Conditions affecting direct gene transfer into rodent muscle in vivo. *Biotechniques* **11**, 474–485.
26. Wells, D. J. (1993) Improved gene transfer by direct plasmid injection associated with regeneration in mouse skeletal muscle. *FEBS Lett.* **332**, 179–182.
27. Danko, I., Fritz, J. D., Latendresse, J. S., Herweijer, H., Schultz, E., and Wolff, J. A. (1993) Dystrophin expression improves myofiber survival in mdx muscle following intramuscular plasmid DNA injection. *Hum. Mol. Genet.* **2**, 2055–2061.
28. Furth, P. A., Shamay, A., Wall, R. J., and Hennighausen, L. (1992) Gene transfer into somatic tissues by jet injection. *Anal. Biochem.* **205**, 365–368.
29. Ledley, F. D. (1994) Non-viral gene therapy. *Curr. Opin. Biotechnol.* **5**, 626–636.
30. Amara, R. R., Villinger, F., Altman, J. D., Lydy, S. L., O'Neil, S. P., Staprans, S. I., et al. (2001) Control of a mucosal challenge and prevention of AIDS by a multiprotein DNA/MVA vaccine. *Science* **292**, 69–74.
31. Mahato, R. I., Kawabata, K., Takakura, Y., and Hashida, M. (1995) In vivo disposition characteristics of plasmid DNA complexed with cationic liposomes. *J. Drug Target.* **3**, 149–157.
32. Kawabata, K., Takakura, Y., and Hashida, M. (1995) The fate of plasmid DNA after intravenous injection in mice: involvement of scavenger receptors in its hepatic uptake. *Pharm. Res.* **12**, 825–830.
33. Liu, D. and Knapp, J. E. (2001) Hydrodynamics-based gene delivery. *Curr. Opin. Mol. Ther.* **3**, 192–197.
34. Liu, F., Song, Y., and Liu, D. (1999) Hydrodynamics-based transfection in animals by systemic administration of plasmid DNA. *Gene Ther.* **6**, 1258–1266.
35. Zhang, G., Budker, V., and Wolff, J. A. (1999) High levels of foreign gene expression in hepatocytes after tail vein injections of naked plasmid DNA. *Hum. Gene Ther.* **10**, 1735–1737.
36. Acsadi, G., Dickson, G., Love, D. R., Jani, A., Walsh, F. S., Gurusinghe, A., et al. (1991) Human dystrophin expression in mdx mice after intramuscular injection of DNA constructs. *Nature* **352**, 815-818.

37. Zhang, G., Budker, V., Williams, P., Subbotin, V., and Wolff, J. A. (2001) Efficient expression of naked dna delivered intraarterially to limb muscles of nonhuman primates. *Hum. Gene Ther.* **12**, 427–438.

38. Budker, V., Zhang, G., Knechtle, S., and Wolff, J. A. (1996) Naked DNA delivered intraportally expresses efficiently in hepatocytes. *Gene Ther.* **3**, 593–598.

39. Zhang, G., Vargo, D., Budker, V., Armstrong, N., Knechtle, S., and Wolff, J. A. (1997) Expression of naked plasmid DNA injected into the afferent and efferent vessels of rodent and dog livers. *Hum. Gene Ther.* **8**, 1763–1772.

40. Armaleo, D., Ye, G. N., Klein, T. M., Shark, K. B., Sanford, J. C., and Johnston, S. A. (1990) Biolistic nuclear transformation of Saccharomyces cerevisiae and other fungi. *Curr. Genet.* **17**, 97–103.

41. Klein, T. M., Fromm, M., Weissinger, A., Tomes, D., Schaaf, S., Sletten, M., and Sanford, J. C. (1988) Transfer of foreign genes into intact maize cells with high-velocity microprojectiles. *Proc. Natl. Acad. Sci. USA* **85**, 4305–4309.

42. Wang, Y. C., Klein, T. M., Fromm, M., Cao, J., Sanford, J. C., and Wu, R. (1988) Transient expression of foreign genes in rice, wheat and soybean cells following particle bombardment. *Plant Mol. Biol.* **11**, 433–439.

43. Burkholder, J. K., Decker, J., and Yang, N. S. (1993) Rapid transgene expression in lymphocyte and macrophage primary cultures after particle bombardment-mediated gene transfer. *J. Immunol. Methods* **165**, 149–156.

44. Lo, D. C., McAllister, A. K., and Katz, L. C. (1994) Neuronal transfection in brain slices using particle-mediated gene transfer. *Neuron* **13**, 1263–1268.

45. Fynan, E. F., Webster, R. G., Fuller, D. H., Haynes, J. R., Santoro, J.C., and Robinson, H.L. (1993) DNA vaccines: protective immunizations by parenteral, mucosal, and gene-gun inoculations. *Proc. Natl. Acad. Sci. U S A* **90**, 11478–11482.

46. Roy, M. J., Wu, M. S., Barr, L. J., Fuller, J. T., Tussey, L. G., Speller, S., et al. (2001) Induction of antigen-specific CD8+ T cells, T helper cells, and protective levels of antibody in humans by particle-mediated administration of a hepatitis B virus DNA vaccine. *Vaccine* **19**, 764–778.

47. Neumann, E., Schaefer-Ridder, M., Wang, Y., and Hofschneider, P. H. (1982) Gene transfer into mouse lyoma cells by electroporation in high electric fields. *EMBO J.* **1**, 841–845.

48. Chu, G., Hayakawa, H., and Berg, P. (1987) Electroporation for the efficient transfection of mammalian cells with DNA. *Nucleic Acids Res.* **15**, 1311–1326.

49. Knutson, J. C. and Yee, D. (1987) Electroporation: parameters affecting transfer of DNA into mammalian cells. *Anal. Biochem.* **164**, 44–52.

50. Presse, F., Quillet, A., Mir, L., Marchiol-Fournigault, C., Feunteun, J., and Fradelizi, D. (1988) An improved electrotransfection method using square shaped electric impulsions. *Biochem. Biophys. Res. Commun.* **151**, 982–990.

51. Yang, T. A., Heiser, W. C., and Sedivy, J. M. (1995) Efficient in situ electroporation of mammalian cells grown on microporous membranes. *Nucleic Acids Res.* **23**, 2803–2810.

52. Melkonyan, H., Sorg, C., and Klempt, M. (1996) Electroporation efficiency in mammalian cells is increased by dimethyl sulfoxide (DMSO). *Nucleic Acids Res.* **24**, 4356–4357.

53. Johnson, P. G., Gallo, S. A., Hui, S. W., and Oseroff, A. R. (1998) A pulsed electric field enhances cutaneous delivery of methylene blue in excised full-thickness porcine skin. *J. Invest. Dermatol.* **111**, 457–463.

54. Titomirov, A. V., Sukharev, S., and Kistanova, E. (1991) In vivo electroporation and stable transformation of skin cells of newborn mice by plasmid DNA. *Biochem. Biophys. Acta* **1088**, 131–134.

55. Heller, R., Jaroszeski, M., Atkin, A., Moradpour, D., Gilbert, R., Wands, J., and Nicolau, C. (1996) In vivo gene electroinjection and expression in rat liver. *FEBS Lett.* **389**, 225–228.

56. Aihara, H. and Miyazaki, J. (1998) Gene transfer into muscle by electroporation in vivo. *Nat. Biotechnol.* **16**, 867–870.

57. Rols, M. P., Delteil, C., Golzio, M., Dumond, P., Cros, S., and Teissie, J. (1998) In vivo electrically mediated protein and gene transfer in murine melanoma. *Nat. Biotechnol.* **16**, 168–171.

58. Dujardin, N., Van Der Smissen, P., and Preat, V. (2001) Topical gene transfer into rat skin using electroporation. *Pharm. Res.* **18**, 61–66.

59. Wu, M. H., Liebowitz, D. N., Smith, S. L., Williams, S. F., and Dolan, M. E. (2001) Efficient expression of foreign genes in human CD34(+) hematopoietic precursor cells using electroporation. *Gene Ther.* **8**, 384–390.

60. Lohr, F., Lo, D. Y., Zaharoff, D. A., Hu, K., Zhang, X., Li, Y., et al. (2001) Effective tumor therapy with plasmid-encoded cytokines combined with in vivo electroporation. *Cancer Res.* **61**, 3281–3284.

61. Mitragotri, S., Blankschtein, D., and Langer, R. (1996) Transdermal drug delivery using low-frequency sonophoresis. *Pharm. Res.* **13**, 411–420.

62. Wyber, J. A., Andrews, J., and D'Emanuele, A. (1997) The use of sonication for the efficient delivery of plasmid DNA into cells. *Pharm. Res.* **14**, 750–756.

63. Tata, D. B., Dunn, F., and Tindall, D. J. (1997) Selective clinical ultrasound signals mediate differential gene transfer and expression in two human prostate cancer cell lines: LnCap and PC-3. *Biochem. Biophys. Res. Commun.* **234**, 64–67.

64. Maresca, G., Summaria, V., Colagrande, C., Manfredi, R., and Calliada, F. (1998) New prospects for ultrasound contrast agents. *Eur. J. Radiol.* **27 Suppl 2**, S171–178.

65. Lawrie, A., Brisken, A. F., Francis, S. E., Cumberland, D. C., Crossman, D. C., and Newman, C.M. (2000) Microbubble-enhanced ultrasound for vascular gene delivery. *Gene Ther.* **7**, 2023–2027.

66. Huber, P. E. and Pfisterer, P. (2000) In vitro and in vivo transfection of plasmid DNA in the Dunning prostate tumor R3327-AT1 is enhanced by focused ultrasound. *Gene Ther.* **7**, 1516–1525.

67. Alexakis, T., Boadi, D. K., Quong, D., Groboillot, A., O'Neill, I., Poncelet, D., and Neufeld, R. J. (1995) Microencapsulation of DNA within alginate microspheres

and crosslinked chitosan membranes for in vivo application. *Appl. Biochem. Biotechnol.* **50**, 93–106.

68. Han, S., Mahato, R. I., Sung, Y. K., and Kim, S. W. (2000) Development of biomaterials for gene therapy. *Mol. Ther.* **2**, 302–317.

69. Godbey, W. T., Wu, K. K., and Mikos, A. G. (1999) Tracking the intracellular path of poly(ethylenimine)/DNA complexes for gene delivery. *Proc. Natl. Acad. Sci. USA* **96**, 5177–5181.

70. Tabata, I., Atomi, Y., Mutoh, Y., and Miyashita, M. (1990) Effect of physical training on the responses of serum adrenocorticotropic hormone during prolonged exhausting exercise. *Eur. J. Appl. Physiol. Occup. Physiol.* **61**, 188–192.

71. Hedley, M. L., Curley, J., and Urban, R. (1998) Microspheres containing plasmid-encoded antigens elicit cytotoxic T-cell responses. *Nat. Med.* **4**, 365–368.

72. Hedley, M. L., Strominger, J. L., and Urban, R. G. (1998) Plasmid DNA encoding targeted naturally processed peptides generates protective cytotoxic T lymphocyte responses in immunized animals. *Hum. Gene Ther.* **9**, 325–332.

73. Wang, D., Robinson, D. R., Kwon, G. S., and Samuel, J. (1999) Encapsulation of plasmid DNA in biodegradable poly(D, L-lactic-co-glycolic acid) microspheres as a novel approach for immunogene delivery. *J. Control Release* **57**, 9–18.

74. Luo, D., Woodrow-Mumford, K., Belcheva, N., and Saltzman, W. M. (1999) Controlled DNA delivery systems. *Pharm. Res.* **16**, 1300–1308.

75. Wu, G. Y. and Wu, C. H. (1988) Receptor-mediated gene delivery and expression in vivo. *J. Biol. Chem.* **263**, 14621–14624.

76. Midoux, P. and Monsigny, M. (1999) Efficient gene transfer by histidylated polylysine/pDNA complexes. *Bioconjug. Chem.* **10**, 406–411.

77. Perales, J. C., Ferkol, T., Beegen, H., Ratnoff, O. D., and Hanson, R. W. (1994) Gene transfer in vivo: Sustained expression and regulation of genes introduced into the liver by receptor-mediated uptake. *Proc. Natl. Acad. Sci. USA* **91**, 4086–4090.

78. Nishikawa, M., Takemura, S., Takakura, Y., and Hashida, M. (1998) Targeted delivery of plasmid DNA to hepatocytes in vivo: optimization of the pharmacokinetics of plasmid DNA/galactosylated poly(L-lysine) complexes by controlling their physiochemical properties. *J. Pharmacol. Exp. Ther.* **287**, 408–415.

79. Diebold, S. S., Kursa, M., Wagner, E., Cotton, M., and Zenke, M. (1999) Mannose polyethyleneimine conjugates for targeted DNA delivery into dendritic cells. J. Biol. Chem. **274**, 19087–19094.

80. Wagner, E., Zenke, M., Cotten, M., Beug, H., and Birnstiel, M. L. (1990) Transferrin-polycation conjugates as carriers for DNA uptake into cells. *Proc. Natl. Acad. Sci. USA* **87**, 3410–3414.

81. Kircheis, R., Schuller, S., Brunner, S., Ogris, M., Heider, K. H., Zauner, W., and Wagner, E. (1999) Polycation based DNA complexes for tumor-targeted gene delivery in vivo. *J. Gene Med.* **1**, 111–120.

82. Ferkol, T., Kaetzel, C. S., and Davis, P. B. (1993) Gene transfer into respiratory epithelial cells by targeting the polymeric immunoglobulin receptor. *J. Clin. Invest.* **92**, 2394–2400.

83. Li, S., Tan, Y., Viroonchatapan, E., Pitt, B. R., and Huang, L. (2000) Targeted gene delivery to pulmonary endothelium by anti-PECAM antibody. *Am. J. Physiol. Lung Cell Mol. Physiol.* **278**, 504–511.

84. Ross, G. F., Morris, R. E., Ciraolo, G., Huelsman, K., Bruno, M., Whitsett, J. A., et al. (1995) Surfactant protein A-polylysine conjugates for delivery of DNA to airway cells in culture. Hum. Gene Ther. **6**, 31–40.

85. Wagner, E., Plank, C., Zatloukal, K., Cotten, M., and Birnstiel, M. L. (1992) Influenza virus hemagglutinin HA-2 N-terminal fusogenic peptides augment gene transfer by transferrin-polylysine-DNA complexes: Toward a synthetic virus-like gene-transfer vehicle. *Proc. Natl. Acad. Sci. USA* **89**, 7934–7938.

86. Nishikawa, M., Yamauchi, M., Morimoto, K., Ishida, E., Takakura, Y., and Hashida, M. (2000) Hepatocyte-targeted in vivo gene expression by intravenous injection of plasmid DNA complexed with synthetic multi-functional gene delivery system. *Gene Ther.* **7**, 548–555.

87. Tinsley-Bown, A. M., Fretwell, R., Dowsett, A. B., Davis, S. L., and Farrar, G. H. (2000) Formulation of poly(D,L-lactic-co-glycolic acid) microparticles for rapid plasmid DNA delivery. *J. Control Release* **66**, 229–241.

88. Lunsford, L., McKeever, U., Eckstein, V., and Hedley, M. L. (2000) Tissue distribution and persistence in mice of plasmid DNA encapsulated in a PLGA-based microsphere delivery vehicle. *J. Drug Target.* **8**, 39-50.

89. Hsu, Y. Y., Hao, T., and Hedley, M. L. (1999) Comparison of process parameters for microencapsulation of plasmid DNA in poly(D,L-lactic-co-glycolic) acid microspheres. *J. Drug Target* **7**, 313-323.

90. Denis-Mize, K. S., Dupuis, M., MacKichan, M. L., Singh, M., Doe, B., O'Hagan, D., et al. (2000) Plasmid DNA adsorbed onto cationic microparticles mediates target gene expression and antigen presentation by dendritic cells. *Gene Ther.* **7**, 2105–2112.

91. Walter, E., Dreher, D., Kok, M., Thiele, L., Kiama, S. G., Gehr, P., and Merkle, H. P. (2001) Hydrophilic poly(DL-lactide-co-glycolide) microspheres for the delivery of DNA to human-derived macrophages and dendritic cells. *J. Control Release* **76**, 149–168.

12

Gene Delivery to Mammalian Cells by Microinjection

Robert King

1. Introduction

Microinjection—that is, the direct-pressure injection of a solution into a cell through a glass capillary—is an effective and reproducible method for introducing exogenous material into cells in culture. The method has been in existence almost as long as there have been microscopes to observe the process. Barber *(1,2)* first described the technique that forms the basis for today's microinjection applications. Described here is a simple method for microinjecting individual adherent cells in culture. This application may be used for microinjecting the myriad mammalian cell types that may be encountered in this field.

2. Materials

2.1. Cells

The best cells for microinjecting will be large, easily adherent, with a pronounced nucleus, giving them a tall aspect. For these reasons, PtK1, MDBK, or CHO, all available from ATCC (Manassas, VA), are appropriate for learning good microinjection technique. In theory, any mammalian cell can be injected in the manner described here, although some types provide more challenges than others. Contractile cells such as muscle often change shape rapidly in response to being injected (particularly when calcium is present in the medium), and cells that do not lay flat when cultured may need to be held in place with a second, "holding" micropipet.

From: *Methods in Molecular Biology, vol. 245:*
Gene Delivery to Mammalian Cells: Vol. 1: Nonviral Gene Transfer Techniques
Edited by: W. C. Heiser © Humana Press Inc., Totowa, NJ

2.2. Micropipet

1. Glass capillary tubing for fabricating micropipet (e.g., Sutter Instrument Company, Novato, CA, Cat. no. BF100-78-10).
2. Micropipette puller for preparing the glass micropipettes (e.g., Flaming-Brown-type puller, Sutter Instrument Company, Cat. no. P-97) (*see* **Note 1**).

2.3. DNA

Any plasmid containing a cytomegalovirus (CMV) promoter-driven reporter gene that may be assayed in individual cells (e.g., green fluorescent protein [GFP] or β-galactosidase) may be used for monitoring the efficiency of microinjection.

1. Prepare plasmid by standard methods and purify using CsCl-ethidium bromide gradient centrifugation *(3)*.
2. Resuspend the DNA in a buffer-free solution at a concentration of up to 1 mg/mL *(4)*.

2.4. Microinjection Apparatus

1. Microscope (e.g., Olympus IX-51 with phase contrast optics, 10×, 20×, 40×, objectives, 10× eyepieces, and three-plate moving stage [Olympus America Inc., Melville, NY]) (*see* **Note 2**).
2. Micromanipulator (e.g., XenoWorks micromanipulator, Bio-Rad Laboratories, Hercules, CA, Cat. no. 165-2802) (*see* **Note 3**).
3. Microinjector (e.g., XenoWorks "digital" microinjector, Cat. no. 165-2805 [110/120 volt] or Cat. no. 165-2806 [220/240 volt] [Bio-Rad Laboratories]) (*see* **Note 4**).

3. Methods

3.1. Cell Preparation

Inoculate the cells onto coverslips (*see* **Note 5**) in 35- or 60-mm tissue-culture plates. Use the appropriate growth medium for the cells (e.g., Dulbecco's Modified Eagle's Medium; DMEM) buffered with 10–20 mM HEPES, pH 7.5 (*see* **Note 6**). Inoculate the cells at a low density so that there is minimal contact between adjacent cells.

3.2. Micropipet Preparation

Immediately before use, fabricate a number of micropipets from 1-mm outer diameter, thin-walled capillary glass with an inner filament (the filament aids filling the very tip of the micropipet with injection solution) using a micropipet puller according to the manufacturer's directions.

3.3. DNA Preparation

1. Centrifuge the purified DNA solution at 10,000–15,000g for 10 min prior to loading the supernatant into the micropipet.
2. Using a 27G needle attached to a 1-mL syringe, load a few microliters of DNA solution through the back-end of the micropipet.
3. Fit the micropipet to the microinjector and lower the micropipet tip into the culture medium. Too long a delay between loading the micropipet and immersing the tip in the medium may cause the liquid in the tip of the micropipet to evaporate, resulting in blockage.

3.4. Microinjection Apparatus Set-Up

1. Ensure that the micromanipulator is set with all three axes at the center of their movement ranges. This ensures that one does not run out of travel during an experiment.
2. Place the dish containing the cell-covered coverslips in the center of the microscope stage, select the lowest power objective available, and focus the microscope on the cells. The plane of focus now corresponds with the bottom of the dish.
3. Rack the microscope objective up a few millimeters so that the plane of focus of the microscope is now a little above the cells.
4. Insert a microinjection pipet into the micropipet holder. Place the holder in the clamp of the micromanipulator and, using the pitch adjustment on the clamp, align the micropipet so that its tip projects into the optical axis of the microscope. Take great care not to touch the micropipet tip against anything—it will break very easily.
5. Look down the microscope while using the micromanipulator joystick to position the micropipet tip in the center of the field of view. Bring the micropipet tip into focus by moving the Z-axis of the micromanipulator, **not** the focus control of the microscope. The micropipet tip should now be in the medium and a few millimeters above the cells. By focusing the microscope up and down, one can view the micropipet tip or the cells.
6. Slowly and very carefully, begin to rotate the Z-axis of the micromanipulator, lowering the micropipet tip toward the cells. Follow the tip down toward the cells with the focus of the microscope.
7. As the cells come into focus, stop lowering the micropipet tip; if the tip touches the coverslip, it can break.
8. Set the microinjector to provide a burst of injection pressure for as long as the footswitch is held down (so-called "continuous-flow" injection) (*see* **Note 7**). Also, adjust the balance pressure so that a slight trickle of solution is always coming from the micropipet tip.

9. Prior to injecting any cells, try a test injection to ensure that the DNA solution is flowing by pressing the microinjector footswitch (*see* **Note 8**).

3.5. Injection Procedure *(see Note 9)*

1. Identify a target cell within the field. Microinjection is a concert of three simultaneous actions:
   a. Twisting the Z-axis control of the micromanipulator joystick, lowering the micropipet tip toward the cell;
   b. Movement of the joystick so that the micropipet penetrates the cell (*see* **Fig. 1**) (*see* **Note 10**); and
   c. Depression of the footswitch of the microinjector to inject the DNA into the cell. Ideally, a small quantity of no more than a few percent of the cell's volume will be introduced and will be visible as a tiny shockwave passing through the cytoplasm.
2. Immediately after injection, reverse the direction of movement of the joystick to withdraw the micropipet and release the footswitch.

4. Notes

1. A micropipet puller is required to fabricate consistent microinjection pipets. The Flaming-Brown-type puller (P-97, Sutter Instrument Company) is a

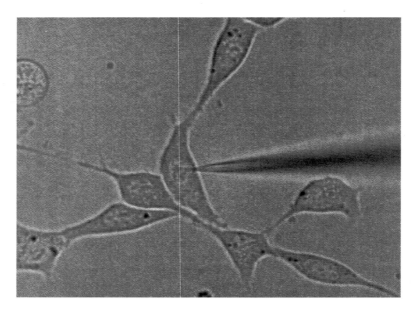

Fig. 1. View through the microscope of a microinjection micropipet entering from the right and projecting down at an approx 45° angle to impale a cultured adherent cell.

sophisticated device that can be calibrated to ensure consistency from one micropipet to the next and produces excellent micropipets for injecting adherent cells. The ideal micropipet will have an internal opening at the tip of less than one micron and have a gentle taper from the tip to the un-pulled section of the capillary.

2. Microscope: In general, an inverted microscope, one whose objective lenses focus beneath a fixed stage, is best suited to microinjection. Although microinjection *can* be performed on an upright microscope, the short working distance between the objective and the specimen make setting the micropipet very difficult. Other good examples of inverted microscopes suited to microinjection are the Nikon TE 2000 (Melville, NY), the Leica DM-IRE2 (Deerfield, IL), and Zeiss Axiovert 200 (Thornwood, NY). For injecting a monolayer of cultured cells, phase-contrast optics produce the best image.

3. Micromanipulator: There are many micromanipulators available for positioning the microinjection pipet, from simple mechanical reduction mechanisms to computer-controlled motorized systems that can interface with the microinjector and microscope. The best micromanipulator will be joystick-operated, comfortable to use over long periods, mechanically stable, and able to move smoothly and responsively.

4. Microinjector: A good microinjector will have the capability to finely adjust the injecting pressure, be able to apply pressure (and therefore inject solution) in discrete pulses or in a continuous stream, and have a high-pressure "clean" function for clearing blockages (which are extremely common) from the tip of the micropipet.

5. Eppendorf Scientific (Westbury, NY) markets "CELLocate," a coverslip with an indexed grid of 175- or 55-micron squares allowing easy identification of injected cells.

6. Carbonate-buffered medium will turn basic upon exposure to the air during microinjection and so should not be used. Most cells will tolerate being injected at room temperature, but if the process is likely to take more than an hour or so the dish should be returned to the incubator periodically. Alternatively, a microscope stage incubator or warm stage insert can be used, both of which are available from the microscope vendor.

7. The precise injection pressure required will depend on a number of factors: the inner diameter of the micropipet tip will determine the flow characteristics and the concentration of the DNA will affect its viscosity, the more viscous the solution, the greater pressure will be required. Trial and error will determine the correct pressure. Start with an injection pressure of about 20–30 psi and slowly increase from there. In general, it is better to work with a higher pressure and spend less time with the microinjection pipet inside the cell (*see* **Note 9**).

8. Under phase-contrast optics the fluid released from the micropipet should be visible as a stream of differing contrast. If no stream is noticeable, try to blow a small fragment of unattached cell around the dish. If no fluid flow can be detected, then the micropipet tip may be blocked. Try cleaning the tip with a quick blast of "clean" pressure and by raising the micropipet tip out of the medium—breaking the surface meniscus can dislodge foreign material.

9. In general, the success of this type of microinjection is highly dependent on operator skill and experience. With practice, quite consistent volumes may be delivered to each cell. The following factors should be borne in mind during the injection:

   a. Spend as short a time as possible with the tip of the micropipet inside the cell, no more than 1 or 2 s.
   b. Enter the cell in a straight line along the axis of the micropipet. Sideways movements will tear the cell membrane and affect the cell's ability to repair the membrane after the injection.
   c. Limit the number of injections performed for each micropipet: cytoplasmic material and fragments of plasma membrane will inevitably adhere to the glass and diminish the performance of the micropipet. After a number of injections, the tip will either block completely or the amount of material on the tip will stick to the contents of the cell and destroy the cell membrane when the tip is withdrawn. Most micropipets may be used for injecting 20–30 cells before there is a diminished performance.
   d. An example of solution that may be used to practice microinjection is 0.01% fluorescein-dextran (Molecular Probes) in calcium-free phosphate-buffered saline (PBS). Successfully injected cells may be identified by fluorescence microscopy *(5)*.

10. Looking at the dish from the side, the tip of the micropipet should describe a "J" shape with the bottom point of the "J" corresponding to the target being injected, usually the perinuclear area, which is the thickest part of the cell. Under phase contrast, the penetration of the cell can be seen clearly, since a bright white spot appears on the surface as the micropipet tip touches the cell membrane.

Manufacturer's addresses:
Sutter Instrument Company
51 Digital Drive
Novato, CA 94949 USA
(415) 883 0128

Bio-Rad Laboratories, Inc.
2000 Alfred Nobel Drive
Hercules, CA 94547 USA
(800) 876 3425

Olympus America, Inc.
2 Corporate Center Drive
Melville, NY 11747 USA
(800) 645 8160

Molecular Probes, Inc.
29851 Willow Creek Road
Eugene, OR 97402 USA
(541) 465-8300

References

1. Barber, M. A. (1911) A Technic for the inoculation of bacteria and other substances into living cells. *J. Infect. Dis.* **8**, 348–360.
2. Barber, M. A. (1914) The pipette method in the isolation of single microorganisms and the inoculation of substances into living cells. *Phillipp. J. Sci. B.* **9**, 307–360.
3. Sambrook, J., Fritsch, E.F., and Maniatis, T. (1989) *Molecular Cloning: A Laboratory Manual*, 2nd ed. Cold Spring Harbor Laboratory Press, Cold Spring Harbor, NY.
4. Graessmann, M. and Graessmann, A. (1983) Microinjection of tissue culture cells, in *Methods in Enzymology*, vol. 101, Academic Press Inc., pp. 482–492.
5. Silver, R. B. (1998) Quantitative microinjection of living cells, in *Cells: A Laboratory Manual*, vol. 2 (Spector, D. L., Goldman, R. D. and Leinwand, L. A., eds.), Cold Spring Harbor Laboratory Press, Cold Spring Harbor, NY, pp. 831–832

13

Delivery of DNA to Cells in Culture Using Particle Bombardment

William C. Heiser

1. Introduction

Numerous chemical, physical, and viral methods have been developed for delivering genes to mammalian cells in culture. Physical methods of gene transfer have the advantage of transporting the genetic material directly into the cytoplasm or nucleus of the cell, bypassing the need for specific membrane requirements for transport into the cell. Several different instrument designs have been described for biolistic delivery of genes *(1–5)*, however, only two of these, the PDS-1000/He and the Helios Gene Gun, are commercially available. Although both instruments may be used for delivery of DNA into mammalian cells in vitro, in this chapter I describe protocols using the PDS-1000/He. In experiments comparing gene expression in cells bombarded using both instruments, expression is routinely two- to three-fold higher in cells bombarded using the PDS-1000/He (unpublished results).

2. Materials

2.1. Cell Culture

1. Cell-culture medium: e.g., Dulbecco's Modified Eagle's Medium (DMEM) or Hams F-12 for adherent cells, or RPMI-1640 for suspension cells. Supplement medium with serum and antibiotics as required for each cell type (Invitrogen/GIBCO, Carlsbad, CA).

From: *Methods in Molecular Biology, vol. 245:*
Gene Delivery to Mammalian Cells: Vol. 1: Nonviral Gene Transfer Techniques
Edited by: W. C. Heiser © Humana Press Inc., Totowa, NJ

2. Trypsin/ethylenediaminetetraacetic acid (EDTA): 0.05% Trypsin, 0.53 m*M* EDTA in phosphate-buffered saline (PBS) (Invitrogen/GIBCO).

2.2. Plasmids

Amplify plasmids according to standard methods *(6)* and purify by CsCl gradient centrifugation or by commercial methods (e.g., Quantum Prep or Aurum columns, Bio-Rad Laboratories, Hercules, CA). For assaying transient gene expression in tissue-culture cells, use a plasmid that expresses the firefly (*Photinus pyralis*) luciferase gene from the cytomegalovirus (CMV) immediate early promoter.

2.3. Particle Bombardment

1. PDS-1000/He with 1.6-μm microcarriers, stopping screens, macrocarriers, and rupture disks (Bio-Rad).
2. 100% and 70% ethanol (prepare 70% ethanol fresh daily).
3. 50% glycerol (sterilize by filtering or autoclaving).
4. 2.5 *M* CaCl$_2$.
5. 0.1 *M* Spermidine (free base; Sigma, St. Louis, MO, Cat. no. S-0266) (*see* **Note 1**).
6. Helium (Grade 4.5; tank pressurized at least 200 psi above bombardment pressure).
7. Platform vortexer (e.g., Genie-2, Scientific Industries, Bohemia, NY).
8. Ultrasonic water bath (e.g., Branson, Model 1510, Danbury, CT).
9. Vacuum source capable of evacuating the bombardment chamber of the PDS-1000/He to 15-in Hg vacuum (vacuum systems found in most research laboratories are capable of accomplishing this).

2.4. Luciferase Assay

1. Luciferase Lysis Buffer: 100 mM phosphate buffer, pH 7.8, 1% Triton X-100, 2 m*M* EDTA, 1 m*M* dithiothreitol (DTT). Store at −20°C; stable for 1 yr.
2. Reagent A: 3.0 m*M* ATP, 15 m*M* MgSO$_4$, 10 m*M* DTT, 30 m*M* Tricine, pH 7.8. Store at −20°C; stable for 1 yr.
3. D-Luciferin (potassium salt; Biosynth, Naperville, IL, Cat. no. L-8220). Prepare a 10X solution in distilled water at 2 m*M* and store in aliquots at −70°C; this is stable for at least 1 yr. Dilute the 10X solution to 0.2 m*M* with distilled water and store at 4°C for up to 1 mo.
4. Firefly Luciferase (Promega, Madison, WI, Cat. no. E-1701). The specific activity of this solution provides a means of converting the luminometer reading of relative light units (RLU) to the amount of luciferase in the cell extract. Prepare dilutions of luciferase in Luciferase Lysis Buffer contain-

ing 1 mg/mL bovine serum albumin (BSA) and 15% glycerol (750 μL of Luciferase Lysis Buffer, 100 μL of μg/μL BSA, 150 μL of glycerol).

3. Methods

The first three sections describe the preparation of the cells, the microcarriers and macrocarriers, and the PDS-1000/He for cell bombardment. Coordinating the timing for each of these steps will come with experience. Those items that can be prepared well ahead of time include the microcarriers (**Subheading 3.2.1.**), the macrocarriers (**Subheading 3.2.2., step 1**), and sterilizing the fixed nest for the PDS-1000/He (**Subheading 3.3., step 1**). Begin preparing the cells (**Subheading 3.1.**) 1–2 d prior to bombardment. On the day of the bombardment, clean the bombardment chamber of the PDS-1000/He and adjust the helium regulator (**Subheading 3.3. steps 2** and **3**), prepare the suspension cultures (**Subheading 3.1.2., steps 2–5**), precipitate the DNA onto the microcarriers, and load the macrocarriers (**Subheading 3.2.2., step 2** and **Subheading 3.2.3., steps 1–8**). Note that the loaded macrocarriers should be used within 1 h of loading (**Subheading 3.2.3., step 8**). If many macrocarriers will be prepared, keep the DNA-coated microcarriers in ethanol (**Subheading 3.2.3., step 7**) and load the macrocarriers immediately before bombardment. **Subheading 3.3., steps 4–8** are the final steps to be performed to bombard each plate of cells.

3.1. Preparation of Cells

3.1.1. Adherent Cultures

1. One day prior to bombardment, inoculate log-phase cells into 35 mm plates so that they are 50–75% confluent the following day.
2. On the day of the experiment, aspirate the medium from the plates and bombard the cells as described in **Subheading 3.3**.

3.1.2. Suspension Cultures

1. Two days prior to bombardment, dilute the cells so that they are in mid-log phase on the day of bombardment.
2. On the day of the experiment, determine the cell density by counting the cells in a hemocytometer.
3. Harvest the cells by transferring them to a sterile 50-mL disposable centrifuge tube. Pellet the cells by centrifugation at 400*g* for 6–8 min at room temperature.
4. Aspirate and discard the supernatant. Resuspend the cell pellet at a density of 2–5 × 10^7 cells/mL.

5. Immediately prior to bombardment, dispense 20 µL of the cell suspension in a circle 22-mm in diameter in the center of a 35-mm sterile Petri dish (*see* **Note 2**).
6. Bombard the cells as described in **Subheading 3.3**.

3.2. Preparation of Microcarriers and Macrocarriers

3.2.1. Microcarrier Preparation

1. Weigh 30 mg of 1.6 µm gold microcarriers in a 1.5-mL microfuge tube. Add 1 mL of 70% ethanol. Vortex for 5 min. Incubate for 15 min.
2. Centrifuge for 5 s. Remove the supernatant and discard.
3. Wash the microcarriers three times with water:
   a. Add 1 mL of sterile water.
   b. Vortex for 1 min.
   c. Allow the particles to settle for 1 min.
   d. Centrifuge for 2 s.
   e. Remove the supernatant and discard.
4. Resuspend the microcarriers at 60 mg/mL by adding 500 µL of sterile 50% glycerol (*see* **Note 3**).

3.2.2. Macrocarrier Preparation

Prepare one macrocarrier for each plate of cells to be bombarded.

1. Rinse each macrocarrier in isopropanol; insert one macrocarrier into a macrocarrier holder. Autoclave the prepared macrocarrier holders.
2. Shortly before loading the DNA-coated microcarriers onto the macrocarrier, place a prepared macrocarrier holder in a desiccated dish, macrocarrier-side up (*see* **Note 4**).

3.2.3. Precipitating DNA onto Microcarriers and Loading the Macrocarriers

This section describes preparing microcarriers at a DNA/Au ratio of 1.6 with 0.5 mg of microcarriers per macrocarrier (*see* **Note 5**).

1. Vortex the microcarriers prepared in **Subheading 3.2.1.** for 5 min. If necessary, sonicate 2–4 s in an ultrasonic water bath (*see* **Note 3**).
2. Remove 50 µL (3 mg) of microcarriers to a 1.5-mL microfuge tube (*see* **Notes 6** and **7**).
3. While vortexing, add the following in order to the microcarriers:
   a. 5 µL plasmid (1 µg/µL).
   b. 50 µL $CaCl_2$ (2.5 *M*).
   c. 20 µL Spermidine (0.1 *M*).
4. Continue vortexing for 2 min.

5. Centrifuge for 2 s. Remove and discard the supernatant, being careful not to disturb the pellet.
6. Add 140 µL of 70% ethanol down the side of the tube without disturbing the pellet. Remove the liquid and discard.
7. Add 140 µL of 100% ethanol down the side of the tube without disturbing the pellet. Remove the liquid and discard.
8. Add 48 µL of 100% ethanol. Gently resuspend the pellet by tapping the side of the tube several times, then by vortexing at low speed for several seconds.
9. Remove six 6 µL aliquots and transfer them to the center of a prepared macrocarrier (**Subheading 3.2.2.**) (*see* **Notes 6** and **8**). Allow the ethanol to evaporate and use these loaded macrocarriers for bombardment within 1 h.

3.3. Preparation of the PDS-1000/He and Cell Bombardment

1. Wrap the fixed nest in aluminum foil (*see* **Note 9**). For each plate of cells to be bombarded, place one stopping screen in a glass Petri dish. Autoclave the fixed nest and stopping screens.
2. On the day of bombardment, clean the bombardment chamber and target shelf using 70% ethanol and a clean paper towel. Allow to air dry.
3. Adjust the helium regulator so that the output pressure is 200 psi over the rupture disk pressure rating (*see* **Note 10**).
4. Using forceps, place a rupture disk into the rupture disk retaining cap and tighten it onto the end of the gas acceleration tube.
5. Place a stopping screen and a loaded macrocarrier (prepared in **Subheading 3.2.3.**) into the fixed nest; attach the macrocarrier retaining cap. Insert the fixed nest immediately below the rupture disk retaining cap in the bombardment chamber.
6. Uncover the tissue-culture dish with the cells (from **Subheading 3.1.1.** or **3.1.2.**) and place it on the target shelf at the closest level (~3 cm) below the fixed nest.
7. Close the chamber door, draw a vacuum to 15 in Hg, and bombard the cells.
8. Release the vacuum from the bombardment chamber, remove the cells, add 1.5 mL of media, and return the cells to the incubator.
9. Assay transient gene expression in cells 24–48 h postbombardment.

3.4. Luciferase Assay

1. Aspirate the media from the plates.
2. Add 1.0 mL of Luciferase Lysis Buffer (4°C).
3. Incubate the plates on a tray of ice for 10–15 min; rock the plates every 5 min.
4. Tap the plates and rock back and forth to mix the cell extract (*see* **Note 11**).
5. Assay a sample of the cell extract in a luminometer (*see* **Note 12**).

6. Convert the RLU of luciferase to the amount (ng) of luciferase in the cell extract by preparing a standard curve using purified luciferase with a known specific activity.

4. Notes

1. Order spermidine in 1 g units. Add 5.8 mL of sterile water to prepare a 1 M (10X) solution (the total volume should be 6.9 mL) and distribute into 1.5-mL microfuge tubes leaving a small amount of space for expansion upon freezing. Store at $-70°C$. Thaw a tube when needed, dilute 1:10 with sterile water, and distribute the 0.1 M solution into 0.5-mL microfuge tubes. Store at $-70°C$. Thaw the 0.5-mL tubes as needed for daily use; discard any unused material.

2. Draw a 22-mm diameter circle on paper to use as a template under the 35-mm plate. Spread the cells in the 22-mm diameter area in the dish using a sterile genomic pipet tip.

3. Microcarriers may be stored in glycerol for up to 1 mo. After vortexing, freshly prepared microcarriers should remain in suspension without any visible, agglomerated, sand-like particles settling out at the bottom of the microfuge tube. Upon prolonged storage, microcarriers become agglomerated and cannot be dispersed either by vortexing or sonicating.

4. Prepare desiccated dishes by sprinkling Drierite into a 60-mm Petri dish to cover about one-half of the bottom of the dish. Place a sterile filter paper on the Drierite and the loaded macrocarrier holder on the filter paper; keep the lid on the Petri dish. The Petri dish provides a semi-sterile environment and the Drierite reduces the drying time of the microcarrier/DNA solution.

5. Several variables are at play in optimizing gene delivery *(3,7)*. In order to express a gene, a gold particle must enter the cell. Therefore, sufficient numbers of particles must be delivered to penetrate a sufficient number of cells. However, cells are killed when penetrated by too many particles. In addition, the amount of DNA that may be coated onto the surface of the gold particles is limited to about 2 µg/mg gold (i.e., a DNA to gold ratio (DNA/Au) of 2 µg plasmid/mg gold). **Figure 1** shows the results of increasing the amount of DNA coated onto the gold microparticles. In one experiment, the amount of microparticles was kept constant at 0.25 mg/shot. In these samples, luciferase expression increased with increasing DNA delivered up to 0.42 µg/shot (DNA/Au = 1.67). Further increasing the DNA in the coating procedure did not result in a further increase in luciferase expression because the gold appears to have a limited number of binding sites that have been saturated. In the other experiment, the ratio of DNA to gold was kept constant at 1.67 µg plasmid/mg gold. In this case, luciferase expression peaked at 0.42 µg/shot. Delivering additional DNA (and microparticles be-

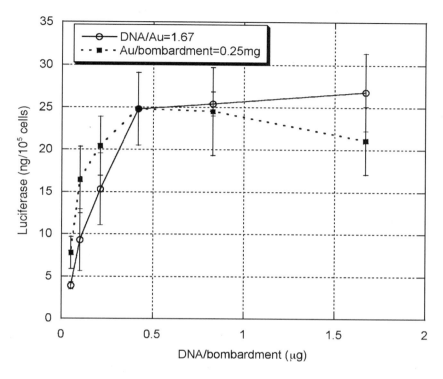

Fig. 1. Effect of the amount of DNA on luciferase expression in CHO cells. CHO cells approx 70% confluent in 35-mm plates were bombarded at 1100 psi and Target Level 1 with 1.6-μm gold particles prepared with increasing amounts of a luciferase plasmid expressed from the CMV immediate early promoter. In one case the DNA/Au ratio was kept constant at 1.67. In the other case the amount of gold per bombardment was kept constant at 0.25 mg. Luciferase expression was determined 24 h postbombardment and expressed as a function of the cell number at the time of bombardment. The values are averages of three replicates; the vertical bars represent the standard deviation.

cause the DNA-to-gold ratio is constant) resulted in no further increase in luciferase expression because of a reduced number of cells in the culture (measured by protein/plate; data not shown). In a third experiment, the amount of DNA was kept constant at 0.42 μg/shot (**Fig. 2**). In these samples, luciferase expression increased with increasing amount of gold delivered up to 0.25 mg/shot. Further increasing the amount of gold in the sample, while reducing the DNA/Au ratio below 1.67, did not result in a further increase in luciferase expression because the additional gold particles result in more cell killing.

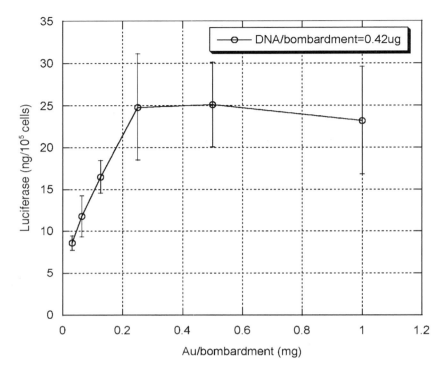

Fig. 2. Effect of the amount of gold per bombardment on luciferase expression in CHO cells. CHO cells approx 70% confluent in 35-mm plates were bombarded at 1100 psi and Target Level 1 with 1.6-μm gold particles prepared with increasing amounts of a luciferase plasmid expressed from the CMV immediate early promoter. Luciferase expression was determined 24 h postbombardment and expressed as a function of the cell number at the time of bombardment. The values are averages of three replicates; the vertical bars represent the standard deviation.

6. When removing aliquots of microcarriers, vortex the tube continuously at an intermediate speed to keep the microcarriers resuspended and to maximize uniform sampling.

7. This procedure describes a method for preparing macrocarriers sufficient for six bombardments. If more bombardments will be performed, prepare replicate tubes rather than increasing the volume in one tube. If fewer bombardments will be performed, prepare an amount of gold microcarriers in **steps 2** and **3** sufficient for at least three bombardments, wash with the same volumes of ethanol, and resuspend the pellet in 24 μL of ethanol in **step 6**. Because of the inherent variability in pipetting the gold particles, samples should be assayed at least in triplicate *(3,7)*.

8. Spread the microcarriers evenly in the central ~10-mm portion of the macrocarrier: dispense the entire aliquot in the center of the macrocarrier, immediately draw the solution back into the tip of the pipet, then slowly dispense the microcarriers while moving the pipet tip around the center of the macrocarrier. Use "standard" pipet tips, not "graduated beveled" tips. The former have a slightly wider tip and bore, facilitating DNA loading and spreading and reducing the chance of clogging.

9. Before autoclaving, prepare the fixed nest with the stopping screen at the middle level (the macrocarrier flight distance is ~8 mm) and adjust the gap distance to 3/16 in (4.8 mm). These conditions have been found to be optimum for bombardment of mammalian cells in the PDS-1000/He and have been used almost exclusively for our work at Bio-Rad *(7)*.

10. For mammalian cells, maximum expression occurs at helium pressures between 650 and 1500 psi *(7)*. Although the optimum pressure should be determined for each cell line, using either 1100 or 1350 psi rupture disks generally provides nearly optimal conditions.

11. This procedure lyses the cells sufficiently and removing the cell debris is not necessary. Many procedures call for scraping the cells from the plates, vortexing to disrupt the cells, then pelleting the cell debris in a microfuge. These added steps do not result in higher luciferase activity and, in fact, the vortexing and centrifugation steps may lead to reduced luciferase activity.

12. Using either a tube luminometer or a microplate luminometer, pipet 1–100 µL of cell extract into the luminometer tube or microplate. If the luminometer has a single injector, also pipette 100 µL of Reagent A into the tube and set the injector to deliver 100 µL of luciferin. If the luminometer has dual injectors, set the first injector to deliver 100 µL of Reagent A and the second injector to deliver 100 µL of luciferin. In either case, program the luminometer to capture light output for 10 s immediately after the luciferin is injected. The amount of luciferin in the sample may be determined by assaying known amounts of luciferase and preparing a standard curve.

References

1. Yang, N.-S., Burkholder, J., Roberts, B., Martinell, B., and McCabe, D. (1990) In vivo and in vitro gene transfer to mammalian somatic cells by particle bombardment. *Proc. Natl. Acad. Sci. USA* **87**, 9568–9572.
2. Williams, R. S., Johnston, S. A., Riedy, M., deVit, M. J., McElligott, S. G., and Sanford, J. C. (1991) Introduction of foreign genes into tissues of living mice by DNA-coated microprojectiles. *Proc. Nat. Acad. Sci. USA* **88**, 2726–2730.
3. Sanford, J. C., Smith, F. D., and Russell, J. A. (1993) Optimizing the biolistic process for different biological applications. *Methods Enzymol.* 217, 483–509.
4. Becker, D. J. and MacDougald, O. A. (1998) Transfection of adipocytes by gene gun-mediated transfer. *BioTechniques* **26**, 660–663.

5. Nicolet, C. M. and Yang, N.-S. (2000) The use of particle-mediated gene transfer for the study of promoter activity in somatic tissues, in *Methods in Molecular Biology*, vol. 130, *Transcription Factor Protocols*, (Tymms, M.J., ed.), Humana Press, Totowa, NJ, pp. 103–116.

6. Sambrook, J., Fritsch, E. F., and Maniatis, T. (1989) *Molecular Cloning: A Laboratory Manual*, 2nd ed. Cold Spring Harbor Laboratory Press, Cold Spring Harbor, NY.

7. Heiser, W. C. (1994) Gene transfer into mammalian cells by particle bombardment. *Anal. Biochem.* **217**, 185–196.

14

Delivery of DNA to Skin by Particle Bombardment

Shixia Wang, Swati Joshi, and Shan Lu

1. Introduction

Particle bombardment or particle-mediated DNA-delivery technologies were developed as physical gene-transfer methods for various in vivo, ex vivo, or in vitro applications. The basic concept is to directly deliver naked DNA plasmids into target cells by using accelerated particles as physical carriers. This technology was first established in plant gene-transfer systems and described as "biolistic (biological ballistic)" in 1987 by Sanford and colleagues (1,2). Based on the same concept, helium-driven gene-gun systems have been developed, such as the Accell gene gun by Agracetus, Inc. (3–6) and the Helios gene-gun system commercially available at Bio-Rad Laboratories. In this chapter, the Bio-Rad gene-gun system is used as a model to describe the delivery of DNA to skin by particle bombardment and the parameters affecting such gene transfer.

Particle bombardment as a physical gene-transfer approach employs a high-velocity stream for intracellular delivery of carrier particles, which can be coated with a number of different macromolecules, such as nucleic acids (DNA or RNA), proteins, or peptides. Many fine, coated particles can be transferred into hundreds or even thousands of cells in a single delivery. The efficiency of such gene transfer can be affected by several parameters, including the size and material of particles, the amount of the particles and DNA, the ratio of DNA to the particles, the process of coating DNA onto particles, the acceleration driving force, the distribution of the particles at the targeting site, and the types of targeted cells and tissues. The particles involved must be non-/low toxic, non-/low

From: *Methods in Molecular Biology, vol. 245:*
Gene Delivery to Mammalian Cells: Vol. 1: Nonviral Gene Transfer Techniques
Edited by: W. C. Heiser © Humana Press Inc., Totowa, NJ

reactive, and subcellular-sized (0.5–5 micron) spheres with sufficient density to penetrate the skin. Pure gold beads in the desired size range have been commonly used because of their chemical and physical properties and commercial availability. Purified plasmids are precipitated onto the gold beads. The DNA/gold bead complex is coated around the inside of Teflon tubing, which is then cut into cartridges. The cartridges are loaded into the gene gun and the coated beads are accelerated into target cells or tissues. Parameters that lead to optimized performance are described and discussed in this chapter.

Skin as an anatomical site normally encounters many exogenous antigens and contains a wide range of specialized cells in the epidermal layer capable of eliciting immune responses *(7–9)*. The epidermal cells in a wide range of mammalian species can be transfected efficiently *(10,11)*. Therefore, skin is an easily accessible site for particle-mediated DNA delivery to trigger immune responses *(7)*. In the past decade, gene-gun delivery of DNA to cells in the epidermis has been successfully used in many DNA vaccine studies *(3,12–16)*. The proteins or antigens encoded by the DNA are synthesized in the transfected cells and are processed similarly as the other newly synthesized proteins. Once antigens are expressed, they follow the rules of antigen processing and presentation. Although the DNA can only be delivered locally via the particle bombardment method, both humoral and cellular immune responses against the specific antigen encoded can be elicited *(17–19)*. Delivery of DNA to skin by particle bombardment has been widely used in DNA immunization and can also be used for specialized gene therapy purposes.

Particle-mediated DNA delivery technology has been effectively applied to transfer genes into different biological systems. A number of advantages can be achieved by particle bombardment, as compared with traditional needle injection techniques. First, this approach can be used to directly transfer genes to a wide range of cells by overcoming physical barriers such as the cell wall and the stratum corneum of the epidermis. Transfer is independent of cell types, ligands/receptors, and cell-surface markers or molecules. Second, co-delivering multiple genes or components at the same sites or to the same cells can be achieved. Therefore, particle bombardment delivery can be used to study the interactions between different gene products before they are released from the cells in which they are expressed. Third, gene-gun inoculation has been considered the most efficient DNA vaccination approach in terms of DNA amount needed. Many studies have demonstrated that inoculations of a few micrograms of DNA by gene gun could raise excellent immune responses in small animals such as rabbits *(20–22)*, mice *(23–25)*, and guinea pigs *(26)* and even nonhuman primates *(17,27,28)*. Thus, particle bombardment is an effective method in a wide variety of biological systems, such as gene transfer to mammalian cells in vivo and in vitro *(29–31)*, gene therapy applications *(31–35)*, and DNA vaccination in experimental animal subjects *(3,12,30)* against a number of pathogens, such as hepatitis B surface antigen *(36–39)*, malaria *(40–42)*, HIV *(13,28,43)*, and rabies *(27,44,45)*.

Although it is a very effective DNA-delivery system, delivery of DNA to skin by particle bombardment does have its limitations. First of all, it needs expensive devices and reagents such as a gene gun and gold particles. Second, the gene gun may selectively induce very high humoral rather than cellular immune responses *(27,46,47)* and also antibody isotype biases *(47)*. It was shown that intramuscular injection of influenza HA antigen-expressing DNA vaccines to mice generated predominantly Th1 (IgG2a) responses, whereas gene-gun inoculation with the same DNA vaccines raised predominantly Th2 (IgG1) responses *(47,48)*.

2. Materials

2.1. Preparation of DNA-Coated Gold Beads

1. Plasmid DNA expressing the gene of interest in TE (10 mM Tris–HCl, pH 8.0, 1 mM ethylenediaminetetraacetic acid [EDTA]) solution at a concentration of 1–5 μg/μL, store at −20°C.
2. Gold beads of 0.5–5 μm diameter. For most animal species, 1 μm gold beads should be used as the first step (Bio-Rad Laboratories, Hercules, CA).
3. 100 mM nuclease-free spermidine (free base) (Cat. no. S-0266, Sigma, St. Louis, MO).
4. 2 M CaCl$_2$ solution, sterile (nuclease-free).
5. Dehydrated ethyl alcohol, 200 proof (such as Cat. no. ET107, Spectrum, Gardena, CA).
6. 2.0-mL Eppendorf tubes.
7. Vortexer.
8. Microcentrifuge.
9. P20, P200, and P1000 Pipettemen and tips.
10. 22-mL glass scintillation vials with Teflon caps.
11. Parafilm.

2.2. Preparation of Tubing Coated With DNA/Gold Beads

1. DNA-coated gold beads prepared as described in **Subheading 3.1.**
2. Tefzel tubing, outer diameter 0.127-in, inner diameter 0.093 in (Bio-Rad Laboratories).
3. Tubing prep station (Bio-Rad Laboratories).
4. Vortexer.
5. Sonication water bath.

6. Compressed nitrogen gas (N$_2$) (grade 4.8) with nitrogen regulator.
7. Timer.
8. 10-mL syringes with flexible adaptor tubing.
9. Tubing cutter (Bio-Rad Laboratories).
10. 22-mL glass scintillation vials.
11. Desiccant capsules.

2.3. Delivery of the Gold Beads Coated With DNA Plasmids

1. Gene gun (Bio-Rad Laboratories).
2. Compressed helium gas (grade 4.5) with helium regulator.
3. Gold bead-coated cartridges prepared in **Subheading 3.2.**
4. Anesthetic agents for animals.
5. Animal subjects.
6. Vacuum cleaner.
7. Electric shaver.

3. Methods

3.1. Preparation of DNA-Coated Gold Beads

Prior to coating the gold beads with DNA, the following parameters should be determined for each study: gold bead load ratio per cartridge (GLR), DNA load ratio per cartridge (DLR), DNA/gold beads ratio (DGR), the number of cartridges (shots) to be used per immunization, and the number of immunizations needed. Typically as a starting point, 1 μg of DNA and 0.5 mg of gold beads are loaded onto each cartridge, with the DNA to gold beads ratio as 2 μg to 1 mg (*see* **Table 1** and **Note 1**).

Table 1
Calculation of DNA and Gold Beads

DGR (μg/mg)	GLR (mg/cartridge)	DLR (μg/cartridge)	No. of cartridges	DNA needed (μg)	Gold beads needed (mg)
4	0.5	2	60	120	30
2	0.5	1	60	60	30
1	0.5	0.5	60	30	30
0.5	0.5	0.25	60	15	30
4	1	2	60	240	60
2	1	1	60	120	60
1	1	0.5	60	60	60
0.5	1	0.25	60	30	60

DGR, DNA and gold beads ratio; GLR, Gold beads loading ratio; DLR, DNA loading ratio.

1. Based on the calculations in **Table 1**, weigh the proper amount of gold bead powder into a microfuge tube (*see* **Note 2**).
2. Add 100 μL of 100 m*M* spermidine to the tube containing gold beads (*see* **Note 3**). Vortex for about 30 s at high speed to suspend the gold beads in the spermidine solution. Sonicate the gold/spermidine mixture for 5–10 s to disperse any aggregated gold beads. Make sure the gold beads are completely dispersed.
3. Based on the DGR, add the proper amount of plasmid DNA to the gold-spermidine mixture and promptly vortex on high for 15–30 s. If more than one DNA construct will be coated on the same beads, pre-mix the DNA beforehand.
4. Precipitate the DNA onto the gold beads: Add 200 μL of 2 *M* CaCl$_2$ drop-wise (*see* **Note 4**) to the gold bead-spermidine-DNA mixture, while vortexing at medium speed. The solution should always remain inside the tube while vortexing. Let the mixture stand at room temperature for 2–5 min to settle. When most of the gold beads have settled, centrifuge for 8–10 s in a microfuge at 9,000g and discard the supernatant.
5. Wash the coated gold beads four times with 1 mL of dehydrated absolute ethanol each time. Add the ethanol and vortex briefly to resuspend, being careful not to handle roughly as the DNA will shear. Centrifuge for 8–12 s and remove the supernatant. The beads are sticky initially and only a short spin is needed to pellet them (too long of a centrifugation will make it difficult to resuspend the beads).
6. After the final ethanol wash, add 100% ethanol to resuspend the DNA-coated gold beads at 8.06 mg/mL and transfer the suspension to a 15-mL screw-capped polypropylene tube.

3.2. Preparation of Tubing Coated With DNA/Gold Beads

After precipitating the plasmid DNA onto the gold beads, the gold beads will be coated evenly onto the inner surface of Tefzel tubing. The ideal size of Tefzel tubing is 0.127 in for the outer diameter and 0.093 in for the inner diameter, which fits the cartridge holder and provides sufficient surface for coating the gold beads. After coating, the tubing is cut into 0.5 in cartridges.

1. Insert the Tefzel tubing into the tubing prep station and blow N$_2$ through it at 0.3 L per min (lpm) for 30–60 min to remove any moisture.
2. When the tubing is sufficiently dried, cut it 5–6 cm beyond the right end of the tubing prep station. Re-cap the ends of the remaining roll of tubing. Turn off the N$_2$ gas. Place a 2–3-cm piece of flexible rubber tubing on the tip of a 10-mL syringe. Attach the flexible tubing to the right end of the

Tefzel tubing while the latter tubing is still in the prep station. Pull the Tefzel tubing from the prep station with the syringe attached.

3. Suspend the tube of gold beads by briefly immersing in a sonicating water bath. Place the free end of the Tefzel tubing into the gold suspension and, using the syringe, draw the gold beads into the tubing until the tubing is about two-thirds filled.

4. Pull about 5 cm of air into the end of the tubing after the gold. Wipe excess gold from the outside of the tubing and then gently insert the tubing back into the prep station. *Do not* seat the tubing end into the nitrogen port seal in the left end of the station. Re-cap the remaining gold suspension in the vial to avoid evaporation and moisture.

5. Let the gold settle out of suspension in the tubing for 10 min, then slowly draw the ethanol from the tubing with the syringe at about 5 cm per second. Remove the syringe from the tubing and discard the ethanol. If the waste ethanol still contains gold, increase the settling time with the next preparation.

6. Push the tubing into the seal of the nitrogen port. Turn the prep station on to rotate the tubing at 20 rpm and allow the gold to spread over the inside of the tubing for 1 min. Slowly turn the N_2 to 0.4 lpm and dry the tubing for 5 min as it continues to rotate. The color of the gold will become bright metallic as it dries.

7. Cut the tubing into half-inch cartridges using the tubing cutter. Put the cartridges into a small vial, seal the vial with parafilm, and store at $-20°C$. Be sure to label both the vial and the cap with the DNA information and the date of preparation (*see* **Note 5**). If tubing is to be stored for later use, add a desiccant capsule to the vial. Replace the razor blade in the tubing cutter and wipe away any gold from the cutter box after each batch (paying special attention to the tubing insertion hole).

3.3. Delivery of the Gold Beads Coated With DNA Plasmids

The epidermal layer of skin is a desirable site for gene-gun delivery of DNA plasmids. The gene gun delivers the majority of the gold beads into the epithelial layer, rather than to the stratum corneum or to the relatively acellular layer underlying the dermal tissues, without excessive tissue damage. Optimal delivery parameters for each animal species is determined in part by the skin thickness, the accelerating pressure, and the size of the gold beads. The depth of gold bead delivery should be optimized for each animal species. The examples in **Table 2** show the common parameters used in our studies that affect the delivery of gold beads into skin at the proper depth. **Figure 1** demonstrates the gold bead distribution in mouse skin as a function of helium pressure. The gold beads were evenly distributed at the epidermal layer by optimized delivery with the

Table 2
Parameters for Gold Bead Delivery into Skin of Various Animal Species

Species	Delivery sites	Gold beads size (μm)	Helium pressure (psi)
Guinea pig	Abdomen	0.5–1.5	300–350
Monkey (Rhesus macaques)	Abdomen, inner thigh	1–3	375–400
Mouse	Abdomen	0.5–1.5	300–350
Rabbit	Abdomen	0.5–1.5	350–400
Rat	Abdomen	0.5–1.5	300–350

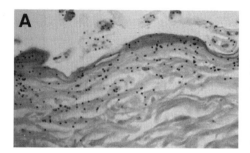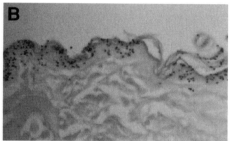

Fig. 1. Distribution of gold beads on mouse dermal tissue delivered by a gene gun with different pressure settings for partical bombardment. (**A**) 350 psi. (**B**) 200 psi. The gold beads can be seen as black dots (magnification = ×400).

Bio-Rad gene gun when the helium pressure was 350 psi (pounds per square inch) (**Fig. 1A**). The gold beads were distributed superficially and unevenly in the mouse skin when 200 psi was used (**Fig. 1B**).

1. Preparation of the gene gun: Install a battery and insert a barrel liner into the gene gun. Connect the gene gun to the helium tank regulator by the helium hose. Open the helium tank main valve and adjust the regulator to the proper output pressure. Install an empty cartridge holder and fire an empty shot to pressurize the system.
2. Loading cartridges: Insert the cartridges into the cartridge holder, then place the cartridge holder back into the gene gun.
3. Preparation of animals: Anesthetize the animals following the specific procedure for the species under study based on the IACUC-approved protocol at

each institution. The anesthesia should allow the animals to sleep for 30–60 min depending on the need. Shave the target area and vacuum the fur away.

4. Place the gene gun over the target site and fire the gene gun, then move the cylinder to the next cartridge position. Be sure to understand the numbering system on the cartridge holder in order to track how many shots have been fired.

4. Notes

1. In order to have the best DNA-delivery result, every step involved in particle bombardment should be optimized. The following are especially important in success of gene-gun delivery:

   a. The diameter of the gold beads and the helium pressure. Because the thickness of skin varies in different animal species, both parameters can affect the depth required to deliver the majority of the gold beads into the epidermal layer. See **Table 2** for suggested starting conditions delivering DNA to skin in various species.

   b. Moisture/humidity is very damaging to the quality of the gold bead-coated tubing. Thus, the ethanol and tubing should be moisture-free. If the humidity is high, it may be necessary to dry the tubing for a longer time.

   c. The gold beads should be evenly distributed inside the tubing. In order to achieve the best DNA-delivery result, the gold bead-coated tubing with any of the following appearance may not be good to use:

      i.   The gold beads are lined on one side of the tubing;

      ii. The gold beads are distributed as stripes in the tubing; or

      iii. The color of the gold bead-coated tubing is very faint or light.

2. If 0.5 mg of gold beads is loaded into each cartridge with DGR of 2, 30 mg of gold is needed for 60 cartridges. For each preparation, the number of cartridges should be limited to 30–80 shots for easy handling and optimum results. Separate preparations should be made if a larger number of shots are needed.

3. If the volume of the DNA solution to be used in **Subheading 3.1.**, **step 3** is between 100 and 200 μL, add 200 μL of 100 mM spermidine to the gold beads.

4. The volume of $CaCl_2$ should be twice the volume of spermidine.

5. Ideally, the cartridges should be made immediately after the gold beads are coated with DNA. However, the gold bead suspension may be stored at −20°C for a few weeks. To store the beads, seal the cap of the suspension with parafilm to prevent the ethanol from absorbing water from the air, which will reduce the performance of the tube-coating process.

References

1. Klein, T., Wolf, E., Wu, R., and Sanford, J. (1987) High velocity microprojectiles for delivering nucleic acids into living cells. *Nature* **327**, 70–73.
2. McCabe, D. M., Swain, E., Martinell, B., and Christou, P. (1988) Stable transformation of soybean (glycine max) by particle acceleration. *Biotechnology* **6**, 923–926.
3. Fynan, E. F., Webster, R. G., Fuller, D. H., Haynes, J. R., Santoro, J. C., and Robinson, H. R. (1993) DNA vaccines: protective immunizations by aperteral, mucosal, and gene-gun inoculations. *Proc. Natl. Acad. Sci. USA.* **90**, 11478–11482.
4. Pertmer, T. M., Eisenbraum, M. D., McCabe, D. M., Prayaga, S. K. Fuller, D. H., and Haynes, J. R. (1995) Gene gun-based nucleic acid immunization: elicitation of humoral and cytotoxic T lymphocyte responses following epidermal delivery of nanogram quantities of DNA. *Vaccine* **13**, 1427–1430.
5. Eisenbraum, M. D., Fuller, D. H., and Haynes, J. R. (1993) Examination of parameters affecting the elicitation of humoral immune responses by particle bombardment-mediated genetic immunization. *DNA Cell Biol.* **12**, 791–797.
6. Macklin, M. D., Drape, R. J., and Swain, W. F. (1998) Preparations for particle-mediated gene transfer using the Accell gene gun. *Methods Mol. Med.* **29**, 297–303.
7. Kupper, T. S. and Groves, R. W. (1995) The interleukin-1 axis and cutaneous inflammation. *J. Invest. Dermatol.* **105**, 62S–66S.
8. Jakob, T., Ring, J., and Udey, J. R. (2001) Multistep navigation of Langerhans/dendritic cells in and out of the skin. *J. Allergy Clin. Immunol.* **108**, 688–696.
9. Pober, J. S., Kluger, M. S., and Schechner, J. S. (2001) Human endothelial cell presentation of antigen and the homing of memory/effector T cells to skin. *Ann. NY Acad. Sci.* **941**, 12–25.
10. Sun, W. H., Burkholder, J., Paller, A. S., Ershler, W. B., and Yang, N. S. (1995) Use of epidermis as a target for in vivo cytokine gene therapy. *J. Invest. Dermotol.* **104**, 596.
11. Larregina, A. T., Watkins, S. C., Erdos, G., Spencer, L. A., Storkus, W. J., Beer Stolz, D., and Falo, L. D., Jr. (2001) Direct transfection and activation of human cutaneous dendritic cells. *Gene Ther.* **8**, 608–617.
12. Fuller, D. H. and Haynes, J. R. (1994) A qualitative progression in HIV type I glycoprotein 120-specific cytotoxic cellular and humoral immune responses in mice receiving a DNA based glycoprotein-120 vaccine. *AIDS Res. Hum. Retroviruses* **10**, 1433–1441.
13. Lu, S., Arthos, J., Montefiori, D. C., Yasutomi, Y., Manson, K., Mustafa, F., et al. (1996) Simian immunodeficiency virus DNA vaccine trial in macaques. *J. Virol.* **70**, 3978–3991.
14. Lu, S., Santoro, J. C., Fuller, D. H., Haynes, J. R., and Robinson, H. L. (1995) Use of DNAs expressing HIV-1 Env and noninfectious HIV-1 particles to raise antibody responses in mice. *Virology* **209**, 147–154.
15. Lu, S., Wyatt, R., Richmond, J. F., Mustafa, F., Wang, S., Weng, J., and Robinson, H. L. (1998) Immunogenicity of DNA vaccines expressing human immunodeficiency virus type 1 envelope glycoprotein with and without deletions in the V1/2 and V3 regions. *AIDS Res. Hum. Retroviruses* **14**, 151–155.

16. Robinson, H. L., Montefiori, D. C., Johnson, R. P., Manson, K., Kalish, M. L. Lifson, J. D., et al. (1999) Neutralizing antibody-independent containment of immunodeficiency virus challenges by DNA priming and recombinant pox virus booster immunizations. *Nat. Med.* **5**, 612–614.

17. McCluskie, M. J., Brazolot Millan, C. L., Gramzinski, R. A., Robinson, H. L. Santoro, J. C., Fuller, J. T., et al. (1999) Route and method of delivery of DNA vaccine influence immune responses in mice and non-human primates. *Mol. Med.* **5**, 287–300.

18. Yoshida, A., Nagata, T., Uchijima, M., Higashi, T., and Koide, Y. (2000) Advantage of gene gun-mediated over intramuscular inoculation of plasmid DNA vaccine in reproducible induction of specific immune responses. *Vaccine* **18**, 1725–1729.

19. Richmond, J. F., Lu, S., Santoro, J. C., Weng, J., Hu, S. L., Montefiori, D. C., and Robinson, H. L. (1998) Studies of the neutralizing activity and avidity of anti-human immunodeficiency virus type 1 Env antibody elicited by DNA priming and protein boosting. *J. Virol.* **72**, 9092–9100.

20. Han, R., Cladel, N. M., Reed, C. A., Peng, X., Budgeon, L. R., Pickel, M., and Christensen, N. D. (2000) DNA vaccination prevents and/or delays carcinoma development of papillomavirus-induced skin papillomas on rabbits. *J. Virol.* **74**, 9712–9716.

21. Han, R., Cladel, N. M., Reed, C. A., Peng, X., and Christensen, N. D. (1999) Protection of rabbits from viral challenge by gene gun-based intracutaneous vaccination with a combination of cottontail rabbit papillomavirus E1, E2, E6, and E7 genes. *J. Virol.* **73**, 7039–7043.

22. Richmond, J. F., Mustafa, F., Lu, S., Santoro, J. C., Weng, J., O'Connell, M., et al. (1997) Screening of HIV-1 Env glycoproteins for the ability to raise neutralizing antibody using DNA immunization and recombinant vaccinia virus boosting. *Virology* **230**, 265–274.

23. Yoshida, A., Nagata, T., Uchijima, M., and Koide, Y. (2001) Protective CTL response is induced in the absence of CD4+ T cells and IFN-gamma by gene gun DNA vaccination with a minigene encoding a CTL epitope of Listeria monocytogenes. *Vaccine* **19**, 4297–4306.

24. Price, B. M., Galloway, D. R., Baker, N. R., Gilleland, L. B., Staczek, J., and Gilleland, H. E., Jr. (2001) Protection against Pseudomonas aeruginosa chronic lung infection in mice by genetic immunization against outer membrane protein F (OprF) of P. aeruginosa. *Infect. Immun.* **69**, 3510–3515.

25. Tahtinen, M., Strengell, M., Collings, A., Pitkanen, J. Kjerrstrom, A., Hakkarainen, K., et al. (2001) DNA vaccination in mice using HIV-1 nef, rev and tat genes in self-replicating pBN-vector. *Vaccine* **19**, 2039–2047.

26. Barnett, S. W., Rajasekar, S., Legg, H., Doe, B., Fuller, D. H., Haynes, J. R., et al. (1997) Vaccination with HIV-1 gp120 DNA induces immune responses that are boosted by a recombinant gp120 protein subunit. *Vaccine* **15**, 869–873.

27. Lodmell, D. L., Parnell, M. J., Bailey, J. R., Ewalt, L. C., and Hanson, C. A. (2001) One-time gene gun or intramuscular rabies DNA vaccination of non-human primates: comparison of neutralizing antibody responses and protection against rabies virus 1 year after vaccination. *Vaccine* **20**, 838–844.

28. Fuller, D. H., Murphey-Corb, M., Clements, J., Barnett, S., and Haynes, J. R. (1996) Induction of immunodeficiency virus-specific immune responses in rhesus monkeys following gene gun-mediated DNA vaccination. *J. Med. Primatol.* **25**, 236–241.

29. Yang, N. S., Burkholder, J., Roberts, B., Martinell, B., and McCabe, D. M. (1990) In vivo and in vitro gene transfer to mammalian somatic cells by particle bombardment. *Proc. Natl. Acad. Sci. USA* **87**, 9568–9572.

30. Tang, D. C., Devit, M., and Johnston, S. A. (1992) Genetic immunization is a simple method for eliciting an immune response. *Nature* **356**, 152–154.

31. Lin, M. T., Pulkkinen, L. Uitto, J., and Yoon, K. (2000) The gene gun: current applications in cutaneous gene therapy. Int. *J. Dermatol.* **39**, 161–170.

32. Sun, W. H., Burkholder, J., Sun, J., Culp, J., Turner, J., Lu, X. G., et al. (1995) In vivo cytokine gene transfer by gene gun suppresses tumor growth in mice. *Proc. Natl. Acad. Sci. USA* **92**, 2889–2893.

33. Mahvi, D. M., Sheehy, M. J., and Yang, N. S. (1997) DNA cancer vaccines: a gene gun approach. Immunol. *Cell. Biol.* **75**, 456–460.

34. Seigne, J., Turner, J., Diaz, J., Hackney, J., Pow-Sang, J., Helal, M., et al. (1999) A feasibility study of gene gun mediated immunotherapy for renal cell carcinoma. *J. Urol.* **162**, 1259–1263.

35. Wang, J., Murakami, T., Hakamata, Y., Ajiki, T., Jinbu, Y., Akasaka, Y., et al. (2001) Gene gun-mediated oral mucosal transfer of interleukin 12 cDNA coupled with an irradiated melanoma vaccine in a hamster model: successful treatment of oral melanoma and distant skin lesion. *Cancer Gene Ther.* **8**, 705–712.

36. Roy, M. J., Wu, M. S., Barr, L. J., Fuller, J. T., Tussey, L. G., et al. (2000) Induction of antigen-specific CD8+ T cells, T helper cells, and protective levels of antibody in humans by particle-mediated administration of a hepatitis B virus DNA vaccine. *Vaccine* **19**, 764–778.

37. Schirmbeck, R. and Reimann, J. (2001) Modulation of gene-gun-mediated Th2 immunity to hepatitis B surface antigen by bacterial CpG motifs or IL-12. *Intervirology* **44**, 115–123.

38. Swain, W. E., Heydenburg Fuller, D., Wu, M. S., Barr, L. J., Fuller, J. T., et al. (2000) Tolerability and immune responses in humans to a PowderJect DNA vaccine for hepatitis B. *Dev. Biol.* **104**, 115–119.

39. Macklin, M. D., McCabe, D., McGregor, M. W., Neumann, V., Meyer, T., Callan, R., et al. (1998) Immunization of pigs with a particle-mediated DNA vaccine to influenza A virus protects against challenge with homologous virus. *J. Virol.* **72**, 1491–1496.

40. Belperron, A. A., Feltquate, D., Fox, B. A., Horii, T., and Bzik, D. J. (1999) Immune responses induced by gene gun or intramuscular injection of DNA vaccines that express immunogenic regions of the serine repeat antigen from Plasmodium falciparum. *Infect. Immun.* **67**, 5163–5169.

41. Yoshida, S., Kashiwamura, S. I., Hosoya, Y., Luo, E., Matsuoka, H., Ishii, A., et al. (2000) Direct immunization of malaria DNA vaccine into the liver by gene gun protects against lethal challenge of Plasmodium berghei sporozoite. Biochem. *Biophys. Res. Commun.* **271**, 107–115.

42. Degano, P., Schneider, J., Hannan, C. M., Gilbert, S. C., and Hill, A. V. (1999) Gene gun intradermal DNA immunization followed by boosting with modified vaccinia virus Ankara: enhanced CD8+ T cell immunogenicity and protective efficacy in the influenza and malaria models. *Vaccine* **18**, 623–632.

43. Vinner, L., Nielsen, H. V., Bryder, K., Corbet, S., Nielsen, C., and Fomsgaard, A. (1999) Gene gun DNA vaccination with Rev-independent synthetic HIV-1 gp160 envelope gene using mammalian codons. *Vaccine* **17**, 2166–2175.

44. Lodmell, D. L., Ray, N. B., Ulrich, J. T., and Ewalt, L. C. (2000) DNA vaccination of mice against rabies virus: effects of the route of vaccination and the adjuvant monophosphoryl lipid A (MPL). *Vaccine* **18**, 1059–1066.

45. Lodmell, D. L., Ray, N. B., and Ewalt, L. C. (1998) Gene gun particle-mediated vaccination with plasmid DNA confers protective immunity against rabies virus infection. *Vaccine* **16**, 115–118.

46. Boyle, C. M. and Robinson, H. L. (2000) Basic mechanisms of DNA-raised antibody responses to intramuscular and gene gun immunizations. *DNA Cell Biol.* **19**, 157–165.

47. Robinson, H. L. (1999) DNA vaccines: basic mechanism and immune responses (Review). *Int. J. Mol. Med.* **4**, 549–555.

48. Feltquate, D. M., Heaney, S., Webster, R. G., and Robinson, H. L. (1997) Different T helper cell types and antibody isotypes generated by saline and gene gun DNA immunization. *J. Immunol.* **158**, 2278–2284.

15

Biolistic Transfection of Cultured Organotypic Brain Slices

A. Kimberley McAllister

1. Introduction

Transfection of postmitotic neurons remains one of the most challenging goals in the field of gene delivery. For many cell types, gene transfer using the various methods described in this book has become a routine tool for studying gene regulation and function. However, postmitotic neurons are not easily transfected by most gene-delivery methods. Conventional techniques to transfect neurons in intact tissue are problematic because they are often unreliable or too time-consuming.

Recently, biolistic transfection has proven successful in transfecting postmitotic neurons in organotypic brain slices *(1,2)*. Biolistics is a physical method of transfection *(3–5)*. In biolistics, micron-sized gold particles are coated with plasmid DNA and then accelerated at high speeds toward target tissue. Cells penetrated by a gold particle have a high likelihood of becoming transfected. For the biolistic device discussed in this protocol (the Helios gene-gun system produced by Bio-Rad), high-pressure helium accelerates the gold particles into the tissue. Biolistic transfection is efficient, reliable, and does not require advanced molecular biological facilities for its application.

To study the signals that regulate neural development, we use organotypic brain-slice cultures *(1,2)*. Organotypic slice preparations are a powerful experimental system as they preserve the three-dimensional architecture and local environment of neurons while allowing easy access for experimental manipula-

From: *Methods in Molecular Biology, vol. 245:*
Gene Delivery to Mammalian Cells: Vol. 1: Nonviral Gene Transfer Techniques
Edited by: W. C. Heiser © Humana Press Inc., Totowa, NJ

tion and observation *(6)*. Interface cultures, in particular, preserve the three-dimensionality of the tissue slice *(6)*. To prepare this culture type, slices are placed on culture inserts so that the top surface of the slice is exposed to the incubator atmosphere while the lower surface contacts the culture medium *(6)*. Organotypic, interface slices remain healthy for weeks and continue to differentiate, forming layer-specific connections *(7,8)*. There are several methods for cutting brain slices: the tissue slicer, the tissue chopper, or the vibratome. The method that produces the healthiest slices for culturing organotypic slices of young cerebral cortex is the tissue slicer, a device designed by Lawrence C. Katz *(9)* and distributed by SD Instruments. This method of slicing is extremely fast; the time from brain removal to placing slices in the incubator is only a few minutes.

Here, we provide information on optimizing the parameters for biolistic transfection of neurons in cultured, organotypic brain slices. Biolistic transfection is a relatively simple and effective method for transfecting neurons in brain slices, resulting in high levels of expression of reporter proteins in transfected cells. Once optimized, biolistic transfection is a reliable and efficient method for studying gene function in many cell types, including postmitotic neurons. Neurons can be transfected immediately following slice preparation. In general, biolistics consists of three steps: coating gold particles with DNA, preparing the cartridges, and accelerating the particles into the tissue.

2. Materials

2.1. Slice Cultures

1. Agarose 1000 (Invitrogen Corp. Carlsbad, CA, Cat. no. 10975027).
2. Basal Media Eagle (BME; Invitrogen Corp., Cat. no. 21010046).
3. Hank's Balanced Salt Solution (HBSS; Invitrogen Corp., Cat. no. 24020117).
4. Horse Serum heat-inactivated, (Hyclone, Logan UT, Cat. no. SH30074.02).
5. HEPES Buffer Solution (1 *M* solution; Invitrogen Corp., Cat. no.15630080).
6. Penicillin-streptomycin liquid (Invitrogen Corp., Cat. no. 15070063).
7. One glass Petri dish (painted black to ease visualization of opaque brain slices), autoclave.
8. HEPES (Invitrogen Corp., Cat. no. 11344-033).
9. Paint brushes, size 000, rinsed in EtOH; autoclave.
10. Pasteur pipets: cut thin tips off and place pipet bulb on cut side so slices can be sucked into large-bore end; autoclave.
11. Tissue slicer (9; SD Instruments, Grants Pass, OR): prepare winders according to procedure enclosed with apparatus. To sterilize, rinse winders with 95% EtOH and expose to UV for 10 min each side.

12. Sterile tissue-culture inserts, 0.4-μm pore size (Millicell CM Inserts Cat. no. PICM 03050, Millipore Corp., Bedford, MA or Falcon Cat. no. 3090, Fisher Scientific Co., Pittsburgh, PA).
13. Sterile tissue culture plates (Falcon, Cat. no. 1146).
14. Sterile disposable tissue culture filter units (*see* **Note 1**; 0.2-μm filter size; Nalgene Nalge Nunc International, Rochester, NY, Cat. no. 156-4020).
15. Fine scissors (Fine Science Tools, Foster City, CA, Cat. no.14063-09), autoclave
16. Blunt forceps (Fine Science Tools, Cat. no. 11006-12), autoclave.
17. Fine forceps (Fine Science Tools, Cat. no. 11251-35), autoclave.

2.2. Solutions

All solutions should be made in a sterile hood, unless indicated otherwise.

1. Culture medium for slice cultures: 50% BME, 25% HBSS, 25% Horse serum, 33 mM D-glucose, 10 mM HEPES (from a 1 M solution stock), 100 U/mL Penicillin-streptomycin. Sterilize culture medium by running through a sterile filter unit (*see* **Note 1**). Maintain 100 mL of media on ice for slicing and 400 mL of media in a 5% CO_2/ 37°C incubator for feeding cultures.
2. HEPES ACSF: 140 mM NaCl, 5 mM KCl, 1 mM $MgCl_2$ 24 mM D-glucose, 10 mM HEPES, 1 mM $CaCl_2$. pH solution to 7.2, sterile filter using a filter unit, and store at 2–4°C (*see* **Note 2**).
3. 1.85% Agarose: Weigh enough agarose to make a 1.85% solution (weight to volume). Add to water. Dissolve by heating agarose solution in a microwave. Pour into a large glass petri dish (150 × 15 mm) allow to solidify, cover, and store at 2–4°C for up to 2 wk. UV sterilize before use.
4. 1 M $CaCl_2$ solution: 1 M solution dissolved in distilled H_2O. This solution does not need to be sterile.
5. PVP solutions: Dissolve 20 mg of PVP in 1 mL of fresh 100% EtOH to make a stock solution. Then, dilute 20 μL of PVP stock into 8 mL of fresh 100% EtOH (final concentration 0.05 mg/mL) (*see* **Note 3**).
6. Spermidine solution: Dilute spermidine (free base, Sigma, St. Louis, MO, Cat. no. S-0266) to a final concentration of 0.05 M in sterile distilled H_2O. Store the dilute solution in a refrigerator for up to 6 mo.

2.3. Biolistic Transfection

1. Bio-Rad Helios gene-gun system (Bio-Rad, Hercules, CA, Cat. no. 165-2431): includes gene gun, tubing prep station, optimization kit, helium hose, and high-pressure helium regulator (purchase the regulator with the lowest possible pressure gauge).
2. 5% $CaCl_2$.

3. 100% EtOH (fresh bottle).
4. Gold, 1.6 μm (Bio-Rad).
5. Compressed helium (Grade 4.5 [99.995%] or higher); placed beside the laminar flow hood.
6. Compressed nitrogen (Grade 4.8 (99.998%) or higher); with a regulator that registers a maximum of 30 psi; placed near and connected to the tubing prep station.
7. Nylon Mesh (Small Parts Cat. no. CMN-90-D, Clear Lakes, FL).
8. Polyvinylpyrrollidone (PVP; Bio-Rad).
9. Razor blades.
10. Gold-coat tubing (Bio-Rad).

2.4. Plasmids

1. Supercoiled plasmid DNAs can be purified using commercial plasmid isolation kits, such as Qiagen spin miniprep kits (Qiagen, Cat. no. 27104) or Bio-Rad midiprep kits (Bio-Rad, Cat. no. 732-6120) (*see* **Note 4**).

2.5. Animals

Cultured organotypic slices can theoretically be made from the brain of any mammal. However, this protocol is optimized for culturing slices of cerebral cortex from newborn ferrets and rats or mice. There are multiple vendors for these animals and they cannot be purchased before obtaining approval from your institutional IACUC committee. The most common vendor for ferrets is Marshall Farms, North Rose, NY; and for mice and rats is Charles River Laboratories, Wilmington, MA.

3. Methods
3.1. Slice Cultures
3.1.1. Preparation of Slicer

1. Place slice culture medium (Solution 1), HEPES ACSF (Solution 2), and agarose (Solution 3) on ice.
2. Place dissection microscope, black Petri dish, 250-mL beaker, surgical instruments, culture dishes and inserts, brushes, slicer in hood. Wash everything that is not sterile with 70% EtOH, and expose to UV light for 20 min.
3. Cut a slab of agarose to fit the slicer, place on removable slicer component, and place both on ice.
4. Place winder into slicing apparatus.

3.1.2. Preparation of Slice Cultures

1. Wear gloves, wipe hands with 70% EtOH often, and wear a face mask.

2. Anesthetize the animal, according to your IACUC protocol.
3. Rinse the animal's head with 70% EtOH and decapitate with razor blade.
4. Transfer the head into the hood and place on paper towel.
5. Remove the brain and place in a black petri dish filled with ice-cold HEPES ACSF. Using ice-cold solutions enhances slice health and makes the tissue more solid and easier to handle.
6. Viewing under the dissecting microscope, remove pia from desired brain region using fine forceps and separate the appropriate block of tissue.
7. Place the tissue onto the agarose and insert into the slicer (follow instructions from SD Instruments).
8. Activate slicer spring. The winder will slice tissue into 400 μm-thick slices (or a different thickness, depending on winder preparation as described in Slicer instructions from SD Instruments).
9. Unscrew winder from slicer and remove the tissue with winder.
10. Gently place the tissue and winder into a clean Petri dish with fresh, ice-cold HEPES ACSF (Solution 2).
11. Carefully separate the slices with a paintbrush, touching the brain tissue as little as possible.
12. Place culture inserts into the culture plates.
13. Suck the slices into the large-bore end of a Pasteur pipet and drop them onto an insert.
14. Using a paintbrush, gently separate and flatten slices so they can't grow together.
15. Using the thin-bore side of a Pasteur pipet, remove as much HEPES ACSF as possible from the filter without touching the slices.
16. Place 1 mL of culture medium (Solution 1) under each insert; the medium should wet the insert without floating the slices. Remove any bubbles from beneath the insert.
17. Maintain slices in a 5% CO_2/37°C incubator.
18. Change the medium every 3 d.

3.2. Biolistic Transfection

The details for this procedure are clearly described in the Helios Gene Gun System Instruction Manual from Bio-Rad Laboratories. Several modifications to these basic procedures are detailed below that enhance transfection reliability for organotypic slice cultures (8,9).

3.2.1. Preparation of DNA-Coated Gold

1. Prepare fresh PVP solutions (Solution 5).
2. In a 1.5-mL microcentrifuge tube, mix 0.0125 g of gold (1.6 μm in diameter; *see* **Note 5**) and 100 μL spermidine solution (Solution 6).

3. Vortex and sonicate for 5 s (*see* **Note 6**).
4. Add 25 µg of DNA and vortex for 5 s.
5. Add 100 µL of 5% $CaCl_2$ drop-wise, while vortexing.
6. Allow the gold and DNA to precipitate for 10 min, rotating the tube periodically.
7. To pellet the gold, spin for 15 s at maximum speed in a microcentrifuge.
8. Remove the supernatant and discard.
9. Wash the gold particles three times with 1 mL of fresh 100% EtOH. Between each wash, spin for 5 s at maximum speed.
10. Resuspend the gold particles in 200 µL of dilute PVP solution (Solution 5) and transfer to a 15-mL centrifuge tube.
11. Rinse all of the gold particles out of the microcentrifuge tube with 800 µL dilute PVP solution (Solution 5) and transfer to a 15-mL centrifuge tube.
12. Add 2 mL of dilute PVP solution (Solution 5) to a final volume of 3 mL of gold particles in PVP solution. This solution is the DNA/PVP/EtOH mixture.

3.2.2. Preparation of Biolistic Cartridges

1. Insert a 30-in length of gold-coat tubing into the tubing prep station and allow nitrogen to flow through for 30 min at 0.35 l per minute (lpm).
2. Turn off the nitrogen with the flowmeter on the prep station.
3. Remove the tubing from the tubing prep station. Attach a syringe to one end of the tubing (using the adapter tubing provided by Bio-Rad).
4. Vortex the gold suspension and invert the tube twice.
5. Pull 3 mL of the DNA/PVP/EtOH mixture into the open end of the purged gold-coat tubing using the syringe. Fill the tubing carefully, avoiding bubbles, until the DNA/PVP/EtOH mixture is 2 in from the syringe.
6. Keeping the tubing horizontal and the syringe attached, insert the loaded tubing into the tubing prep station.
7. Allow the gold to settle for 5 min, keeping the syringe attached.
8. Slowly and steadily remove the PVP/EtOH solution using the syringe, leaving the settled gold undisturbed. Detach the syringe and discard the solution (*see* **Note 7**).
9. Immediately turn the tubing 180° using the switch on the prep station, allowing the gold to coat the tubing for 5–15 s.
10. Turn the switch on the prep station to ON to rotate the tubing for 30 s.
11. While the tubing is rotating, open the valve on the flowmeter to allow 0.35 lpm nitrogen to dry the gold onto the inside of the tubing for 5 min.
12. Turn the motor OFF. Remove the tubing.
13. Cut off any pieces of tubing not uniformly coated with gold. Use the tubing cutter (provided with the system by Bio-Rad) to cut the tubing into 0.5-in cartridges.

14. Store the cartridges in a vial (supplied by Bio-Rad) with a dessicator pellet, label, and wrap with parafilm. Cartridges can be stored for 3 mo at 4°C.

3.2.3. Transfecting Slices

1. Attach a piece of 2 × 2 inch nylon mesh with tape to the barrel liner, covering all of the exposed parts (*see* **Note 8**). Sterilize this mesh and the cartridge holders by exposure to UV light for 30 min.
2. Insert an empty cartridge holder and a 9V battery into the Helios Gene Gun.
3. Using the special helium hose provided by Bio-Rad, attach the gene gun to the regulator (provided with the Bio-Rad Biolistic system) on the helium tank. To pressurize the helium hose and the reservoirs in the gun, fire a few shots at 100 psi.
4. Load the gold-containing cartridges into the cartridge holders (provided by Bio-Rad) and load the holder into the gene gun.
5. Attach the barrel liner with the nylon mesh to the gene gun.
6. Using sterilized forceps, place a filter insert containing slice cultures onto a slab of agarose (presterilized by UV, Solution 3).
7. Quickly, position the gene gun above the slices at a 90° angle to the slices and bring the end of the barrel flush with the top of the filter insert.
8. Shoot the gold into the slices at a pressure of 75–110 psi (*see* **Note 9**).
9. Quickly replace the inserts into their original medium and return to the incubator.
10. Close the helium tank, release the pressure from the gene gun, and detach the gene gun from the helium tank.
11. Remove the cartridge holder and discard the used cartridges.
12. Check for successful transfection by visualizing the gold in the slices using a dissection microscope. The gold particles should be evenly distributed in the slice to a depth of 300 µm. Subsequently, transfected cells should be visualized by visualizing the reporter gene (i.e., by visualizing EGFP with fluorescence microscopy; *see* **Notes 9** and **10**).

4. Notes

1. Nalgene filters are the best choice as they have minimal detergent in their filters. The use of filters from other companies, including Corning, can lead to unhealthy slices, possibly owing to detergent in their filters.
2. This protocol is optimized for organotypic slices from the ferret visual cortex. Cortical slice cultures from rat or mouse are more difficult to keep healthy. Healthy slices are more often obtained when NaCl is replaced with equimolar sucrose in the HEPES ACSF (dissection solution).
3. The PVP solution must be made fresh each day from a newly opened bottle of EtOH (to avoid any water at the last step in the cartridge-coating pro-

cedure). An alternative to using a new bottle each time is to aliquot a fresh bottle of EtOH and freeze the aliquots at −20°C.

4. DNA amplified with the specified kits is usually pure enough to facilitate reliable transfections. However, if transfection is unsuccessful, it is possible that the DNA might be contaminated.

5. For cortical and hippocampal neurons, 1.6 μm gold particles transfect the greatest number of healthy neurons. It is possible that smaller diameter cells will be optimally transfected with smaller gold particles, which are also available from Bio-Rad.

6. The concentration of gold particles in this protocol corresponds to half the Bio-Rad-recommended concentration. In cortical slices, the Bio-Rad-recommended amount of gold causes too much tissue damage. The concentration recommended here is optimal for transfection of cortical slices.

7. The Bio-Rad manual suggests that the PVP/EtOH solution be removed with a peristaltic pump. This is unnecessary as the solution can easily be removed with the syringe without disturbing the gold.

8. The nylon mesh reduces the detrimental effects of the helium shock wave. This is critical to the success of biolistic transfection of organotypic cortical slices.

9. A number of factors affect the success of biolistic transfection. The number of neurons transfected using biolistics depends on the amount of gold per shot, the efficiency of the expression construct, and the helium pressure. These parameters must be optimized for each tissue type, and controls must be designed to ensure that biolistic transfection does not damage the slices. For every tissue, there is a threshold gold concentration that will result in the largest number of transfected cells without tissue damage. Increasing the gold concentration above that threshold can cause extensive tissue damage in brain slices; decreasing the gold concentration too far will result in low numbers of transfected cells. Similarly, there is an optimal helium pressure for each tissue type that will provide the greatest penetration of the slices without tissue damage. Pressures above that threshold produce a helium shock wave that kills cells in slices; pressures below the threshold prohibit transfection. Because a few cells can be transfected even in cases with significant tissue damage, the simple observation of transfected cells is not sufficient to ensure the health of the tissue or cells. The nylon mesh described above minimizes this shock wave, but the helium pressure must still be optimized and careful controls performed to ensure that biolistics is not damaging the transfected cells.

10. For organotypic brain slices, the most important parameter for transfection success is neuronal health. If slices are unhealthy, levels of transfection will

be low. More often than not, if no transfected cells are obtained, it is the health of the cultures that is the problem.

References

1. Lo, D. C., McAllister, A. K, and Katz, L. C. (1994) Neuronal transfection in brain slices using particle-mediated gene transfer. *Neuron* **13**, 1263–1268.
2. McAllister, A. K., Lo, D. C., and Katz, L. C. (1995) Neurotrophins regulate dendritic growth in developing visual cortex. *Neuron* **15**, 791–803.
3. Klein, T. M., Wolf, E. D., Wu, R., and Sanford, J.C. (1987) High-velocity microprojectiles for delivering nucleic acids into living cells. *Nature* **327**, 70–73.
4. Johnston, S. A. (1990) Biolistic transformation: microbes to mice. *Nature* **346**, 776–770.
5. Sanford, J. C., Smith, F. D., and Russell, J. A. (1993) Optimizing the biolistic process for different biological applications. *Meth. Enzymol.* **217**, 483–509.
6. Stoppini, L., Buchs, P. A., and Muller, D. (1991) A simple method for organotypic cultures of nervous tissue. *J. Neurosci. Methods* **37**, 173–182.
7. Yamamoto, N., Yamada, K., Kurotani, T., and Toyama, K. (1992) Laminar specificity of extrinsic cortical connections studies in coculture preparations. *Neuron* **9**, 217–228.
8. Dantzker, J. L. and Callaway, E. M. (1998) The development of local, layer-specific visual cortical axons in the absence of extrinsic influences and intrinsic activity. *J. Neurosci.* **18**, 4145–4154.
9. Katz, L. C. (1987) Local circuitry of identified projection neurons in cat visual cortex brain slices. *J. Neurosci.* **7**, 1223–1249.

16

Efficient Electroporation of Mammalian Cells in Culture

Peter A. Barry

1. Introduction

Electroporation exploits the fact that high-voltage electrical fields can temporarily disrupt the structural integrity of cell membranes (*1*). When an electrical pulse is delivered to cells placed between two electrodes, pores develop in the membrane within 3 milliseconds (ms) and increase in size up to 120 nm by 20 ms (*3*). The presence of the pores is transient, and they reseal within a few seconds. During this brief window of time, it is possible to introduce a wide variety of macromolecules into mammalian tissue culture cells. Electroporation is broadly applicable to multiple cell types, highly efficient, reproducible, and appropriate for both transient and stable transfections.

However, electroporation conditions are cell line-dependent. Similar to chemically mediated transfection techniques, optimal conditions must be established for each cell line. This chapter is designed to provide the investigator with a reliable protocol that enables a high level of transfection efficiency and gene expression. There are many variables that can influence the success of electroporation. These include the state of cell growth at the time of electroporation, the amount of transfecting DNA, size of the capacitor used to store the charge, and the voltage of the electrical charge delivered to the cells. Although the focus will be on transient gene expression, the strategy can also be extended to stable transfection.

From: *Methods in Molecular Biology, vol. 245:*
Gene Delivery to Mammalian Cells: Vol. 1: Nonviral Gene Transfer Techniques
Edited by: W. C. Heiser © Humana Press Inc., Totowa, NJ

2. Materials

2.1. DNA

1. Purify plasmids for electroporation by standard techniques *(7)* and determine concentration by optical density (OD) at 260 nm. Adjust DNA concentration to 0.5–1.0 μg/mL. Aliquot DNA samples and store at −80°C.
2. Purify genomic DNA by sodium dodecyl sulfate (SDS)/proteinase K lysis, phenol-chloroform extraction, and ethanol precipitation *(7)*. Adjust DNA concentration to 0.5–1.0 μg/mL. Aliquot DNA samples and store at −80°C.

2.2. Electroporation

1. Gene Pulser II System (Bio-Rad Laboratories, Hercules, CA).
2. 0.4 cm electroporation cuvet (Bio-Rad Laboratories).
3. RPMI 1640 (Invitrogen, Carlsbad, CA).

2.3. Analysis of Electroporation

Vectastain Elite ABC Kit and DAB Substrate Kit (Vector Laboratories, Burlingame, CA).

3. Methods

3.1. Preparation of Cells

1. Subdivide adherent or suspension cell cultures 16–24 h prior to day of electroporation (Day −1, **Table 1**). Approximately 4×10^6 cells will be required per electroporation sample. Adjust plating density with normal growth medium so that the cells will be in logarithmic phase of growth at time of electroporation (Day 0, **Table 1**; *see* **Notes 1** and **2**).
2. On day of electroporation, trypsinize adherent cell cultures. Pellet trypsinized or suspension cell cultures in a table top centrifuge (5 min at 500*g*). Wash cells two times with phosphate-buffered saline (PBS).
3. Resuspend cells in RPMI 1640 media (without supplements); count viable cells and adjust the cell density to 1.3×10^7 cells per mL. Maintain cells at room temperature (*see* **Note 3**).

3.2. Electroporation

1. Add DNA to bottom of labeled cuvets. Use a maximum of 10 μg (10–20 μL) of DNA per cuvet. Maintain at room temperature (*see* **Note 4**).
2. Working with three to four cuvets at a time, add 0.3 mL of resuspended cells to the bottom of each cuvet. Gently tap cuvet to mix DNA and cells. Avoid introducing air bubbles. Maintain at room temperature.

Table 1
Protocol for Electroporation

Phase	Day	Procedure
Pre-electroporation	−1	Subdivide cells the day before electroporation so that cells are in the logarithmic phase of growth at time of electroporation the next day.
Electroporation cells	0	Harvest cells, wash twice in PBS, and resuspend cells in RPMI 1640 media (no supplements) at a density of 1.3×10^7 cells per mL. Add 0.3 mL of resuspended cells per 4 mm cuvet. Maintain at room temperature.
DNA		Use 5–10 µg of super-coiled DNA per cuvet (10–20 µL volume).
Conditions		To be determined (see text). Work with three to four cuvets at one time. Add DNA to cuvet, gently tap to mix, but avoid introducing air bubbles. Pulse each cuvets, tapping the cuvet after pulsing to redistribute the cells. Keep cells at room temperature. After pulsing the batch of cuvet, immediately transfer contents of cuvette to a T25 flask or well of a six-well plate containing normal growth media.
Post-electroporation	+2	Either harvest cells for transient assay, immuno-cytochemistry, Western blot, RNA analysis. Alternatively, place cells in selective media.

3. Place a cuvet into the cuvet holder of the Gene Pulser. Adjust capacitor and voltage settings (*see* **Note 5**). Pulse cuvet, remove, and tap gently to mix electroporated cells. Record Time Constant of decay. Maintain cuvet at room temperature until the batch of three to four cuvets has been pulsed.

4. Remove cells from cuvet and place into normal growth medium within a T25 flask or a single well of a six-well plate. Rinse cuvet with growth medium and pool with remainder of cells. Place flask in a 37°C/5% CO_2 humidified incubator.

5. Complete electroporation of cells, working with three to four cuvets at a time.

6. Maintain cells at 37°C for 24–48 h (Day +2, **Table 1**) then assay transient gene expression. Note that the optimal time for expression will need to be empirically determined and will depend on both the cell line and the gene being delivered.

3.3. Analysis of Electroporation

1. Determine viable cell number. This can be accomplished using a variety of techniques, such as staining cells with a vital dye and counting cells in a hemocytometer.
2. For transient transfections, harvest cells for enzymatic assays, RNA, protein, or immunocytochemistry, according to recommended protocols (*see* **Note 6**). For stable transfections, remove growth medium, and replace with appropriate selective medium.

4. Notes

1. Maintain stocks of cells in relation to the size of the experiment; 4×10^6 cells are recommended per DNA sample or electroporation condition as a starting cell number. The number of cells can be adjusted downward if the sensitivity of the experimental assay is sufficient.
2. Optimal electroporation requires that cells be in an active state of growth at the time they are prepared for pulsing. If cells approach stationary phase, electroporation efficiency drops (data not shown). Cells can be subdivided more than 1 d in advance as long as the plating density is accordingly reduced.
3. RPMI 1640 works exceedingly well as an electroporation medium, even for those cells normally cultured in a different growth medium. No supplements (e.g., serum, penicillin/streptomycin) are added to the RPMI for electroporation. Other basal growth media should also work well. Optimal capacitor and voltage settings may be slightly different from those observed with RPMI 1640 owing to differences in formulations. It should be emphasized that cells are returned to their normal growth medium after pulsing.
4. There is an upper limit to the amount of DNA that should be used for electroporation; 10–20 μg results in a high efficiency of transfection and strong gene expression. Use of greater amounts can reduce cell viability and gene expression (data not shown).
5. Optimal electroporation conditions should be empirically determined for each cell line. Key factors that will influence the interpretation of the results will be the level of gene expression, efficiency of gene uptake, and cell viability after electroporation. A typical experiment involves slight adjustments in the capacitor size and voltage using a constant amount of

Table 2
Optimization of Electroporation Conditions for Jurkat Cells (960 µF)

Voltage	Time constant (ms)	CAT Activity (cpm)
100	38	6,120
150	32	595,600
200	34	2,171,840
250	31	2,185,100
300	30	162,320

DNA. This is illustrated for Jurkat cells (a human T-lymphoid cell line) (**Table 2**, only voltage adjustments are presented).

6. As an example, Jurkat cells were transfected with plasmids containing the chloramphenicol acetyltransferase (CAT) gene under the transcriptional control of the HIV-1 promoter (pHIVLTR/CAT) and the HIV-1 tat gene driven by the SV40 promoter (pSV/tat) *(2)*. Cells were analyzed at 48 h for CAT activity *(5)*. The results demonstrate that pulsing Jurkat cells with 200 and 250 volts and the 960 µF capacitor produced the highest level of CAT expression. A similar approach has been used to identify efficient electroporation conditions for a variety of mammalian cells (**Table 3**).

The level of transgene expression is generally a function of the efficiency of DNA uptake by electroporated cells. Electroporation conditions that result in high levels of gene expression (e.g., 200 volts, **Table 2**) are

Table 3
Optimal Electroporation Conditions for Mammalian Cells[a]

Cell line	Capacitor (µF)	Voltage (kilovolts)
Chinese hamster ovary (CHO)	960	0.25
CEMx174	960	0.2
GCT	960	0.2
HeLa	500	0.3
HuT78	960	0.25
L929	500	0.35
MRC-5	960	0.3
U373	960	0.3
BALB/c	500	0.4
Cos7	960	0.3
Vero	960	0.3

[a]In all cases, electroporation is in 0.4-cm cuvets with 0.3 mL of cells.

associated with the highest percentage of cells that express input DNA. This can be readily assayed by electroporation of a reporter plasmid for which the presence of the expressed protein product can be assayed by a simple fluorescent assay (e.g., enhanced green fluorescent protein; EGFP) *(4)* or immunocytochemical stain (e.g., human cytomegalovirus [HCMV] immediate-early [IE] 1 protein *[2]*).

a. For transfection analysis by EGFP expression, transfect cells with a plasmid-expressing EGFP under the transcriptional control of a constitutive promoter, such as that of the SV40 early promoter. If the investigator has access to an inverted fluorescent microscope, the EGFP-expressing cells can be visualized directly in the tissue culture flask without prior fixation. Otherwise, cells should be plated on chamber slides (Fisher Scientific, Pittsburgh, PA) after pulsing. Two days after electroporation, fix cells for 30 min in 2% paraformaldehyde (in PBS) then extensively wash them in PBS. Mount slides with a cover-slip using Fluoromount G (Electron Microscopy Sciences, Fort Washington, PA) as a mounting medium. Visualize EGFP-expressing cells under fluorescent light using single-by-pass filters (Omega Optical, Brattleboro, VT). Compare the number of fluorescent cells per microscopic field under fluorescent light with the number of cells under visible light. Alternative reporter plasmids can be substituted for EGFP. For example, β-galactosidase has the advantage of an enzymatic stain that can be detected with visible light *(6)*.

b. For immunocytochemical detection of gene expression, plate cells either on multiwell dishes (e.g., 96-well) or chamber slides. After 44–48 h (Day +2) wash cells twice with PBS and fix cells. The choice of fixative and antibody staining parameters depends on the particular antibody preparation. For HCMV IE1 protein, fix cells in cold methanol:acetone (1:1) for 30 min. Remove fixative and allow plates to dry. Rehydrate cells by three washes in PBS and then incubate in 3% hydrogen peroxide-distilled water. After washing in PBS, incubate cells for 2 h (25°C) in anti-HCMV IE1 (clone MAb810, Chemicon, Temecula, CA), diluted 1:1,000 (in PBS-0.1% Tween-20, PBS-T). Following three washes in PBS-T, incubate cells in secondary antibody (biotinylated goat anti-rabbit, 1:800; Vector Laboratories) for 1 h at room temperature. Wash the slides three times in PBS-T, and add avidin-biotin complex-peroxidase (ABC) (Vector), followed by diaminobenzidine (DAB) (Vector) as a substrate until there is sufficient color development. Depending on the level of gene expression, intense staining may be observed within 5 min at room temperature. Terminate the reaction by rinsing the cells in distilled water. Visualize cells using a microscope.

For many of the cell types listed in **Table 3**, optimal electroporation conditions can result in 50–75% of the viable cells expressing the transgene (data not shown). Conversely, inefficient conditions (e.g., 100 volts, **Table 2**) will result in a very low percentage of cells expressing input DNA.

There is a trade-off between efficiency of electroporation and cell viability. Conditions that result in the majority of cells expressing input DNA are usually accompanied by large increases in cell death. Rates of cell mortality can range between 25 and 75% of the pulsed cells (data not shown). Too extreme a set of conditions (such as 300 volts for Jurkat cells, **Table 2**) results in a very high level of cell killing, ultimately reducing the level of CAT activity. Determining what constitutes "optimal" electroporation conditions will depend, in part, on the experimental goal. If the investigator wishes to compare expression between two or more plasmids, it probably is not necessary to achieve the highest electroporation efficiency. As a result, a number of cells lower than 4×10^6 could be used for pulsing. Conversely, if there is a need to transfect the majority of cells, then the pulsing parameters that establish the highest efficiency should be identified.

References

1. Andreason, G. L. and Evans, G. A. (1988) Introduction and expression of DNA molecules in eukaryotic cells by electroporation. *Biotechniques* **6,** 650–660.
2. Barry, P. A., Pratt-Lowe, E., Peterlin, B. M., and Luciw, P. A. (1990) Cytomegalovirus activates transcription directed by the long terminal repeat of human immunodeficiency virus type 1. *J. Virol.* **64,** 2932–2940.
3. Chang, D. C. and Reese, T. S. (1990) Changes in membrane structure induced by electroporation as revealed by rapid-freezing electron microscopy. *Biophys.* J. **58,** 1–12.
4. Chang, W. L. W., Tarantal, A. F., Zhou, S. -S., Barthold, S. W. and Barry P. A. (2002) Recombinant green fluorescent protein-expressing rhesus cytomegalovirus for studying congenital cytomegalovirus infection in a primate model. *J. Viral.* **76,** 9493–9504.
5. Nordeen, S. K., Green, P. P. I. and Fowlkes, D. M. (1987). A rapid, sensitive, and inexpensive assay for chloramphenicol acetyltransferase. *DNA* **6,** 173–178.
6. Ohki, E. C., Tilkins, M. L., Ciccarone V. C., and Price, P. J. (2001) Improving the transfection efficiency of post-mitotic neurons. *J. Neurosci. Methods* **112,** 95–99.
7. Sambrook, J., Fritsch, E. F., and Maniatis, T. (1989) *Molecular cloning: A laboratory manual,* 2nd ed. Cold Spring Harbor Laboratory Press, Cold Spring Harbor, NY.

17

Delivery of DNA to Skin by Electroporation

Nathalie Dujardin and Véronique Préat

1. Introduction

The easy accessibility and the large area of the skin make it a potential target for gene therapy *(1)* while its immune activity makes it an efficient site of DNA vaccination *(2)*. Despite these potential advantages for the delivery of DNA into the skin, a significant physical barrier impedes the transfer of large molecules, including DNA, into the epidermal cells. First, transdermal transport of molecules is limited by the low permeability of the stratum corneum, the outermost layer of the skin. Only potent lipophilic low molecular-weight (<500) drugs can be delivered by passive diffusion at therapeutic rates. Hence, the transdermal penetration of hydrophilic and/or high molecular-weight molecules, including DNA, requires the use of methods to enhance skin permeability and/or to provide a driving force acting on the permeant. Both chemical (e.g., penetration enhancer) and physical (e.g., iontophoresis, electroporation, or sonophoresis) methods have been used but their efficacy remains low for the transport of macromolecules through the stratum corneum. Besides the stratum corneum, a second barrier to the penetration of DNA in the target cells of the skin is the cell membrane.

Electroporation consists of applying short high-voltage pulses that create aqueous pathways across lipid bilayers. It has been shown to (1) increase the transport of molecules, including macromolecules, through the stratum corneum *(3–5)*; and (2) enhance the permeability of the skin and the subcutaneous tissue *(6–7)*. Hence, electroporation could be a promising method to deliver genes into

From: *Methods in Molecular Biology, vol. 245:*
Gene Delivery to Mammalian Cells: Vol. 1: Nonviral Gene Transfer Techniques
Edited by: W. C. Heiser © Humana Press Inc., Totowa, NJ

the skin either when the DNA is applied topically on the skin (topical delivery) or when the DNA is injected in the skin (intradermal delivery).

This chapter describes the method to deliver DNA into skin in vivo using electroporation to enhance the permeability of stratum corneum or skin cells following topical or intradermal delivery of a plasmid.

1.1. Transdermal Delivery by Electroporation

Transdermal gene delivery consists in topically applying DNA on the skin before electroporation *(8–10)*. Owing to the barrier function of the skin, the transfection efficiency is rather low and is mainly restricted to the epidermis.

1.2. Intradermal Delivery Using Electroporation

Intradermal gene delivery consists of injecting DNA either intradermally or subcutaneously. Pulses are applied to enhance the transfection efficiency *(11–16)*.

The various protocols of electroporation investigated in DNA delivery into skin after topical and intradermal delivery are summarized in **Tables 1** and **2**, respectively.

2. Materials
2.1. Transdermal Delivery (10)

1. Animal model: 2-mo-old hairless male rats (Iffa Credo, France) housed in standard cages at room temperature on a 12 h light and 12 h dark cycle, with access to standard laboratory food and water ad libitum (*see* **Note 1**).
2. Anesthesia: diethylether (*see* **Note 2**).
3. Instrumentation: pulse generator that produces rectangular waves such as PA-4000 (Cyto Pulse Sciences, Columbia, MD). During a pulse, the electrical behavior is measured with an oscilloscope (Model 54602B, Hewlett-Packard, Belgium) (*see* **Note 3**).
4. Electrodes: a skin-fold is made with a clip containing two small reservoirs (100 µL) and platinum electrodes of 1 cm² (99.99% purity, Aldrich Chemie, Belgium). The electrodes are placed at the outer surface of each reservoir at a distance of 6 mm including skin-fold (*see* **Note 4**).
5. Plasmid: the plasmid pEGFP-N1 (4.7 kb) (CLONTECH, Germany) encoding the green fluorescent protein (GFP) under the transcriptional control of a cytomegavirus (CMV) promoter is used as a marker of the transfection efficiency (*see* **Note 5**).
6. Plasmid preparation: GigaPrep Kit for plasmid preparation (Qiagen, Inc., The Netherlands).

Table 1
Topical DNA Delivery in Skin by Electroporation

Pulses conditions[a]	DNA	Electrodes	Model	Results	References
3X—120V—10 or 20 ms (ED) (pressure 9 min)	pM-MuLV-SV-*LacZ* (40 μg)	Plate caliper electrodes (1 cm² each)	Hairless mice xenograft	*LacZ* expression mainly in the dermis (370 μm depth) 200–700 cells transfected /mm²	(8,9)
10X—1000V—100 μs (SW) 10X—335V—0.5 ms (SW) 10X—335V—5 ms (SW)	pEGFP-N1 (50 μg)	Platinum electrodes with reservoir and a clip	Hairless rats	Decrease in expression within 1 wk Expression in epidermis only	(10)

ED, Exponential decay pulse; SW, square wave pulse.
[a]Number of pulses – Voltage – Pulse duration.

Table 2
DNA Delivery in Skin by Electroporation After Intradermal Delivery

Pulses conditions[a]	DNA	Electrodes	Model	Results	References
2X—400–600 V/cm— 100–300 µs (2–12.5 µg)	pSV3neo Neomycin resistance pHEB4 (10–12 µg) E1A region of adeno-virus 2 to immortalize cellsi	Two flat stainless-steel (2 cm^2)	Newborn mice → fibroblast culture	Transfection efficiency: $1.2–7 \times 10^{-4}$ Cell immortalisation	(11)
10X—252–378 V — 5 ms (SW)	pUT531 (pSV40-βgal)	Two flat stainless-steel (2-cm length, 4.2 mm distance)	Murine melanoma	Maximum β-gal positive cells and β-gal activity at 336 V	(12)
6X—1750 V/cm— 100 µs	p-EGFP-C1 (100 µg) pND2lux (100 µg)	Two rows of seven pins (7 mm)	Pig	EP enhances luciferase expression 20–40 fold over free DNA injection GFP expression in dermis	(13)
4X + 4X perpendicularly with or without 1 expo-nentially enhanced pulse (EEP)	pDND2Lux (100 µg)	Two flat-plate electrodes with caliper	Mice	Luciferase expression: 8X, 100 V/cm—20 ms < 8X, 1500 V/cm, 50 µs < 8X, 750 V/cms— 50 µs + 1X— 40–750 V/cm— 20 ms EEP	(15)

8X—1500 V/cm—100 µs or 8X—100 V/cm—20 ms (SW)	pCMV-Luc (100 µg) pIRES-mu IL-12 (50 µg)	Two flat stainless-steel plates with forceps; Three parallel wires with 1- or 2-mm gap	Mice	Luciferase expression: 1500V—100 µs >100 V—20 ms; EP: 10-fold enhancement of IFN-γ plasma level 2 mm > 1 mm > flat electrodes	*(16)*
6X—1750 V/cm—100 µs	PND2lux, pEGFP, lacZ PCMV-HBS(s)	Calliper plate Pin	Mice Yorkshire pigs	Luciferase expression: 83-fold increase with EP; Distribution of gene expression: dermis; Antibody response enhanced by EP Shift from Th2 to mixed Th1/Th2 response with EP	*(14)*

EEP, exponentially enhanced pulse; EP, electroporation; SW, square wave.
aNumber of pulses – Voltage – Pulse duration

7. Confocal Laser Scanning Microscope: Biorad MRC 1020 confocal unit equipped with an argon-krypton laser and mounted on a Zeiss Axiovert 135M-inverted microscope (Bio-Rad Laboratories, Hercules, CA).
8. Scotch Crystal Tape (3M, St. Paul, MN).

2.2. Intradermal Delivery (16)

1. Animal model: female C57bl/6 mice 9–10 wk old (*see* **Note 1**).
2. Anesthesia: inhalation of mixture of 3% isoflurane and 97% oxygen (*see* **Note 2**).
3. Instrumentation: a pulse generator that produces rectangular waves such as a PA-4000 (Cyto Pulse Sciences) or a T820 (BTX, San Diego, CA) (*see* **Note 3**).
4. Electrodes: Two flat stainless steel plates (0.5×0.5 cm) mounted on the end of forceps (*see* **Note 4**).
5. Plasmid: plasmid pCMV-Luc$^+$ containing the firefly luciferase cDNA under the control of the CMV promoter, and pIRES-mu IL-12, containing the murine IL-12 cDNA under the control of the CMV promoter (*see* **Note 5**).
6. Plasmid preparation: MegaPrep Kit for plasmid preparation (Qiagen).
7. Electromyography paste: Spectra 360 electrode gel (Parker Laboratories, Orange, NJ)
8. Tissumizer (Tekmar, Cincinnati, OH).
9. Homogenization buffer: (50 mM K_3PO_4, 1 mM ethylenediaminetetraacetic acid [EDTA], 1 mM diothiothreitol [DTT], 10% Glycerol and Reporter Lysis buffer (Promega, Madison, WI).
10. Luminometer: Microtiter plate luminometer (Dynex Technologies, Chantilly, VA).
11. Luciferase detection: assay luciferase expression using a Luciferase Detection Kit with CoA-luciferin as a substrate (Promega).
12. Interferon-γ (IFN-γ) ELISA Kit to assay IL-12 activity (R&D Systems, Minneapolis, MN).

3. Methods

3.1. Transdermal Delivery (see **Table 1**)

1. Anesthetize the animal by placing it in a chamber charged with diethylether. When anesthetized, place its nose in a diethylether atmosphere if it must be anesthetized longer than a few minutes (*see* **Note 1**).
2. Strip the abdominal skin 20 times with Scotch crystal tape (*see* **Note 6**).
3. Clamp a fold of the abdominal skin between the two reservoirs and electrodes using the clip.

4. Dilute the plasmid in 10 mM phosphate buffer, pH 7.4 (50 µg/100µL) (*see* **Note 5**). Fill the cathodal compartment with the plasmid solution (100 µl). Place 100 µl of phosphate buffer in the anodal compartment.
5. Immediately apply ten 100 µs pulses with a field strength of 1600V/cm (1000V for an electrode gap of 6.25 mm) (*see* **Note 7**).
6. Mark the treatment site by applying an alcoholic crystal violet solution (0.1%) around the electroporated area.

3.2. Intradermal Delivery (see Table 2)

1. Anesthetize the animal by placing it in an induction chamber infused with a mixture of 3% isoflurane and 97% oxygen for several minutes. After it is anesthetized, fit with a standard rodent mask and continue delivering 2.5% isoflurane in oxygen (*see* **Note 2**).
2. Dilute the plasmid in sterile saline to the appropriate concentration (100 µg/50 µL for pCMV-Luc$^+$, 50 µg/50µL for pIRES-mu IL-12).
3. Inject the plasmid intradermally into the left flank of the mice with a 31-gauge needle (*see* **Note 8**).
4. Place the plate electrodes on each side of the injected area, and pinch the skin containing the injection area between two plate electrodes. Use electrocardiography paste to assure sufficient contact between the electrodes and skin.
5. Administer eight 100 µs pulses at a nominal field strength of 1500V/cm 2 min after the injection (*see* **Note 7**).
6. Mark the treated area with a permanent-marking pen.

3.3. Detection of Gene Expression in the Skin (see Note 5)

3.3.1. Localization of GFP Expression (10)

1. After 1–7 d, preferably after 48 h, kill the rat by cervical dislocation and remove the skin in the electroporated zone with scissors and forceps touching only the border of the sample.
2. Place the skin sample in a sample holder consisting of two microscopy slides separated by pieces of microscopy slides glued on the slides with nail varnish. Add a few drops of phosphate-buffered saline (PBS) in the sample holder (7).
3. Observe immediately by Confocal Laser Scanning Microscopy using a filterblock, which selects the 488-nm laser line to illuminate the specimen and transmits emitted light with a wavelength in the 522–535 nm range. Perform optical sectioning parallel to the skin surface (xy planar section), at different focal planes (z) below the surface of the skin (20–100 µm, depending on the depth of penetration).

3.3.2. Quantification of Reporter Gene Expression (see *Note 5*)

3.3.2.1. LUCIFERASE *(16)*

1. Forty-eight hours after plasmid delivery, kill the rat by cervical dislocation and remove the treated skin sample with scissors.
2. Homogenize in homogenization buffer and Reporter Lysis buffer using a Tissumizer.
3. Quantify luciferase activity in extracts using a Luciferase Detection Kit and a luminometer. Prepare a skin sample from an untreated area as a control. Subtract this background luminescence from each treated sample.
4. Report luciferase expression as total Relative Light Units (RLU) per sample volume.

3.3.2.2. IL-12 *(16)*

1. Draw blood (200–500 µL) from the tail vein at multiple time points after treatment. Centrifuge blood for 3 min at 5000 rpm at 4°C. Collect serum and store at −20°C.
2. Measure IFN-γ plasma levels to evaluate the efficacy of IL-12 expression using an ELISA kit as described by the manufacturer.

4. Notes (*see* Tables 1 and 2)

1. Animals Species. Delivery of DNA to skin by electroporation is in the preclinical phase. Most of the studies have been performed in rodents (rats and mice) but experiments with pigs and macaques have been done as well *(13–14)*. For transdermal delivery, the use of hairless rodents or pig is recommended. Xenografts of human skin can also be transplanted onto nude mice *(9)*.
2. Anesthesia. Different methods can be used to anesthetize the animals: (a) diethylether inhalation, (b) an inhalation of mixture of 3% isoflurane and 97% oxygen, (c) intraperitoneal injection of pentobarbital (60 mg/kg, iv) subcutaneous injection of ketamine hydrochloride (22–50 mg/kg) with or without xylazine (2.5–6.5 mg/kg) and/or acepromazine maleate (0.75 mg/kg). In humans, high voltage pulses have been applied to skin treated with a local anesthetic.
3. Electroporation Device. The electroporation device can deliver either exponentially decaying pulses or square pulses. Exponentially decaying pulses provide a longer driving force for the transport the DNA, but their duration depends on the electrical properties of the skin and electroporation environment *(17)*. Hence, most of the recent work has been performed with square wave pulses to get a better reproducibility of the pulse parameters. For routine work, any square wave generator can be used. For optimization

of the pulsing conditions, more versatile device should be selected. The electrical parameters of the pulses must be controlled with an oscilloscope to monitor the voltage and duration of the pulses and preferably to measure skin resistance during and after pulse administration.

4. Electrodes (*see* **Tables 1** and **2**). Different materials and design of electrodes have been used for topical and intradermal delivery of DNA in skin by electroporation. For topical delivery, a skin-fold is made with a clip containing a small reservoir (100 µL) and platinum electrodes *(10)* or caliper brass electrodes *(8–9)*. For intradermal delivery, three electrode designs have been used. Two flat-plate stainless steel electrodes coated with electrocardiography paste to obtain a good electrical contact are placed around the injection site making a fold *(12,14–16)*. Three parallel wires with a 2-mm gap between the wires placed over the injected area, which must be electroporated without fitting the skin between the electrodes, have also been used *(16)*. Alternatively, needle electrodes can be employed. These electrodes consist of two rows of seven, 7-mm pins (1 × 5.4 mm gaps), which penetrate 2.5 mm into the animal's skin *(13–14)*

5. Plasmids (*see* **Tables 1** and **2**). When delivered topically or intradermally with electroporation to enhance the transfection efficiency, DNA is delivered as a naked plasmid. CMV-promoter-driven plasmids are usually employed. Depending on the purpose of the experiment, plasmids encoding, different types of genes can be used: (a) reporter genes for the localization or the quantification of gene expression (b) therapeutic genes for the treatment of local or systemic disease, or (c) genes for DNA vaccination whose protein is capable of generating an immune response.

Plasmids used: A list of plasmids investigated in the delivery of DNA in skin by electroporation is summarized in **Tables 1** and **2**. pGFP can be used to localize gene expression in the skin by confocal microscopy *(10)* or by microscopy of skin cryosections *(14)*. LacZ (e.g. pNDβ-gal), is often used as a reporter gene to localize gene expression by immunohistochemistry or staining with a substrate, 5-bromo-4-chloro-3-indolyl-β-D-galactoside (X-Gal), which is hydrolyzed by the enzyme β-galactosidase to generate galactose and soluble indoxyl molecules that in turn are converted into insoluble indigo *(8–9,12)*. Enzymatic activity can also be quantified *(12)*. Luciferase (e.g., pND-lux) is the most sensitive quantitative reporter gene used. Its enzymatic activity in the skin can be measured with a luminometer *(13,15–16)*.

Genes encoding therapeutic proteins have also been investigated. Expression is usually monitored by measuring the protein concentration in the plasma by ELISA *(16)*.

Taking advantage of the immune activity of the skin *(2)*, intradermal immunization with DNA vaccines (e.g., plasmid encoding the hepatitis B surface antigen) have been performed. The IgG titers are measured in the plasma by ELISA. Whether a Th1 or Th2 response is induced can be assessed by measuring IL4 and IFN-γ plasma level or the IgG1/IgG2a ratios.

Formulation of the plasmid: The plasmid is diluted with water, saline, or buffer. Typically, 50–100 µg DNA dissolved in 50–100 µL are injected or applied on the skin. The naked plasmid can be mixed with a competitive nuclease inhibitor (e.g., aurintricarboxilic acid) to enhance its expression *(13)*.

Kinetics of expression: Gene expression is usually quantified 24–48 h after DNA delivery at the time of maximum expression. A time course of gene expression can be performed to determine the duration of gene expression and to determine when the maximum level of expression occurs (usually, 1–2 d). Although expression after topical delivery to the epidermis is usually limited to 7–10 d owing to epidermis turnover, longer expression may be achieved with viral gene delivery or by delivery of DNA to the dermis *(18)*.

6. Skin Treatment for Transdermal DNA Delivery. Because the target cells for gene expression are in the epidermis, the stratum corneum can be partly removed to reduce the main barrier to DNA permeation. This may be achieved by abrasion, brushing, or tape stripping, which removes all or part of the stratum corneum depending on the number of strips used *(10)*. Pretreatment of the skin with 70% isopropyl alcohol can act as an absorption promoter and decrease DNA degradation by bacterial DNase *(8)*.

7. Electroporation Protocols. The efficacy of gene delivery and transdermal transport is influenced by the characteristics of the pulses: duration (µs-ms), number, frequencies, and waveform *(6,17,19)*. The optimization of gene transfer by electroporation after local DNA injection has been extensively studied in different tissues, including the skin, by varying pulse voltage (or electric field) and pulse duration *(6,12,13)*. The number of pulses is limited to 4–10 pulses. Two types of pulsing conditions have been compared: (a) low-voltage, long pulses, typically 100–200 V/cm, 10–20 ms; and (b) high-voltage, short pulses, typically 1000–1750 V/cm, 100 µs. In contrast to muscle, high-voltage pulses seem to be more efficient than low voltage pulses for gene transfection of skin *(15,16)*.

8. Intradermal Injection. To verify that the injection is an intradermal injection, that is, it is localized in the dermis, add black India ink to the plasmid. After removing the skin, the ink localization can be observed visually or histologically on vertical cryostat section of the skin stained with hematoxylin-eosin.

References

1. Khavari, P. (1997) Therapeutic gene delivery to the skin. *Mol. Med. Today,* **3,** 533–538.

2. Babiuk, S., Baca-Estrada, M., Babiuk, L. A., Ewen, C., and Foldvari, M. (2000) Cutaneous vaccination : the skin as an immunologically active tissue and the challenge of antigen delivery. *J. Control. Rel.* **66**, 199–214.

3. Banga A. K., and Prausnitz M. R. (1998) Assessing the potential of skin electroporation for the delivery of protein and gene-based drugs. *TIBTECH* **16**, 408–412.

4. Regnier, V., De Morre, N., Jadoul, A., and Préat, V. (1999) Mechanisms of phosphorothioate oligonucleotide delivery by skin electroporation. *Int. J. Pharm.* **184**, 147–156.

5. Lombry, C., Dujardin, N., and Préat,V. (2000) Transdermal delivery of macromolecules using skin electroporation. *Pharm. Res.* **17**, 32–37.

6. Mir, L., Bureau, M. F., Gehl, J., Rangara, R., Rouy, D., Caillaud, J. M., et al. (1999) High-efficiency gene transfer into skeletal muscle mediated by electric pulses *Proc. Natl. Acad. Sci.*, **96,** 4262–4267.

7. Regnier, V., and Préat, V. (1998) Localization of a FITC-labelled phosphorothioate oligodeoxynucleotide in the skin after topical delivery by iontophoresis and electroporation. *Pharm. Res.* **15**, 1596–1602

8. Zhang, L., Li, L., Hofmann, G. A., and Hoffman, R. (1996) Depth-targeted efficient gene delivery and expression in the skin by pulsed electric fields: an approach to gene therapy of skin aging and other diseases. *Biochem. Biophys. Res. Commun.* **220**, 633–636.

9. Zhang, L., Li, L., An, A., Hoffman, R. M., and Hofmann, G. A. (1997) In vivo transdermal delivery of large molecules by pressure-mediated electroincorporation and electroporation: a novel method for drug and gene delivery. *Bioelectrochem. Bioenerget* **42**, 283–292.

10. Dujardin, N., Van Der Smissen, P., and Préat, V. (2001) Topical gene transfer into rat skin using electroporation. *Pharm. Res.* **18**, 61–66.

11. Titomirov, A. V., Sukharev, S., and Kistanova, E. (1991) In vivo electroporation and stable transformation of skin cells of newborn mice by plasmid DNA. *Biochim. Biophys. Acta* **1088**, 131–134.

12. Rols, M.-P., Delteil, C., Golzio, M., Dumond, P., Cros, P., and Teissié, J. (1998) In vivo electrically mediated protein and gene transfer in murine melanoma. *Nat. Biotechnol.* **16**, 168–171.

13. Glasspool-Malone, J., Somiari, S., Drabick, J. J., and Malone, R. W. (2000) Efficient nonviral cutaneous transfection, *Mol. Ther.* **2**, 140–146.

14. Drabick J. J., Glasspool-Malone J., Somiari S., King A., and Malone R. W. (2001) Cutaneous transfection and immune responses to intradermal nucleic acid vaccination are significantly enhanced by in vivo electroporation. *Mol. Ther.* **3**, 249–255.

15. Lucas, M. L., Jaroszeski, M. J., Gilbert, R., and Heller, R. (2001) In vivo electroporation using an exponentially enhanced pulse: a new waveform. *DNA Cell Biol.* **20**, 183–188.

16. Heller, R. Schultz, J., Lucas, M. L., Jaroszeski, M. J., Heller, L. C., Gilbert, R.A., et al. (2001) Intradermal delivery of interleukin-12 plasmid DNA by in vivo electroporation. *DNA Cell Biology* **20**, 21–26.

17. Vanbever R., Leboulangé E., and Préat V. (1996) Transdermal delivery of fentanyl using electroporation. I. Influence of electrical factors. *Pharm. Res.* **13,** 559–65.

18. Préat,V., and Dujardin, N. (2001) Topical delivery of nucleic acids in the skin. *STP Pharm. Sciences,* **11,** 57–68.

19. Vanbever R., Pliquett U., Préat V., and Weaver J.C. (1999) Comparison of the effect of short high voltage and long medium pulses on skin electrical transfer. *J. Control. Rel.* **60,** 35–47.

18

In Vivo DNA Electrotransfer in Skeletal Muscle

Guenhaël Sanz, Saulius Šatkauskas, and Lluis M. Mir

1. Introduction

Appropriate electric pulses can reversibly permeabilize living cells both in vitro and in vivo. Since the pioneering work of E. Neumann and colleagues *(1)*, cell electroporation (also often termed cell electropermeabilization) has become the most frequent method for cell transfection. Indeed, it easily applies in vitro to bacteria, yeast, animal, and plant cells. In vivo, the use of the DNA electrotransfer method is rapidly expanding because of its simplicity and efficiency. First attempts to transfer DNA in vivo to muscle cells were published in 1998 *(2,3)*. This led in 1999 to an extended study on the determination of optimal conditions for DNA electrotransfer in skeletal muscle in mice, rats, rabbits, and even primates using a reporter gene *(4)*. The same year, the first "therapeutic" gene (the erythropoïetin gene) was transferred to mouse muscle in vivo by Rizzuto and colleagues *(5)*. In the last 2 years, several publications have used the conditions described in **ref.** *(4)* and have shown the wide applicability of this method for the electrotransfer of a large number of genes in different muscles *(6–8)*. DNA electrotransfer to skeletal muscle could lead to broad potential applications in the therapeutic field (metabolic disorders correction, vaccination, systemic secretion of angiogenic or antiangiogenic factors, etc.) and for physiological, pharmacological, and developmental studies.

These applications of DNA electrotransfer (now also termed electrogenetherapy) in the skeletal muscle are sustained by several advantages of this approach:

From: *Methods in Molecular Biology, vol. 245:*
Gene Delivery to Mammalian Cells: Vol. 1: Nonviral Gene Transfer Techniques
Edited by: W. C. Heiser © Humana Press Inc., Totowa, NJ

1. Electrotransfer drastically increases the efficiency of intramuscular gene transfer, resulting in a two-log enhancement of gene expression.
2. Electrotransfer strongly decreases variability between individual subjects. This variability is one of the largest impediments to the use of intramuscular gene injection and the delivery of the electric pulses overcomes this problem.
3. Electrotransfer results in increased time of expression of the electrotransferred genes. This allows the possibility of long-term disease correction by a repeated but infrequent treatment.

The principles of electrotransfer to muscle have now been determined with precision using combinations of high-voltage, short duration (HV) electric pulses and of low-voltage, long duration (LV) electric pulses *(9,10)*. The recent results of our team *(10)* clearly demonstrate that in vivo DNA transfer with electric pulses is a four-step process:

1. Injection and distribution of the DNA in the tissue.
2. Target cell permeabilization (HV pulse).
3. Probably an improved DNA distribution in the permeabilized tissue (*see* **Note 1**).
4. DNA transfer facilitated by DNA electrophoresis in the tissue (LV pulse).

In this chapter we describe in detail the experimental protocol for DNA electrotransfer to skeletal muscle in the mouse leg using the gene coding for luciferase as a reporter gene. The notes provide suggestions to extend the method to other experimental situations.

2. Materials

2.1. Mice

Various animal models can be used to examine electroporation of muscle tissue. We have used the immunocompetent C57Bl/6 black mice (R. Janvier, Orléans, France, or IFFA Credo, L'Isle d'Arbesle, France) that are 8 wk of age or older. Maintain mice in animal housing facilities under conventional conditions. Feed them with usual laboratory diet and water given *ad libitum*. Allow the mice to adapt at least 1 wk in their new animal housing facilities before starting experimentation. Many other strains of mice, rats, and rabbits (*see* **Note 2**) may be used following the protocols described here (see in particular **Subheading 3.3.** and **Note 11**).

2.2. DNA

Prepare plasmids using standard procedures. If possible, use preparations containing a high percentage (70–80%) of supercoiled DNA (*see* **Note 3**), and no RNA detectable by gel electrophoresis. DNA can be prepared using kits for

DNA purification providing endotoxin-free material (*see* **Note 4**). For luciferase detection, the plasmid used was the pXL3031 (pCMV-Luc+) containing the cytomegalovirus promoter (nucleotides 229-890 of pcDNA3, Invitrogen, Carlsbad, CA) inserted upstream of the coding sequence of the modified cytosolic luciferase gene.

2.3. Chemicals

For the electrotransfer itself, it is necessary to have the following:

1. Alcohol 70% for disinfecting and for wetting the shaved legs (muscles beneath the skin become more visible).
2. Electrocardiography paste, or any other conductive gel. Put some amount of paste in a small Petri dish so it is easy to deposit some gel on the electrodes by dipping them in the gel.
3. Anaesthetics like Ketamine (Ketalar, Panpharma Parke Davis, Courbevoie, France) and Xylazine (Rompun, Bayer Pharma, Puteaux, France).

For the whole of the experiment, it is also necessary to have, strictly depending on the electrotransferred gene, all the chemicals necessary to detect the product resulting from the expression of this gene.

For muscle homogenization and evaluation of luciferase expression these products are:

4. Cell Culture Lysis reagent (Promega, Charbonnières, France): dilute 10 mL of reagent with 40 mL of distilled water and add one tablet of Protease inhibitor cocktail (Boehringer Mannheim, Mannheim, Germany).
5. Luciferase Assay Substrate (Promega).

2.4. Equipment for Intramuscular DNA Injection

Hamilton syringes (50 µL, Cat. no. 1705RN, or 100 µL, Cat. no. 1710RN, Hamilton, Bonaduz, Switzerland) with needles (16 mm, 26S gauge, pst 4, taper N, Cat. no. 80497/00, Hamilton). These syringes permit precise delivery of very small volumes and allow direct intramuscular injection through the skin.

2.5. Electric Pulses Generator

PS15 electropulser (Jouan, Nantes, France): a square wave electric-pulse generator permitting adjustable voltage, pulse length, and pulse frequency (*see* **Note 5**). The electrical parameters of the delivered pulses may be recorded with a digital oscilloscope (Hitachi VC-6025, Tokyo, Japan).

2.6. Electrodes

The most convenient external electrodes consist of two parallel flat stainless steel rectangles, with rounded inferior edges to avoid any possible tip effect at the level of the contact between the electrodes and the skin (*see* **Note 6**). They are held by an insulating template, which sets the electrode gap (usually 6–8 mm), and allows safe electrical connection to the generator PS15. Typical dimensions of the electrode edge that will be applied on the skin are 1 cm width and 0.5 mm thickness. The electrical contact with the skin is ensured by a film of electrocardiography paste spread on the electrode edges and internal surface. Because the pulses generally used for DNA electrotransfer are short, it is possible to disregard electrolysis as a problem. Thus, any type of insulating material combined with metal electrodes will be convenient. It is, however, important that the electrodes are completely parallel in order to get a uniform electrical field distribution.

2.7. Equipment for Muscle Homogenization and Luciferase Activity Determination

1. FastPrep system (FastPrep120, Qbiogene, Inc., Illkirch, France).
2. Berthold Lumat LB 9501 luminometer (EG&G, Berthold, Evry, France).

3. Methods
3.1. Preparation of the DNA

On the day of the experiment, adjust DNA concentration with physiological saline (0.9% NaCl) or PBS, pH 7.2, to 3 µg to 15 µg of DNA in 30 µL. Prepare sufficient DNA for injection into each *tibial cranial* muscle. Microfuge tubes with a wide opening are convenient for preparing these dilutions. Keep the DNA solution at 4°C.

3.2. Preparation of the Electric Pulses Generator

1. Plug in the generator and the oscilloscope and verify the connections.
2. Adjust the pulse duration to 20 ms, and the repetition frequency to 1 Hz.
3. For the first experiment, wait to set the voltage applied after fixing the distance between the electrodes (*see* **Subheading 3.3., step 4**). If the distance between the electrodes is already known, adjust the electric-field strength (that is the ratio of the delivered voltage to the distance between the electrodes) to 200 V/cm. This electric-field strength has been shown to be optimal for DNA electrotransfer in muscles both in mice and in rats (*4*). This field strength is also quite adequate for DNA electrotransfer in rabbit muscle, but appears to be too low in mice younger than 4 wk (unpublished data).

3.3. Anesthesia and Final Preparation of the Animals

All mice included in a given experiment should be of similar age and immunological status and should be treated on the same day in order to have the same physiological conditions for each animal. For each experiment include about 10 *tibialis cranialis* muscles (*see* **Note 7**) in each experimental group, i.e., five mice if you use both legs independently, or 10 mice if only one leg is used. Conduct all experiments following the local and national recommendations for animal experimentation elaborated by the local (or national) Ethics Committees on Animal Care and Experimentation.

1. Prepare an anesthetic mixture of Xylazine (7.5–10 mg/kg) and Ketamine (75–100 mg/kg): add 0.5 mL of Xylazine (2%) to 9.5 mL of Ketamine (10 mg/mL) (*see* **Note 8**). This mixture can be kept at 4°C for 1 wk. Anesthesia with this drug combination generally lasts 1–1 1/2 h, with maximal effect after about 10 min.
2. Inject the anesthesia intraperitoneally: for mice weighing 20 g, inject 200 µL. Weigh a few mice before starting to anesthetize the animals in order to adjust the dose (anesthetic in excess can kill the mice). Note that male mice are more sensitive to the anesthetic than are female mice, and that nude (*nu/nu*) mice are less sensitive to the anesthetic than are most immunocompetent mice.
3. Using an electric shaver, shave the legs over the muscle to be electroporated, and all around, avoiding skin scrapes.
4. Using a caliper, measure the width of the leg of the mice used, in order to adjust the distance between the electrodes and the voltage setting. For mice legs, the distance between the electrodes usually is between 4.2 and 5.3 mm.

3.4. DNA Administration

1. Inject the DNA (3–15 µg in 30 µL saline) (*see* **Note 9**) directly into the *tibialis cranialis* muscle (*see* **Note 7**), and proceed to the electric pulse delivery. Electrotransfer efficacy does not depend on the period between intramuscular DNA injection and electric pulse delivery, provided that this period is shorter than 4 h *(11)*. This result indicates that rapid extracellular degradation of the injected DNA does not occur in muscle tissue. This slow degradation contributes to the homogeneity and predictability of the expression of the electrotransferred genes (*see* **Note 10**) and it makes the overall procedure easier to perform. However, for practical reasons, it is recommended to deliver the electric pulses as soon as DNA injection is performed.

3.5. Preparation of the Electrodes

Electrical contact with the shaved leg skin must be ensured by means of conductive paste.

1. Spread a thin film of electrocardiography paste on the internal face of the electrodes. To improve the contact between the electrodes and the leg skin, put some gel on the leg.
2. Place the electrodes at both sides of leg, at the level of the injection site (*see* **Note 11**).

3.6. Electric Pulse Delivery

At defined times after DNA injection (e.g., 25 s, or 1 min, depending on the experiments), apply the transcutaneous electric pulses through the two stainless steel plate electrodes, placed at each side of the leg (see **Subheading 3.5., step 2**). For that purpose:

1. Hold firmly the mouse in one hand.
2. Hold the electrodes with the other hand.
3. Apply a run of eight electric pulses to the muscle. Avoid changing the electrode position during the run in spite of the muscular contraction occurring at each pulse. Start the run either by pressing a pedal (device available in Jouan generator and in Genetronics Medpulser) or have an assistant to push the pulse button on the generator.

3.7. Analysis of Expression of the Electrotransferred Gene

For analysis of Luciferase activity (*see* **Note 12**), sacrifice the mice either 2 or 7 d after DNA transfer.

1. Sacrifice the animal using CO_2 gas.
2. Remove the skin over the electroporated area.
3. With a pair of small surgical scissors excise the muscle *in toto* and transfer it to an appropriate vessel, depending on the protocol used to detect the expression of the electrotransferred gene (*see* **Note 13**).
4. Using the FastPrep system, homogenize the muscles in 1 mL of Cell Culture Lysis reagent supplemented with Protease inhibitor cocktail.
5. Pellet the cellular debris by centrifugation at 5000g for 15 min at 4°C. Assess luciferase activity in 10 μL of supernatant, using a Berthold Lumat LB 9501 luminometer, by measuring the light produced for 10 s, starting from 1 s after the addition of 50 μL of Luciferase Assay Substrate (Promega) to the supernatant (*see* **Note 14**).
6. Prepare a standard curve over a wide range of concentrations (from 50 pg/mL to 1 μg/mL) using purified firefly luciferase protein (Boehringer Mannheim).
7. Convert relative light units (RLU) to amount of luciferase and express results as pg luciferase per muscle or ng per muscle ± SEM (*see* **Note 15**).

4. Notes

1. It has been reported that improved DNA distribution and increased gene expression can be achieved by pre-injecting (prior to the DNA injection) either a sucrose solution, which forces the generation of spaces between muscle fibers *(12)*; or a hyaluronidase solution, which breaks down components of the extracellular matrix and provides some permeability of the connective tissues without total disruption *(13)*. It has also been reported recently that poloxamer 188 protects muscle fibers from the deleterious effects of the electric pulses, allowing an increase in the expression of the electrotransferred genes *(14)*. Because these recent reports have not yet been duplicated, these modifications are not included in the protocol described in this chapter.

2. If immunodepressed animals are used, e.g., "nude" *nu/nu* mice, keep these immunodeficient mice in sterile cages with filtering covers.

3. It has been reported that using supercoiled DNA results in the maximum level of expression of the electrotransferred genes. However, no systematic studies comparing efficacy of supercoiled, single-nicked, and linear DNA have been performed.

4. If DNA is purified by means of cesium chloride gradients, carefully remove all traces of cesium chloride.

5. Other electric pulses generators can be used. The essential constraint is that the equipment used must be able to deliver square wave pulses, and not exponentially decaying ones. Indeed, it seems necessary to maintain the applied voltage for a given duration in order to achieve electrophoretic transport of the DNA within the tissue (*see* Introduction). Controlling the duration of the pulse at a sustained voltage is mandatory.

6. Plate and needle electrodes are commercially available (e.g., by BTX, San Diego, CA or IGEA, Carpi, Italy). However, in particular for small laboratory animals, designing simple electrodes for your particular need may be a good alternative. Guidelines to elaborate such devices can be found in Jaroszeski et al. *(15)*.

7. In mice, the *tibialis cranialis* is a muscle that is easily discerned beneath the skin. In the leg, the tibial bone is seen as a white line and the *tibialis cranialis* muscle is located besides the tibia, on the external part of the leg. Wetting the leg with alcohol increases the visibility of the tibial bone and of the *tibialis cranialis* muscle if the shaving is not complete.

8. Alternatively, add 0.5 mL of Xylazine (2%) to 2 mL of Ketamine (50 mg/mL) and 7.5 mL of sterile water.

9. The amount of DNA to be injected depends on the strength of the gene promoter and on the gene product level that must be reached to get either a good signal or a therapeutic effect.

10. When the product of the gene is a protein secreted in the blood, quantification of its production can be followed without sacrificing the animal, by taking blood samples at regular times after the DNA electrotransfer and analyzing the presence of this product in the blood.

11. When working with animals larger than mice or rats, or with muscles larger than the murine *tibialis cranialis*, it is also possible to use implanted stainless steel needle electrodes. The efficiency of needle electrodes for DNA electrotransfer to rat muscle is similar to that of external plate electrodes. A description of the procedure for the gluteal muscle is given here.

   a. Prepare needle electrodes from 22G hypodermic needles. These are single-use and easily replaced, held by insulating plates and directly connected to the generator. Other needle-based devices are extensively described in Jaroszeski et al. *(15)*.

   b. Holding the electrode-device, dip the tips of the mounted needles in ink.

   c. Make an incision over the gluteal muscle with surgical scissors; alternatively, reach the muscle by inserting the needle electrodes through the skin.

   d. Insert the needles into the muscle, trying to make the direction of the electrical field perpendicular to the long axis of the fibers.

   e. Administer the run of pulses and withdraw the needles. When comparing pulsed with unpulsed samples, the needles can be inserted on both sides but pulses delivered on one side only.

   f. After sacrificing the animal, look for the ink marks on the muscle. Hold the animal so that the muscle is in more or less the same position as when you performed electroporation.

   g. Cut the muscle between the ink marks with a pair of small surgical scissors.

   h. Take out the sample and transfer to a plastic tube.

12. In spite of the large reduction in variability conferred by DNA electrotransfer (in comparison to the injection of naked DNA), there is still some variability. In the case of products secreted into the blood stream, it is possible to reduce this variability by transferring the DNA to both legs. This will also result in higher levels of the circulating product. In this case, there is no difference if the two injections are given first followed by the two runs of electric pulses, or if each one of the injections is directly followed by the electric pulse delivery. However, it is more practical to inject both legs first and then pulse one leg after the other. Note that electrotransfer allows the fine modulation of foreign gene expression by varying the amount of DNA injected and/or the electric pulse characteristics and/or the volume of tissue exposed to the electric pulses. Indeed, only local effects were obtained, in

agreement with the known local permeabilizing effects of appropriate electric pulses. Adjustment of these parameters allows fine control over gene expression, making it more likely to achieve the desired investigative or therapeutical goals.

13. Histochemical and/or immunohistochemical procedures can also be used to detect electrotransferred gene expression. In this case, remove, fix, and slice the muscles. Appropriate staining, e.g., by indirect immunohistochemistry, can be performed either *in toto* or on the sections.

14. A Walac Victor² luminometer (PerkinElmer Life Sciences, Courtaboeuf, France) can also be used to assess luciferase activity, by integration of the light produced during 1 s.

15. If data is to be expressed in terms of pg or ng of product per gram of tissue, before sacrificing the animal, weigh the tubes that will be used for the muscle, excise the muscle, transfer it to the tube, and re-weigh the tube and muscle to determine the weight of the muscle. Results may also be expressed in pg or ng of product per gram of protein. In this case, the homogenate must be assayed for protein content.

Acknowledgments

We gratefully acknowledge all our colleagues for the work done in collaboration with them and for the fruitful discussions that have improved, step by step, all the protocols described here.

References

1. Neumann, E., Schaefer-Ridder, M., Wang, Y., and Hofschneider, P. H. (1982) Gene transfer into mouse lyoma cells by electroporation in high electric fields. *EMBO J.* **1,** 841–845.

2. Mir, L. M., Bureau, M. F., Rangara, R., Schwartz, B., and Scherman, D. (1998) Long-term, high level in vivo gene expression after electric pulse-mediated gene transfer into skeletal muscle. *C.R. Acad. Sci. III* **321,** 893–899.

3. Aihara, H. and Miyazaki, J. (1998) Gene transfer into muscle by electroporation in vivo. *Nat. Biotechnol.* **16,** 867–870.

4. Mir, L. M., Bureau, M. F., Gehl, J., Rangara, R., Rouy, D., Caillaud, J. M., et al. (1999) High-efficiency gene transfer into skeletal muscle mediated by electric pulses. *Proc. Natl. Acad. Sci. U.S.A* **96,** 4262–4267.

5. Rizzuto, G., Cappelletti, M., Maione, D., Savino, R., Lazzaro, D., Costa, P., et al. (1999) Efficient and regulated erythropoietin production by naked DNA injection and muscle electroporation. *Proc. Natl. Acad.Sci. USA* **96,** 6417–6422.

6. Bettan, M., Emmanuel, F., Darteil, R., Caillaud, J. M., Soubrier, F., Delaere, P., et al. (2000) High-level protein secretion into blood circulation after electric pulse-mediated gene transfer into skeletal muscle. *Mol. Ther.* **2,** 204–210.

7. Kreiss, P., Bettan, M., Crouzet, J., and Scherman, D. (1999) Erythropoietin secretion and physiological effect in mouse after intramuscular plasmid DNA electrotransfer. *J. Gene Med.* **1,** 245–250.
8. Vilquin, J. T., Kennel, P. F., Paturneau-Jouas, M., Chapdelaine, P., Boissel, N., Delaere, P., et al. (2001) Electrotransfer of naked DNA in the skeletal muscles of animal models of muscular dystrophies. *Gene Ther.* **8,** 1097–1107.
9. Bureau, M. F., Gehl, J., Deleuze, V., Mir, L. M., and Scherman, D. (2000) Importance of association between permeabilization and electrophoretic forces for intramuscular DNA electrotransfer. *Biochim. Biophys. Acta* **1474,** 353–359.
10. Satkauskas, S., Bureau, M. F., Puc, M., Mahfoudi, A., Scherman, D., Miklavcic, D., and Mir, L. M. (2002) Mechanisms of in vivo DNA electrotransfer: respective contribution of cell electropermeabilization and DNA electrophoresis. *Mol. Ther.* **5,** 133–140.
11. Satkauskas, S., Bureau, M. F., Mahfoudi, A., and Mir, L. M. (2001) Slow accumulation of plasmid in muscle cells: supporting evidence for a mechanism of DNA uptake by receptor-mediated endocytosis. *Mol. Ther.* **4,** 317–323.
12. Davis, H. L., Whalen, R. G., and Demeneix, B. A. (1993) Direct gene transfer into skeletal muscle in vivo: factors affecting efficiency of transfer and stability of expression. *Hum. Gene Ther.* **4,** 151–159.
13. McMahon, J. M., Signori, E., Wells, K. E., Fazio, V. M., and Wells, D. J. (2001) Optimisation of electrotransfer of plasmid into skeletal muscle by pretreatment with hyaluronidase: increased expression with reduced muscle damage. *Gene Ther.* **8,** 1264–1270.
14. Hartikka, J., Sukhu, L., Buchner, C., Hazard, D., Bozoukova, V., Margalith, M., et al. (2001) Electroporation-facilitated delivery of plasmid DNA in skeletal muscle: plasmid dependence of muscle damage and effect of poloxamer 188. *Mol. Ther.* **4,** 407–415.
15. Jaroszeski, M. J., Heller, R., and Gilbert, R. (2000) *Electrochemotherapy, Electrogenetherapy, and Transdermal Drug Delivery.* Humana Press, Totowa, NJ.

19

Electrically Mediated Plasmid DNA Delivery to Solid Tumors In Vivo

Mark J. Jaroszeski, Loree C. Heller, Richard Gilbert, and Richard Heller

1. Introduction

Electroporation is a physical method that delivers plasmid DNA in a variety of tissues. This technology is currently gaining recognition and acceptance in the scientific community as evidenced by the number of studies using in vivo electroporation published each year. One potential application for this method is to deliver DNA to the cells that comprise tumors.

Although electroporetic DNA transfer is a relatively new technology, some very encouraging results have been published *(1–8)*. Enough investigations have been conducted to indicate that this technology can be implemented in a wide variety of tumors *(9–19)*. These investigations also indicate that successful implementation in a particular host/tumor system is dependent on the details of the electroporation method used. More specifically, the electrode and electrical parameters used to facilitate the delivery of DNA to tumor cells are two important factors. One efficient strategy for researchers interested in using electroporation is to first optimize delivery with respect to these two factors in their particular system using DNA coding for reporter molecules before delivering DNA that codes for functional/therapeutic molecules.

This chapter provides a protocol that describes methods for delivering plasmid coding for the reporter molecule luciferase to tumors in two specific host/tumor models. One of these systems is a murine subcutaneous tumor model, and the other is a rat model with an established tumor in the liver. The protocol

From: *Methods in Molecular Biology, vol. 245:*
Gene Delivery to Mammalian Cells: Vol. 1: Nonviral Gene Transfer Techniques
Edited by: W. C. Heiser © Humana Press Inc., Totowa, NJ

addresses both subcutaneous and internal tumors because they are two common locations for experimental tumors. This protocol is intended to serve as a starting point for researchers who wish to apply/optimize electroporetic DNA transfer to a particular host/tumor system.

2. Materials

1. Animals: C57Bl/6 mice that are 7 wk of age or older (Harlan, Indianapolis, IN). Sprague Dawley rats (Harlan) that are 175–200 g in weight. Standard housing and feeding are sufficient (*see* **Note 1**).
2. Tumorigenic Cells: B16-F10 melanoma cells (ATCC CRL-6475, American Type Culture Collection [ATCC], Rockville, MD) grown in McCoy's 5A medium supplemented with 10% (v/v) fetal bovine serum FBS and 90 µg/mL gentamycin. N1S1 rat hepatocellular carcinioma cells (ATCC CRL-1604, ATCC) grown in Swimms S-77 medium supplemented with 4 mM L-glutamine, 0.01% Pluronic F68, 9% (v/v) FBS, and 90 µg/mL gentamycin.
3. Anesthesia: Isoflurane (Abbott Laboratories, North Chicago, IL) and an appropriate vaporizer to administer the inhaled anesthetic.
4. Electroporation Pulse Generator: Direct-current pulse generator that produces rectangular waves such as those available from BTX, Inc. (San Diego, CA), Bio-Rad Laboratories (Hercules, CA), or CytoPulse Sciences (Columbia, MD).
5. Oscilloscope: Philips PM3375 oscilloscope (Philips, Eindhoven, Netherlands) is used as a digital storage device to monitor administered electric pulses. This device or an equivalent instrument is recommended but is not essential.
6. Electrode: Several different electrodes are commercially available (BTX, Inc.). These include penetrating needle type electrodes and parallel-plate electrodes that do not penetrate tissue. Custom electrodes are the choice of many researchers (*see* **Note 2**).
7. Plasmid DNA: Plasmid containing DNA coding for luciferase. pRc/CMV (Invitrogen, San Diego, CA) with a luciferase coding sequence inserted has yielded excellent results.
8. Plasmid Preparation Kits: MegaPrep (Qiagen, Inc., Valencia, CA).
9. Luminometer: MLX microtitre luminometer (Dynex Technologies, Chantilly, VA) or similar device.
10. Homogenization Buffer: 25 mM Tris-HCl, pH 7.8, 2 mM dithiothreitol (DTT), 2 mM ethylenediaminetetraacetic acid (EDTA), pH 8.0, and 10% glycerol. Prepare immediately before use and store on ice.
11. Luciferase Assay Buffer: 25 mM glycylglycine, pH 7.8, 15 mM phosphate buffer, pH 7.8, 15 mM MgSO$_4$, 4 mM EGTA, 2 *m*M ATP, 1 mM DTT, and

100 μM luciferin (PharMingen, San Diego, CA). Prepare immediately before use and store on ice.

12. Luciferase Standard: A commercial firefly luciferase solution (Promega, Madison, WI) in homogenization buffer containing 1 mg/mL bovine serum albumin (BSA). Prepare immediately before use and store on ice.
13. Tissue Homogenizer: Tissuemizer (Tekmar, Cincinnati, OH) or similar.

3. Methods
3.1. Plasmid DNA Delivery
3.1.1. Plasmid DNA Delivery to Subcutaneous Tumors

1. Induce anesthesia by placing mice into a chamber charged with 2% isoflurane in oxygen until the animal is unconscious. This should require about 2 min.
2. Transfer the animal to work table and fit with a standard rodent mask suppled with 2% isoflurane.
3. Shave one flank of the animal using standard animal clippers.
4. Inject 50 µL of saline solution containing 1×10^6 B16 cells subcutaneously into the shaved region of skin. A 1-cc syringe with a 25-gauge needle works well for this type of injection.
5. Monitor animals until fully recovered from anesthesia and return all injected animals to normal housing for a period of time to allow the tumor to grow to the desired size. Typically, 7–10 d are required for tumors to reach a size of 5–7 mm in diameter.
6. Prepare plasmid DNA using the Qiagen, or similar, preparation kits in accordance with the manufacturer's instructions (*see* **Note 3**).
7. Induce anesthesia as described in **step 1** after the tumors have grown to the desired size. Then, inject plasmid DNA directly into each tumor. The optimal concentration of DNA is application dependent. A 50-µL injection volume containing 100 µg of DNA has resulted in high luciferase expression in this model for tumors that are about 6 mm in diameter. These quantities can be varied to optimize delivery for any given tumor. A 25-gauge needle attached to a 1-cc syringe is well suited for intratumor injection.
8. Apply an appropriate electrode to the tumor immediately after injecting DNA (*see* **Note 2**). Then, administer direct current pulses to the tumor using an electroporation power supply. For this animal model, pulses that have electric field strengths of 1500 V/cm and durations of 100 µs work well *(13)*. Multiple pulses are typically applied and can be monitored using an oscilloscope to verify the output of the electroporation-pulse generator. This step can be adapted to optimize delivery for any particular tumor type.

Pulses with electric-field strengths ranging from about 200–1500 V/cm and pulse widths ranging from approx 100 μs to 20 ms have resulted in delivery to tumors.

9. Monitor the mice until they are fully awake, then return them to normal housing conditions to allow time for expression of the luciferase reporter molecule.

3.1.2. Plasmid DNA Delivery to Internal Tumors

1. Induce anesthesia by placing rats into a chamber charged with 3% isoflurane in oxygen until the animal is unconscious. This should require about 2 min.
2. Transfer the animal to a surgical table and fit with a standard rodent mask supplied with 2.5–3% isoflurane.
3. Shave the abdomen of the animal and clean the abdominal area with betadine solution.
4. Surgically expose the median liver lobe by making a 3 cm transverse incision approximately 0.5 cm caudal to the xiphoid process.
5. Apply pressure, gently by hand, to the chest in order to force the median liver lobe out of the incision for treatment.
6. Inject 50 μL (1×10^6) of N1S1 cells subcapsularly into the median hepatic lobe of the rat. A-1 cc syringe with a 30-gauge needle works well for this type of injection.
7. Close the animals by inserting the liver back into the abdomen, suturing the abdominal muscles, and stapling the overlying skin. Return all injected animals to normal housing conditions for 8 d to allow the tumors to grow to a size of 5–7 mm in diameter.
8. Prepare plasmid DNA using the Qiagen, or similar, preparation kits in accordance with the manufacturers instructions (*see* **Note 3**).
9. Repeat **steps 1**, **2**, **4**, and **5** to gain surgical access to the tumors; remove the surgical staples from each animal prior to repeating **step 4**.
10. Inject tumors with plasmid DNA in solution. Fifty μL of solution containing 100 μg of luciferase plasmid produces excellent results in tumors that are approx 100 mm³ in volume (approx 6 mm in diameter). A 30-gauge needle and 1-cc syringe work well for this. The optimal quantity of plasmid is application-dependent, so a broad range of DNA quantities should be investigated for optimizing delivery to other tumor systems. Injecting a volume of plasmid DNA solution that is greater than 50% of the tumor volume can result in rupturing the tumor capsule and/or surface of the liver. This will result in leakage of the solution from the tumor, which will reduce delivery efficiency.

11. Apply an appropriate electrode to the tumor immediately after injecting DNA. Then, administer direct current pulses to the tumor using an electroporation power supply. Penetrating needle electrodes are preferred as they insure that the deep tumor margin receives treatment (*see* **Note 2**). For this model, pulses that have electric-field strengths of 1000–2000 V/cm and durations of 100 µs work well *(12)*. Multiple pulses are typically applied and can be monitored with an oscilloscope to verify the output of the electroporation generator. It is possible to adapt this step to optimize delivery to any particular tumor type by investigating the expression that results from different combinations of field strength and pulse width. Generally, pulse widths in the microsecond range are useful for field strengths that are higher than 750–1000 V/cm. Longer pulse widths, in the millisecond range, tend to be more practical for lower field strengths.

12. Close the animals as described in **step 8** of this section, and house the animals to allow time for luciferase expression.

3.2. Tumor Harvesting and Detection of Luciferase Expression

1. Humanely euthanize animals at a desired time point after delivering the plasmid DNA. Generally expression is present starting 24 h after treatment.

2. Excise the B16 melanomas and place them in tubes (25-mL round-bottom plastic tubes) that have their individual tare weights recorded. Immediately place the samples on dry ice for rapid freezing as each tumor is harvested. Follow the same procedure for N1S1 tumors, with the exception that the liver must be surgically exposed in order to excise the tumor. Skin should be trimmed from the melanomas and normal liver tissue should be trimmed from the N1S1 tumors before placing the samples into the tubes.

3. Weigh each tube and determine the tissue weight by subtracting the tare weight of each tube. Record these weights for use in the analysis. Store the weighed tubes containing the tumors at $-70°C$ until analysis is performed.

4. To prepare cell extracts for luciferase assays, place all tubes on ice. Add 1 mL of homogenization buffer to each tube and then homogenize each sample using a tissue homogenizer. Keep samples on ice after homogenization.

5. Transfer an aliquot of each homogenized sample into one well of a white 96-well plate. A 30 µL aliquot is a good starting point, but this quantity may need to be adjusted for different types of experiments. Also, prepare a standard curve using appropriate quantities of luciferase standard to cover the range of expected luciferase activity in the samples.

6. Analyze for luciferase activity using a microplate luminometer: inject 150 µL of luciferase assay buffer, then measure the light output for 10 s.

7. Express luminometer data in units of pg luciferase per mg of tumor using the data from each sample, the results from the wells that contained the luciferase standards, and the tumor weights.

4. Notes

1. Mice younger that 7 wk of age may present problems as they have very fine hair, so it is difficult to shave the flank area prior to inducing subcutaneous tumors. Rats weighing 175–200 g or greater are recommended because their median liver lobe is thick enough to withstand surgical manipulation. Smaller rats have thinner lobes, which are more difficult to manipulate without tearing.
2. Choosing an appropriate electrode for a particular model system is critical. Currently, two basic styles are commercially available. The first style is essentially two parallel plates mounted on an apparatus, such as the ends of forceps or Vernier calipers, that allows them to be moved apart and together. The plates are simply clamped on two sides of the tumor prior to administering electric pulses. The second commercially available variety employs needles that penetrate tissue. This type of design inserted into the tumor provides greater assurance that the deep tumor margin will be treated. Very simple models of each type are commercially available. Many researchers make their own needle and plate electrodes *(4,7,9–14,16,19,20)* to suit their particular needs.
3. Alternatively, prepared plasmid coding for luciferase can be purchased from commercial sources.

Acknowledgment

This protocol described in this chapter was developed as part of a studies supported by the University of South Florida Department of Surgery and Center for Molecular Delivery.

References

1. Muramatsu, T., Shibata, O., Ryoki, S., Ohmori, Y., and Okumura, J. (1997) Foreign gene expression in the mouse testis by localized in vivo gene transfer. *Biochem. Biophys. Res. Commun.* **233**, 45–49.
2. Aihara, H. and Miyazaki, J. (1998) Gene transfer into muscle by electroporation in vivo. *Nat. Biotechnol.* **16**, 867–870.
3. Suzuki, T., Shin, B., Fujikura, K., Matsuzaki, T., and Takata, K. (1998) Direct gene transfer into rat liver cells by in vivo electroporation. *FEBS Lett.* **425**, 436–440
4. Heller, R., Schultz, J., Lucas, M. L., Jaroszeski, M. J., Heller, L. C., Gilbert, R. A., Moelling, K., and Nicolau, C. (2001) Intradermal delivery of IL-12 plasmid DNA by in vivo electroporation. *DNA Cell. Biol.* **20(1)**, 21–26.

5. Kishida, T., Asada, H., Satoh, E., Tanaka, S., Shinya, M., Hirai, H., et al. (2001) In vivo electroporation-mediated transfer of interleukin-12 and interleukin-18 genes induces significant antitumor effects against melanoma in mice. *Gene Ther.* **8**, 1234–1240.

6. Widera, G., Austin, M., Rabussay, D., Goldbeck, C., Barnett, S. W., Chen M., et al. (2000) Increased DNA vaccine delivery and immunogenicity by electroporation in vivo. *J. Immunol.* **164**, 4635–4640.

7. Vicat, J. M., Boisseau, S., Jourdes, P. Laine, M., Wion, D., Bouali-Benazzouz, R., et al. (2000) Muscle transfection by electroporation with high-voltage and short-pulse currents provides high-level and long-lasting gene expression. *Hum. Gene Ther.* **11**, 909–916.

8. Heller, R., Jaroszeski, M., Atkin, A., Moradpour, D., Gilbert, R., Wands, J., and Nicolau, C. (1996) In vivo gene electroinjection and expression in rat liver. *FEBS Lett.* **389**, 225–228.

9. Niu, G., Heller, R., Catlett-Falcone, R., Coppola, D., Jaroszeski, M., Dalton, W., Jove, R., and Yu, H. (1999) Gene therapy with dominant-negative Stat3 suppresses growth of the murine melanoma B16 tumour in vivo. *Cancer Res.* **59**, 5059–5063.

10. Nishi, T., Kimio, Y., Yanashiro, S., Takeshima, H., Sato, K., Hamada, K., et al. (1996) High-efficiency in vivo gene transfer using intraarterial plasmid DNA injection following in vivo electroporation. *Cancer Res.* **56**, 1050–1055

11. Rols, M. P., Delteil, C., Golzio, M., Dumond, P., Cros, S., and Teissie, J. (1998) In vivo electrically mediated protein and gene transfer in murine melanoma. *Nature Biotechnol.* **16(2)**, 168–71

12. Heller, L. C., Jaroszeski, M. J., Coppola, D., Pottinger, C., Gilbert, R., and Heller, R. (2000) Electrically mediated gene delivery to hepatocellular carcinomas in vivo. *Gene Ther.* **7**, 826–829.

13. Heller, L. C., Pottinger, C., Jaroszeski, M. J., Gilbert, R., and Heller, R. (2000) In vivo electroporation of plasmids encoding GMCSF and interluekin-2 into existing B16 melanomas combined with electrochemotherapy induces long term antitumor immunity. *Melanoma Res.* **10**, 577–583.

14. Bettan, M., Ivanov, M. A., Mir, L. M., Boissiere, F., Delaere, P., and Scherman, D. (2000) Efficient DNA electrotransfer into tumors. *Bioelectrochemistry* **52**, 83–90.

15. Tamura, T., Nishi, T., Goto, T., Takeshima, H., Dev, S. B., Ushio, Y., and Sakata, T. (2001) Intratumoral delivery of interleukin 12 expression plasmids with in vivo electroporation is effective for colon and renal cancer. *Hum. Gene Ther.* **12**, 1265–1276.

16. Wells, J. M. Li, L. H., Sen, A., Jahreis, G. P., and Hui, S. W. (2000) Electroporation-enhanced gene delivery in mammary tumors. *Gene Ther.* **7**, 541–547.

17. Goto, T., Nishi, T., Tamura, T., Dev, S. B., Takeshima, H., Kochi, M., et al. (2000) Highly efficient eletro-gene therapy of solid tumor by using an expression plasmid for the herpes simplex virus thymidine kinase gene. *Proc. Nat. Acad. Sci. USA* **97(1)**, 354–359.

18. Lohr, R., Lo, D. Y., Zaharoff, D. A., Hu, K., Zhang, X., Li, Y., et al. (2001) Effective tumor therapy with plasmid-encoded cytokines combined with in vivo electroporation. *Cancer Res.* **61**, 3281–3284.

19. Yamashita, Y., Shimada, M., Hasegawa, H., Minagawa, R. Rikimaru, T., Hamatsu, T., et al. (2001) Electroporation-mediated interleukin-12 gene therapy for Hepatocellular carcinoma in the mice model. *Cancer Res.* **61**, 1005–1012.
20. Gilbert, R., Jaroszeski, M. J., and Heller, R. (1997) Novel electrode designs for electrochemotherapy. *Biochim. Biophys. Acta* **1334**, 9–14.

20

Hydrodynamic Delivery of DNA

Joseph E. Knapp and Dexi Liu

1. Introduction

The study of gene expression and regulation relies on the introduction of DNA into cells, which can be accomplished by a procedure called transfection. In vitro studies with transfected cells expressing the introduced genes (transgenes) have provided many valuable insights into gene function. However, in vitro transfection methods cannot provide the complete range of responses possible in intact organs and tissues, which have unique anatomy and consist of different cell types. Thus, in vivo studies in whole animals remain a critical element in gene-function studies. The effective and efficient in vivo delivery of DNA into whole animals has presented significant challenges. Although much work continues to be carried out to improve the more commonly used viral and nonviral vectors for gene delivery, recent progress in the application of naked DNA methods (*1–4*, and the chapters in this volume) has offered valuable alternatives. The obvious utility of direct introduction of DNA into tissues without the requirement for a viral vector has stimulated development of several alternative delivery methods. Of these, the hydrodyamics method (*5–8*) offers a simple, convenient means for the expression of significant levels of transgenes in the organs, especially the liver, of mice.

The hydrodynamics method involves the rapid injection of a large volume (approximately the blood volume) of a saline solution of DNA into mice via the tail vein. Although this rather simple procedure can achieve transfer of DNA into several organs, its real utility is its ability to provide a quick and efficient means for a high level transfection of liver hepatocytes. It is believed that the

From: *Methods in Molecular Biology, vol. 245:*
Gene Delivery to Mammalian Cells: Vol. 1: Nonviral Gene Transfer Techniques
Edited by: W. C. Heiser © Humana Press Inc., Totowa, NJ

intravascular pressure developed by the injection provides a driving force to affect the efficient transfer of DNA into the perivascular hepatocytes. Although the effect of the injection of such a large volume of fluid into animals might be expected to cause significant damage, the animals recover physically from the injection in 5–10 min. They continue to thrive and gain weight even after multiple injections. Importantly, no organ damage has been detected histologically and only mild, transient elevations in liver enzyme levels (alanine aminotransferase) have been observed *(5)*. The elevated alanine aminotransferase level returns to normal within 3 d. Depending on the particular plasmid construct, gene expression can reach a maximum between 4–8 h after injection and the transgenes can remain transcriptionally active for more than 6 mo *(7,9–11)*. The hydrodynamics method described here:

1. Is convenient, simple, and efficient.
2. Is applicable to linear, circular, or polymerase chain reaction (PCR) derived DNA.
3. Does not require elaborate equipment or surgical procedures.
4. Does not cause injury to the test animals.

The objective of this chapter is to provide investigators with the information necessary to carry out this procedure and to adapt it to their own needs. The procedure can be easily mastered and does not require the development of special skills. The method provides a rapid, convenient, and valuable tool for gene function analysis, protein expression, and gene therapy studies in whole animals.

2. Materials

1. DNA: prepare by any of the applicable standard methods *(12)* (*see* **Note 1**). Purify plasmid DNA amplified in *Escherichia coli* or other microbial source by either ion-exchange column or CsCl-ethidium bromide gradient centrifugation (*see* **Note 1**). Store DNA as a concentrated solution in saline at −20°C.
2. Saline solution: 0.9% NaCl in water. Filter-sterilize and store in at 4°C. Warm to room temperature prior to use.
3. Three-cc syringe and 27-G1/2 needle (Becton Dickinson, Franklin Lakes, NJ) (*see* **Note 2**).

3. Method

1. Weigh mice to be injected. Calculate the volume of solution to be injected individually for each mouse on the basis of 0.09 mL/g of body weight.
2. To the calculated volume of room temperature saline solution add the appropriate amount of concentrated DNA solution to provide 10–100 µg DNA (*see* **Note 3**).

3. Suitably restrain (*see* **Note 4**) the test animal, locate the tail vein (*see* **Note 5**), and inject the DNA solution over a period of 5 s by means of a sterile, 3-cc syringe and 27 G1/2 needle (*see* **Note 6**).
4. Animals should recover from the injection in 5–10 min and transgene expression can be maximal in 4–8 h post-injection, depending on the promoter, transgene, and regulatory elements present in the construct (*see* **Note 7**).

4. Notes

1. The DNA used can be circular, linearized, or PCR-amplified. For gene expression studies, a promoter at the 5′ end and polyA signal at the 3′ end need to be added to the transgene. DNA containing introns as well as exons can also be used. An upper limit in the size of DNA that is transferable by this method has not been established. We have been successful in transferring DNA as large as 25 Kbp. We have not observed obvious differences between DNA purified by CsCl-ethidium bromide gradient centrifugation or endotoxin-free ion-exchange column methods.
2. We have found that a 3-cc syringe is the most convenient size for work with mice. Although other syringe types can be used, we have routinely employed 3-cc, latex-free disposable syringes. Although we routinely use 27-G1/2 needles, other needles can also be used. Needles of smaller gauge number require a slower push on the syringe plunger to deliver the injection and make the injection more difficult to control. Larger gauge number needles require excessive pressure on the plunger to provide the proper injection rate.
3. Although the upper limit for the amount of DNA that can be delivered is likely in the range of several milligrams, we have found that 10–100 µg are sufficient for routine work in mice. It is best to prepare the solution for injection by diluting a small volume of cold, concentrated DNA solution to the required volume with room temperature saline just prior to injection. In this way, any deleterious effects of injecting a large volume of cold solution are prevented.
4. Although animals may be anesthetized to facilitate administration of the intravenous dose, we have found that simply restraining the animals in a simple, commercially available acrylic tube restrainer is a convenient and efficient alternative.
5. Although with experience tail vein injections become easy and routine, we have found that warming the animals briefly (1–2 min) under an infrared heating lamp enhances visualization of the tail vein and allows for easy insertion of the needle.
6. The injection volume and speed have been optimized and are a balance between efficiency of DNA delivery and safety for the experimental animals. Exceeding either the limit on volume or injection speed will result in significant organ damage in the animals.

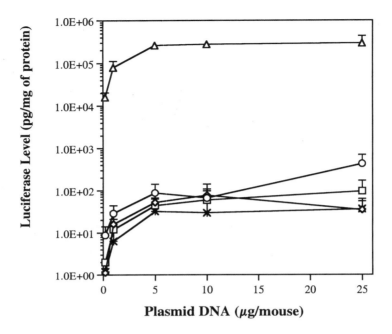

Fig. 1. DNA Dose-dependent luciferase gene expression. Various amounts of plasmid DNA (pCMV-Luc) were injected into each mouse within 5 s. The level of luciferase gene expression was determined 8 h post-injection in liver ($\triangle$), kidney ($\bigcirc$), spleen ($\diamondsuit$), lung ($\square$), and heart (^N_K). Error bars represent s.e.m. from three mice *(5)*.

7. These procedures have been applied to the study of expression of different transgenes *(5–11)* and regulatory elements *(9,10)* for gene expression. **Figure 1** shows the dose-response curve with respect to the amount of plasmid (pCMV-Luc) DNA injected and the level of luciferase gene expression. A significant level of luciferase protein 8 h post-injection was detected in all internal organs, including the lung, heart, liver, spleen, and kidney, when the amount of plasmid DNA injected was as low as 0.2 µg per mouse. The luciferase protein in all examined organs increased with increasing the amount of plasmid DNA injected and reached a saturation level at approximately 5 µg of pCMV-Luc plasmid DNA per mouse. Further increases of the DNA dose up to 25 µg increased the level of transgene expression in kidney by sixfold, but did not result in a significant increase in luciferase protein level in other organs. The amount of luciferase protein expressed at a dose of 5 µg of pCMV-Luc plasmid DNA per mouse is 300 ng per mg of extracted protein from the liver, 0.1 ng from kidney, 0.06 ng from spleen, 0.04 ng from lung, and 0.03 ng from heart. For liver, this level represents

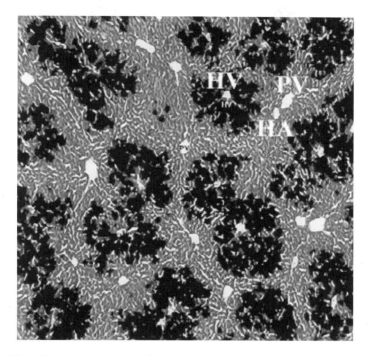

Fig. 2. Histochemical analysis of β-galactosidase gene expression in liver. Ten μg of pCMV-LacZ plasmid DNA were injected into each mouse via tail vein using hydrodynamics procedure. β-galactosidase gene expression was assayed 8 h post-injection. HV, hepatic vein; HA, hepatic artery; PV, portal vein (50×).

45 μg of luciferase per gram of liver in a mouse with a body weight of 18–20 g. More than 1000-fold higher level of transgene expression in the liver than in other organs suggests that the liver is the major site for transgene expression *(5)*.

Figure 2 shows the site and percentage of cells transfected in the liver using a dose of 25 μg of plasmid DNA containing *LacZ* gene (pCMV-LacZ). About 40% of liver cells were identified as β-galactosidase positive. The transfected cells were mainly hepatocytes identifiable by their polygonal shape and round nuclei. The transfected cells are grouped around the hepatic vein and located at the perivenular region in the cell plate.

References

1. Wolff, J. A., Malone, R. W., Williams, P, Chong, W., Acsadi, G., Jani, A., and Felgner, P. L. (1990) Direct gene transfer into mouse muscle in vivo. *Science* **247,** 1465–1468.
2. Yang, S. Y., Hogge, G. S., and MacEwen, G. (1999) Particle-mediated gene delivery:

applications to canine and other large animal systems, in *Nonviral Vectors for Gene Therapy*. (Huang, L., Hung, M. C., Wagner, E., eds.) Academic Press, New York, pp. 171–188.

3. Somiari, S., Malone, J. G., Drabick, J. J., Gilbert, R. A., Heller, R., Jaroszeski, M. J., and Malone, R. W. (2000) Theory and in vivo application of electroporative gene delivery. *Mol Ther* **3**:178–187.
4. Zhang, G., Vargo, D., Budker, V., Armstrong, N., Knechtle, S., and Wolff, J. A. (1997) Expression of naked plasmid DNA injected into the afferent and efferent vessels of rodent and dog livers. *Hum. Gene Ther.* **8**, 1763–1772.
5. Liu, F., Song, Y. K., and Liu, D. (1999) Hydrodynamics-based transfection in animals by systemic administration of plasmid DNA. *Gene Ther.* **6**, 1258–1266.
6. Zhang, G., Budker, V., and Wolff, J. A. (1999) High levels of foreign gene expression in hepatocytes after tail vein injections of naked plasmid DNA. *Hum. Gene Ther.* **10**, 1735–1737.
7. Zhang, G., Song, Y. K., and Liu, D. (2000) Long-term expression of human alpha 1-antitrypsin gene in mouse liver achieved by intravenous administration of plasmid DNA using a hydrodynamics-based procedure. *Gene Ther.* **7**, 1344–1349.
8. Liu, D. and Knapp, J. E. (2001) Hydrodynamics-based gene delivery. *Curr. Opin. Mol. Ther.* **3**, 192–197.
9. Yant, S. R., Meuse, L., Chiu, W., Ivics, Z., Izsvak, Z., and Kay, M. A. (2000) Somatic integration and long-term therapeutic transgene expression in normal and hemophilic mice using plasmid DNA encoding a DNA transposon system. *Nat. Genet* **25**, 35–41.
10. Miao, C. H., Ohashi, K., Patijn, G. A., Meuse, L., Ye, X., Thompson, A. R., and Kay, M. A. (2000) Inclusion of the hepatic locus control region, an intron and untranslated region increases and stabilizes hepatic factor IX gene expression in vivo but not *in vitro*. *Mol. Ther.* **1**, 522–532.
11. Miao, C. H., Thompson, A. R., Loeb, K., and Ye, X. (2001) Long-term and therapeutic-level hepatic gene expression of human factor IX after naked plasmid transfer in vivo. *Mol Ther.* **3**, 947–957.
12. Sambrook, J., Fritsch, E. F., and Maniatis, T. (1989) *Molecular Cloning: A Laboratory Manual*, 2nd ed. Cold Spring Harbor Laboratory Press, Cold Spring Harbor, NY.

21

Naked DNA Gene Transfer in Mammalian Cells

Guofeng Zhang, Vladimir G. Budker, James J. Ludtke, and Jon A. Wolff

1. Introduction

Gene therapy—the goal of which is to cure inheritable and acquired diseases by supplying genetic information to various tissues—is a promising therapy of the new millennium. To date, many strategies have been attempted to cure a disease by adding a foreign gene or correcting a mutation in the genes. The success of these gene-therapy strategies is largely dependent on the development of a vector that delivers and efficiently expresses a therapeutic gene in a specific cell population. Viral vectors are potentially efficient, although nonviral vectors have some advantages in that they are typically less immunogenic and easier to prepare. Direct, nonviral gene transfer into the whole organism remains a desirable goal for gene therapy because it avoids laborious and costly cell culture; the gene could be administered as easily as a drug.

After the demonstration that cationic lipids can mediate the efficient transfer of genes into cells in culture *(1)*, we tried to evaluate the ability of cationic lipids to mediate direct gene transfer into animals *(2)*. In muscle, it was the control group injected with naked mRNA that contained foreign gene expression (chloramphenicol acetyl transferase, CAT). Similar results were subsequently demonstrated using plasmid DNA (pDNA) expression vectors. A new window has been opened since then and many attempts have been made for making vaccines *(3–5)*, and for enhancing intramuscular transfection efficiency by using electroporation *(6–8)* or ultrasound *(9,10)*. In our laboratory we continue to work on in vivo gene transfer by using naked DNA and have created an in-

From: *Methods in Molecular Biology, vol. 245:*
Gene Delivery to Mammalian Cells: Vol. 1: Nonviral Gene Transfer Techniques
Edited by: W. C. Heiser © Humana Press Inc., Totowa, NJ

travascular administration method which produces high levels of foreign gene expression with naked DNA in liver *(11)* and muscle *(12)*.

2. Materials
2.1. Reagents

1. Normal saline: 0.9% sodium chloride in deionized water filtered through a 0.2-μm filter. The normal saline is made in our laboratory and we do not recommend use of commercial solutions because they usually contain preservatives. With large injection volumes, the preservative can cause animal death.
2. Mannitol solution: 15% mannitol in normal saline filtered through a 0.2-μm filter.
3. Ringer's solution: 820–900 mg of NaCl, 25–35 mg of KCl, and 30–36 mg of $CaCl_2$ in 100 mL of deionized water filtered through a 0.2-μm filter.
4. Heparin, 1 u/μL (Elkins-Sinn, Inc., Cherry Hill, NJ).
5. Papaverine (Sigma-Aldrich, St. Louis, MO) 20 mg /mL dissolved in Brij 99 solution (Sigma-Aldrich). Prepare Brij 99 solution by dissolving 60 mg of Brij 99 in 1 mL of deionized water.
6. Endotoxin-free plasmid DNA, commercially prepared (Bayou Biolabs, Harahan, LA or Aldevron, Fargo, ND).
7. Ketamine, xylazine, isofluorane, and metofane for anesthesia.

2.2. Tools

1. Micro Tying Forceps (curved).
2. Forceps (Ewald, serrated).
3. Scissors (114 mm).
4. Needle holder.
5. Micro-vessel-clamps:
   a. Kleinert-Kutz microvessel clip, 1 × 6 mm (Pilling Surgical, Horsham, PA, Cat. no. 65145). This clamp is suitable for blocking all of the vessels in the mouse and for most vessels in the rat.
   b. Curved clamp, 16 mm (Accurate Surgical & Scientific Instruments Corporation, Westbury, NY, Cat. no. ASSI.R3V). This clamp is larger and stronger than the microvessel clip; it is useful for blocking gluteal vessels when delivering plasmid DNA to the rat hindleg through the artery.
   c. Cooley Pediatric Derra Clamp, 7-inch (Pilling Surgical, Cat. no. 354815). This clamp is used for blocking the vena cava downstream from the hepatic vein.
6. Retractors, made from 4.75 cm long giant Gem paper clips. First, straighten the clip, then shape it into a flat-bottomed "U," with about 1.5 cm between

each arm of the "U." Bend the bottom of the "U" at a 90° angle about 1.5 cm from the bottom of the "U." Make another 90° bend about 0.5 cm from the bottom such that the side view of the retractor now looks like a flat-bottomed "J." Connect a strand of suture silk to the end of each arm. The strands can be taped on the table to hold the retractor at the proper position.
7. Fiber optic illuminator.
8. Loop (4-7 X magnification).
9. Suture Silk.
10. Gelfoam, 4 mm × 4 mm squares (Pharmacia and Upjohn Co., Kalamazoo, MI).
11. Cotton-tipped applicators.

3. Methods

The following sections describe methods of delivering pDNA to various organs of the mouse, rat, and monkey. A basic familiarity with the anatomy of these animals is required. Detailed anatomical descriptions of mice and rats can be found in the atlas by Popesko et al. *(13)* and of the monkey in the text by Bast et al. *(14)*.

3.1. Intravascular Delivery of Naked pDNA to Liver

After our laboratory had developed novel transfection complexes of pDNA · and amphipathic compounds and proteins, we sought to deliver them to hepatocytes in vivo *(15,16)*. Our control for these experiments was naked pDNA and we were once again surprised that this control group had the highest expression levels *(11,17)*. We will describe intra-portal, intra-hepatic, and intra-tail vein injection methods separately. Because the intra-bile duct and intra-portal injection methods are similar, they will be described concurrently.

3.1.1. Intra-Portal and Intra-Bile Duct Delivery of pDNA to Mouse Liver

This procedure describes a step-by step method for intra-portal pDNA delivery. See **Note 1** for modifications required for intra-bile duct delivery.

1. Use 25–30 g ICR mice. Anesthetize the animal by intramuscular injection of ketamine (80 mg/kg) and xylazine (2 mg/kg). Metofane or isoflurane inhalation (delivered by placing saturated cotton ball inside a closed container with the mouse) may also be used if necessary.
2. Make a mid-saggital incision from the xiphoid process to the lower one third of the abdomen to open the abdominal cavity. For the purpose of clearly exposing the liver, pull the xiphoid process in the cephalic direction with a hemostat and tape the tail on the table. Pull both sides of the abdominal wall laterally using retractors.

3. Carefully cut the front part of falciform ligament from the front edge to the inferior vena cava to enable clamping of the vena cava and hepatic vein.

4. Hold a curved microvessel clamp using a forceps and place it on the inferior vena cava and hepatic vein through the upper side of the liver. There are several branches of the hepatic vein that feed into the vena cava; place the clamp as close to the liver and as deep as possible. If the clamp does not block all of the hepatic veins, the transfection will not be successful.

5. Immediately after blocking the hepatic vein, inject the portal vein. The portal vein is easily found by pushing the intestine to left side of the abdominal cavity or by putting the intestine to the left and outside of the abdominal cavity. Fix the portal vein using a cotton-tipped applicator. Use a 1-mL syringe with a 30-G needle to inject 1 mL of a 15% mannitol solution containing 2.5 U of heparin and 10–100 μg pDNA into the portal vein at 2–3 ml/min.

6. Place a piece of gelfoam (4 mm × 4 mm) on the injection site before removing the needle; maintain pressure on the gelfoam using the applicator. Bleeding should stop 1–2 min after removal of the needle.

7. Remove the clamp 2 min after DNA injection. Close the abdominal cavity by suturing.

3.1.2. Intra-Hepatic Delivery of pDNA to Liver

This method describes a step-by-step method for intra-hepatic delivery of pDNA to mouse liver. See **Note 2** for modifications required to make the method suitable for use in rats.

1. Anesthetize 25–30 g ICR mice, then open the abdominal cavity as described in **Subheading 3.1.1., steps 1** and **2.**

2. Carefully cut the front part of falciform ligament from the front edge to the inferior vena to enable clamping of the vena cava downstream of the hepatic vein.

3. Separate the inferior vena cava upstream of the renal vein, because there is not enough space in mouse for blocking and injection between the renal vein and liver.

4. Put two clamps on the inferior vena cava both upstream and downstream of the hepatic vein to block blood flow. Carefully push down the liver using a cotton tipped applicator to expose the vena cava; then clamp the vena cava. Quickly push up the liver to expose the upstream part of the inferior vena cava; put a clamp on it upstream of the renal vein (*see* **Note 3**).

5. Use a 1 mL syringe with a 30-G needle to inject 1 mL of 15% mannitol solution containing 2.5 U of heparin and DNA. We usually use approx 100 μg DNA, but have used a range of 10–200 μg (*see* **Note 4**). Insert the needle into the inferior vena cava in the anterograde direction of blood flow

and advance it until the tip of the needle passes over the right renal vein. It may help to bend the needle at a 30–40° angle in order to make it possible to keep the front part of the needle parallel with the vena cava, thereby decreasing the possibility of damaging the back wall of vena cava and causing leakage owing to needle movement during injection.

6. Place a piece of gelfoam on the injection site and apply pressure with a cotton-tipped applicator while removing the needle. Keep the gelfoam in place until the clamps are removed 2 min after injection and the bleeding has stopped.

7. Suture the abdominal cavity closed.

3.1.3. Intra-Tail Vein Delivery of pDNA to Mouse Liver

A method of systemic administration of pDNA solution to the liver in mouse was developed concurrently in two independent laboratories (*18,19*). We will mention our method briefly.

1. Anesthetize the mouse using isofluorane inhalation. To deliver anesthesia, saturate a cotton ball with isofluorane, and place inside a closed container with the mouse. Watch carefully for apnea. Do not inject any drug for anesthesia because the liver that normally catabolizes most anesthesia drugs may be in stroke temporarily after injection of a large volume of solution and toxicity of anesthetics may be much higher. If an animal receives a drug injection for anesthesia, it may have difficulty recovering and in some cases may die.

2. Warm the tail with warm water or a heat lamp to make the vessel dilate. Usually 50–60°C water is suitable.

3. Hold the distal part of the tail using the thumb and ring finger to pull the tail straight. The index and middle fingers hold the root of the tail to block blood flow.

4. Using a 3-mL syringe with a 27-G needle, inject 2–3 mL of Ringer's solution containing pDNA through the tail vein within 7–10 s (*see* **Note 5**).

3.2. Intravascular Delivery of pDNA to Other Internal Organs

Following successful expression in liver, we attempted to transfect various other organs. We have found that the intravascular injection technique is widely applicable as a basic method for gene transfer in vivo. Many types of cells can take up and express pDNA following intravascular delivery; it appears that it is a general characteristic of mammalian cells in vivo. Therefore, if pDNA comes into contact with somatic cells, there is a possibility that it will be taken up and expressed regardless of the delivery method employed. There are large variations from organ to organ with respect to the structure, blood supply, cell biological characteristics, and so on. The optimum conditions for gene transfer will therefore vary from organ to organ. Strategies for optimization are mentioned below and in notes.

3.2.1. Intravascular Injection of pDNA to Mouse Kidney

This section describes a step-by-step method for the intravascular injection of pDNA into the mouse kidney. *See* **Note 6** for modifications required to make the method suitable for rats.

1. Anesthetize 25–30 g ICR mice, then open the abdominal cavity as described in **Subheading 3.1.1**, **steps 1** and **2**.
2. After exposure, separate the inferior vena cava both upstream and downstream of the renal veins. Also separate the inferior phrenic and spermatic vessels.
3. Use two clamps to block the inferior vena cava; place one downstream of the renal vein (between renal vein and liver), and the other upstream and 0.5 cm away from renal vein. Also block the inferior phrenic and spermatic vessels to create high pressure.
4. Prepare a 1-mL syringe containing 0.4–0.6 mL of Mannitol solution containing 2.5 U of heparin and pDNA (usually 10–100 µg) and with a 30 G needle bent at a 30° angle. Insert the needle into the inferior vena cava upstream of the renal vein. Advance the tip close to, but not over, the junction point of the renal vein and the inferior vena cava.
5. Inject the pDNA at a rate of about 1 mL/20 s.
6. Use a piece of gelfoam to stop bleeding and remove clamps 2 min postinjection. No special care is routinely needed after closing the abdominal cavity.

3.2.2. Intravascular Delivery of pDNA to Mouse Intestine

This section describes a step-by-step method for pDNA delivery to mouse intestine. *See* **Note 7** for minor modifications required to make the method suitable for rats.

1. Anesthetize 25–30 g ICR mice, open the abdominal cavity, and expose the intestines as described in **Subheading 3.1.1.**, **steps 1–3**.
2. Locate the superior mesenteric vein by pushing the intestine to left side of the abdomen. Separate the vessel and block the vein just before its juncture with the splenic vein using a microclamp.
3. Insert a 30-G needle into the mesenteric vein in the retrograde direction of blood flow. Inject 1–2 mL of mannitol solution containing 2.5–5 U of heparin and 100 µg/mL of DNA in a 3 mL syringe at high velocity.

4. Place a piece of gelfoam on the injection site before pulling the needle out and maintain pressure using a cotton tipped applicator. Release the clamp 2 min later. Suture the abdominal cavity closed.

3.2.3. Intravascular Injection for Ovary Gene Transfer in Rat

1. Anesthetize the rat by intramuscular injection of ketamine (80–100 mg/kg) and xylazine (2 mg/kg).
2. Make a medial incision about 4 cm long on the lower part of the abdomen and expose the internal organs using retractors.
3. The ovary receives its blood supply from two sources, the ovarian artery and the uterine artery, and there is rich collateral circulation between these two systems. Select one vessel from the ovarian artery, ovarian vein, or uterine vein for delivering DNA solution; the uterine artery is too thin to be used. Separate and clamp all vessels before injection. Place the clamps as close to the ovary as possible, with the exception of the vessel selected for injection, which should be blocked at a point far away from the ovary in order to leave enough space for injection.
4. Inject 0.3–0.5 mL of mannitol containing 100 µg/mL of pDNA at high velocity.
5. Place a piece of gelfoam on the injection site before pulling the needle out and maintain pressure using a cotton-tipped applicator. Release the clamp 2 min later. Suture the abdominal cavity closed.

3.2.4. Intravascular Injection for Testis Gene Transfer in Rat

1. Anesthetize the rat, then open the abdominal cavity and expose the internal organs as described in **Subheading 3.2.3., steps 1** and **2**.
2. DNA may be delivered to the testis by one of two routes:
   a. Block the testicular artery and vein at a position far away from the testis to allow room for injection. Also block the artery and vein of the deferent duct. Using a 1-mL syringe with a 30-G needle, quickly inject 0.5 mL of mannitol solution containing pDNA (100 µg/mL) into the testicular artery or vein.
   b. Clamp all blood outflow and inflow of the testis at a position as close to the testis as possible. Using a 1-mL syringe and a 30-G needle, quickly inject 0.5 mL of mannitol solution containing pDNA (100 µg/mL), into a vessel located on the surface of the testis.
3. Place a piece of gelfoam on the injection site before pulling the needle out and maintain pressure using a cotton tipped applicator. Release the clamp 2 min later. Suture the abdominal cavity closed.

3.3. Intravascular Delivery of pDNA to Skeletal Muscle

The intravascular delivery of naked pDNA to muscle cells is attractive particularly because many muscle groups would have to be targeted for intrinsic muscle disorders such as Duchenne muscular dystrophy. In addition, an intravascular approach would avoid the limited distribution of pDNA through the interstitial space following intramuscular injection. Muscle has a high density of capillaries *(20)* that are in close contact with the myofibers *(21)*. Delivery of pDNA to muscle via capillaries puts the pDNA into direct contact with every myofiber and substantially decreases the interstitial space the pDNA has to traverse in order to access a myofiber. However, the endothelium in muscle capillaries is of the continuous, nonfenestrated type and has low solute permeability, especially to large macromolecules *(22)*. Based on our success with intravascular injection of naked DNA to liver, arterial delivery of DNA to rat muscle was attempted with very positive results. High levels of reporter gene expression were found throughout the hindleg and foot muscles *(12)*. This intra-arterial injection technique has since been applied to large animals such as rabbit, monkey, and dog.

3.3.1. Intra-Artery Delivery of pDNA to Rat Hindleg Muscle

1. Anesthetize a 130–160 g rat by intramuscular injection of ketamine (80–100 mg/kg) and xylazine (2 mg/kg).
2. Make a medial incision from the upper two-thirds to the caudal edge of the abdomen to open the abdominal cavity. Use two retractors, one to pull the right abdominal wall to the right side, and the other to pull the left abdominal wall and internal organs to the left side.
3. Using a cotton-tipped applicator and forceps, push the peritoneum and connective tissues away from the front of iliac artery until it is clearly visualized. Separate the external iliac artery and vein from the surrounding tissue using curved micro tying forceps and place a suture silk around the two vessels. Separate the caudal epigastric vessels (artery and vein), internal iliac vessels, vessels of deferent duct, and caudal gluteal vessels using the same method.
4. Clamp these separated vessels to block blood inflow and outflow of the hindleg. Pay great attention to separating and clamping the gluteal vessels. These vessels are very deep and are not easily visualized. These vessels also have extensive collateral vessels. If damage occurs to these vessels or the blockage is not complete, it will dramatically affect the transfection efficiency. Block the external iliac last, as close to the origin of the artery as possible to allow enough space for injection.

5. Fix the external iliac artery by pulling the two ends of the suture silk that was placed there previously. Using a 3-mL syringe with a 30 G needle, inject 3 mL of normal saline containing 0.5 mg of papaverine into the artery.
6. Five minutes later, insert a 25- or 27-G butterfly needle into the external iliac artery through the original hole used for the papaverine injection. If the butterfly needle is not inserted into the papaverine injection hole, there will be extensive leakage through the hole. Attach a 10-mL syringe to the butterfly needle and inject 10 mL of normal saline containing 500 μg of pDNA (the dose can be varied widely) within 7–15 s.
7. Place a piece of 6 mm × 6 mm gelfoam on the injection site and maintain pressure while removing the needle using a cotton-tipped applicator. Remove the clamps 2 min after injection and maintain the pressure on the gelfoam until all bleeding has stopped. No special care is needed after suturing the abdominal cavity (*see* **Note 8**).

3.2.2. Intra-Arterial Delivery of pDNA to Monkey Skeletal Muscle

1. Anesthetize a 6–14 kg monkey by intramuscular injection of ketamine (10–15 mg/kg im) followed by isofluorane inhalation (2–3% by face mask).
2. For a forearm injection, make a 3 cm long longitudinal incision on the skin along the inside edge of the biceps brachii and 2 cm above the elbow. For a lower-leg injection, make the incision on the upper edge of popliteal fossa.
3. Use a retractor to expose the vessels. Separate the brachial artery or popliteal artery from the surrounding tissue and veins using forceps.
4. Insert a 20-G catheter into the brachial or popliteal artery anterogradely and ligate it in place with a suture (*see* **Note 9**).
5. Block blood flow in the forearm or lower leg using a blood pressure cuff that surrounds the arm or leg proximal to the injection site. Inflate the sphygmomanometer above 300 mm Hg and then inject 30 mL of normal saline containing 5 mg of papaverine into the catheterized vessels.
6. Keep the blood flow blocked; 5 min after papaverine injection, rapidly inject the DNA solution (about 1.5% of body weight) into the vessel using an air pump driven by nitrogen gas. Deliver 120–150 mL of DNA solution into the arm or 150–180 mL into the leg. The injection typically takes 30–60 s (*see* **Note 10**).
7. Place a piece of gelfoam on the injection site and maintain pressure until the catheter is removed and the bleeding has stopped (or, if necessary, until the artery has been repaired). Two minutes after injection, deflate the sphygmomanometer belt and suture the incision.

4. Notes

1. The same procedure is suitable for intra-bile duct injection except that (1) the bile flow is occluded by a clamp at the end of the bile duct to prevent the DNA solution leaking into the duodenum, and (2) the DNA solution is injected into the bile duct retrogradely. When pDNA in 1 mL of hyperosmotic solution was injected into the bile duct, the expression was usually the same or better than intraportal injection. However, the alanine aminotransferase (ALT) level in serum (indicating liver trauma) was higher than with the intraportal delivery method. Very good expression can often be achieved after injecting small DNA volumes using the intra-bile injection method, but there is wide variability. When the DNA solution is delivered through the bile duct without occlusion of the hepatic vein, both the gene expression and ALT levels are lower.

2. The intra-portal, intra-hepatic, and intra-bile duct injection methods have been successfully used in rats and the results are comparable to the mice data. We will not mention the methods separately here because they are identical to those used in mice except that 10 mL of DNA solution, a 25-G butterfly needle, and a 10-mL syringe are used. A modification has also been tried in which a 24-G catheter and tourniquet are used instead of a 25-G butterfly needle and metal clamp. A Rommel tourniquet is put on the inferior vena cava at the position between the renal vein and liver. A catheter is then inserted into the inferior vena cava upstream of the renal vein but passed beyond the tourniquet. When this procedure is followed, DNA solution is delivered only to the liver; no solution is delivered to the kidneys.

3. Several points should be noted for the intra-hepatic injection in mouse. First, the portal vein can be free or clamped when performing the intra-hepatic injection, but if the portal blood flow is blocked, higher expression can be reached. Second, the DNA solution perfuses not only the liver, but also the kidneys because the clamp is located upstream of the renal vein. This is the reason that we advance the needle to pass over the renal vein to cause most of the DNA solution to reach the liver. Third, the order of clamping at the vessels can affect the transfection efficiency. If the first clamp is put on the vena cava downstream of hepatic vein, the liver will be filled up by the blood coming from both the portal system and the lower part of the inferior vena cava system. This causes the pressure in the liver to increase prior to injection, so higher ultimate pressures can be reached, resulting in higher transfection efficiencies. However, if the upstream blocking occurs first, the pressure in the inferior vena cava downstream of the clamp is decreased because the blood coming from the lower part of the body was blocked. In this case the pressure difference between inferior

vena cava and portal system is increased and the blood is quickly drained from the liver, resulting in decreased pressure in the liver and lower transfection efficiencies.

4. Solution volume, injection velocity, and complete occlusion of blood flow are three key factors for achieving successful gene transfer in our intravascular delivery protocol. Alteration of any one of these three factors will substantially affect the transfection efficiency. Combining this surgical approach with improved plasmid vectors has enabled uncommonly high levels of foreign expression in which more than 15 µg of luciferase protein/liver was produced in mice and more than 50 µg of luciferase/liver in rat. Equally high levels of β-galactosidase expression were obtained; more than 5% of hepatocytes had intense blue staining. β-galactosidase expression was rarely seen in other type of cells. Expression of luciferase or β-galactosidase was almost evenly distributed in hepatocytes throughout the entire liver when any of the three routes were injected.

5. The amount of solution volume is variable according to the body weight. Use the following equation to calculate DNA volume: solution volume (mL) = (animal body weight in grams − 5)/10. The amount of pDNA may be varied over a wide range; the maximum we have used is 300 µg per injection and the minimum is 1 ng per injection. The expression is DNA dose-dependent, but we achieve very good expression using 1 ng of luciferase gene per injection.

6. The same protocol can be used for gene transfer to rat kidney (*23*). However, because the renal artery and vein are longer and thicker, a single kidney is usually selected as a target. If the renal vein is selected for gene delivery, the vein must be clamped at the juncture point of renal vein and inferior vena cava. The renal artery can be kept free or blocked. If the artery is selected, both artery and vein should be blocked. It is important to remember that, when the left kidney is selected as target organ, the inferior phrenic and spermatic vessels should be clamped as well.

7. Intravascular gene transfer to rat intestine is somewhat different from that in mice. Usually we select a 5 cm segment of intestine as the target for gene transfer. The two ends of the targeted segment of intestine are clamped, including the intestine itself and the collateral vessels that connect the target segment and the neighboring area. The main trunk of the artery and vein that supply the target or drain blood from the segment should be clamped, as well as any branches that are concerned with the collateral circulation between the segment and neighboring area. DNA solution is injected through the artery anterogradely or through the vein retrogradely. The volume of DNA solution is varied according to the length of selected intestine segment (100 µg for a 5-cm segment).

8. Additionally, our data show that immunosuppression promotes long-term expression following intra-arterial delivery of pDNA to rat muscle *(24)*. Other groups have demonstrated long term gene expression in liver using liver-specific promotors *(25)*. These are exciting advances that hold great promise for use in gene therapy.

9. A modification has been made in large animals for the intra-arterial injection of pDNA (unpublished). A long cannula (F-6) is inserted from the proximal part of the brachial, femoral, or external iliac artery, past the position of the blood pressure cuff to the distal portion of the artery. The blocking and injection are same as that described above. The modified method eliminates leakage because the incision is far away from the injection site and the tourniquet prevents back-flow. This method has produced exciting results in monkey muscle *(24)*. The luciferase expression averages 991 ng/g muscle for the forearm and 1692 ng/g for lower leg 1–2 wk post-injection. In the rat, luciferase expression averages 780 ng/g. X-gal staining produced similar results. The highest percentage of β-galactosidase-positive cells achieved using this method has been 37% and the lowest is about 1%, averaging 6.3% in the forearm and 7.3% in the lower leg muscle.

10. For most organs, if high pressure is obtained inside the vessels, highly efficient gene transfer occurs. It is possible that the high pressure enables pDNA extravasation, and the pDNA is then taken up and expressed by the cells *(26)*. Alternatively, the high pressure may directly affect the membrane of targeted cells so as to enable pDNA uptake. Regardless of the mechanism, the intravascular injection offers an efficient and safe method for gene delivery to many organs.

References

1. Felgner, P. L., Gadek, T. R. Holm, M., Roman, R., Chan, H. W., Wenz, M., et al. (1987). Lipofection: a highly efficient, lipid-mediated DNA-transfection procedure. *Proc. Natl. Acad. Sci USA* **84**, 7413–7417.
2. Wolff, J. A., Malone, R. W., Williams, P. Chong, W. Acsadi, G. Jani, A., and Felgner, P. L. (1990). Direct gene transfer into mouse muscle in vivo. *Science* **247**, 1465–1468.
3. Fonseca, D. P., Benaissa-Trouw, B., van Engelen, M., Kraaijeveld, C. A., Snippe, H., and Verheul, A. F. (2001) Induction of cell-mediated immunity against Mycobacterium tuberculosis using DNA vaccines encoding cytotoxic and helper T-cell epitopes of the 38-kilodalton protein. *Infect. Immun.* **69**, 4839–4845.
4. Nawrath, M., Pavlovic, J., Dummet, R., Schultz, J., Strack, B., Heinrich, J., and Moelling, K. (1999). Reduced melanoma tumor formation in mice immunized with DNA expressing the melanoma-specific antigen gp100/pmel17. *Leukemia* **13**,S48–51.

5. Zheng, B. J., Ng, M. H., He, L. F., Yao, X., Chan, K. W., Yuen, K. Y., and Wen, Y.M. (2001) Therapeutic efficacy of hepatitis B surface antigen-antibodies-recombinant DNA composite in HBsAg transgenic mice. *Vaccine* **19**, 4219–4225.

6. Fewell, J. G., MacLaughlin, F., Mehta, V., Gondo, M. Nicol, F., Wilson, E., and Smith, L. C. (2001) Gene therapy for the treatment of hemophilia B using PINC-formulated plasmid delivered to muscle with electroporation *Mol. Ther.* **3**, 574–583.

7. McMahon, J. M., Signori, E., Wells, K. E., Fazio, V. M., and Wells, D. J. (2001). Optimisation of electrotransfer of plasmid into skeletal muscle by pretreatment with hyaluronidase: increased expression with reduced muscle damage. *Gene Ther.* **8**, 1264–1270.

8. Nakano, A., Matsumori, A., Kawamoto, S., Tahara, H. Yamato, E., Sasayama, S., and Miyazaki, J. I. (2001) Cytokine gene therapy for myocarditis by in vivo electroporation. *Hum. Gene Ther.* **12**, 1289–1297.

9. Amabile, P. G., Waugh, J. M., Lewis, T. N., Elkins, C. J., Janas, W., and Dake, M. D. (2001) High-efficiency endovascular gene delivery via therapeutic ultrasound. *J. Am. Coll. Cardiol.* **37**, 1975–1980.

10. Lawrie, A., Brisken, A. F., Francis, S. E., Cumberland, D. C., Crossman, D. C., and Newman, C.M. (2000) Microbubble-enhanced ultrasound for vascular gene delivery. *Gene Ther.* **7**, 2023–2027.

11. Budker, V., Zhang, G. Knechtle, S., and Wolff, J. (1996) Naked DNA delivered intraportally expresses efficiently in hepatocytes. *Gene Ther.* **3**, 593–598.

12. Budker, V., Zhang, G., Danko, I., Williams, P., and Wolff, J. (1998) The efficient expression of intravascularly delivered DNA in rat muscle. *Gene Ther.* **5**, 272–276.

13. Popesko, P., Rajtova, V., and Horak, J. (1992) *Atlas of the Anatomy of Small Laboratory Animals*, vol. 2. Wolfe Publishing, Ltd., London.

14. Bast, T. H. (1969) T*he Anatomy of the Rhesus Monkey (Macaca mulatta).* Hafner Publishing Co., New York.

15. Budker, V., Gurevich, J., Hagstrom, Bortzov, F., and Wolff, J. A. (1996) pH-sensitive, cationic liposomes: A new synthetic virus-like vector. *Nat. Biotechnol.* **14**, 760–764.

16. Budker, V., Hagstrom, J. E., Lapina, O., Eifrig, D., Fritz, J., and Wolff, J. A. (1997) Protein/amphipathic polyamine complexes enable highly efficient transfection with minimal toxicity. *Biotechniques.* **23**, 139, 142–137.

17. Zhang, G., Vargo, D., Budker, V., Armstrong, N., Knechtle, S., and Wolff, J. A. (1997) Expression of naked plasmid DNA injected into the afferent and efferent vessels of rodent and dog livers. *Hum. Gene Ther.* **8**, 1763–1772.

18. Liu, F., Song, Y., and Liu, D. (1999) Hydrodynamics-based transfection in animals by systemic administration of plasmid DNA. *Gene Ther.* **6**, 1258–1266.

19. Zhang, G., Budker, V., and Wolff, J. A. (1999). High levels of foreign gene expression in hepatocytes after tail vein injections of naked plasmid DNA. *Hum. Gene Ther.* **10**, 1735–1737.

20. Browning, J., Hogg, N., Gobe, G., and Cross, R. (1996) Capillary density in skeletal muscle of Wistar rats as a function of muscle weight and body weight. *Microvasc. Res.* **52**, 281–287.

21. Lee, J. and Schmid-Schonbein, G. W. (1995) Biomechanics of skeletal muscle capillaries: hemodynamic resistance, endothelial distensibility, and pseudopod formation. *Ann. Biomed. Eng.* **23**, 226–246.

22. Taylor, A. and Granger, D. (1984) Exchange of macromolecules across the microcirculation, in *Handbook of Physiology: The Cardiovascular System Microcirculation* (eds. Geiger, S., Renkin, E., and Michel, C.) pp. 467–520

23. Kuemmerle, N. B., Lin, P. S., Krieg, R. J., Jr., Lin, K. C., Ward, K. P., and Chan, J. C. (2000) Gene expression after intrarenal injection of plasmid DNA in the rat. *Pediatr. Nephrol.* **14**, 152–157.

24. Zhang, G., Budker, V., Williams, P., Subbotin, V., and Wolff, J. A. (2001) Efficient expression of naked DNA delivered intraarterially to limb muscles of nonhuman primates. *Hum. Gene Ther.* **12**, 427–438.

25. Miao, C. H., Ohashi, K., Patijn, G. A., Meuse, L., Ye, X., Thompson, A. R., and Kay, M. A. (2000) Inclusion of the hepatic locus control region, an intron, and untranslated region increases and stabilizes hepatic factor IX gene expression in vivo but not in vitro. *Mol. Ther.* **1**, 522–532.

26. Budker, V., Budker, T., Zhang, G., Subbotin, V., Loomis, A., and Wolff, J. A. (2000) Hypothesis: naked plasmid DNA is taken up by cells in vivo by a receptor-mediated process. *J. Gene Med.* **2**, 76–88.

22

Microparticle Delivery of Plasmid DNA to Mammalian Cells

Mary Lynne Hedley and Shikha P. Barman

1. Introduction

Polypeptides with therapeutic benefit and those for which a cellular and humoral immune response is desirable can be produced by injection of plasmid DNA into muscle (for review, *see* **ref.** *[1]*). The safety and efficacy of the DNA-based approach has resulted in several clinical trials with plasmids encoding cytokines, alloantigens, cancer antigens, and antigens from pathogens *(2–8)*, but in most cases the results from these studies have been less than inspiring. Data collected to date would suggest that the success of this method requires higher levels of transfection and sustained expression than are achievable with plasmid in the absence of a delivery system. Furthermore, the efficacy of DNA vaccines will depend on the presentation of DNA encoded antigen by activated professional antigen-presenting cells (APCs) such as Langerhans cells, macrophages, and dendritic cells (DCs) *(1)*.

Given the results from current studies with unformulated DNA in humans, the rationale for formulating plasmid DNA in microparticles includes achieving sustained, therapeutic concentrations of a polypeptide, and targeting DNA uptake by phagocytic cells with particles less than 10 μm in diameter *(9–11)*. Advantages to the encapsulation approach include the potential for controlled release of DNA from particles, elimination of frequent dosing regimens, and protection of DNA from serum components. The selection of biodegradable, biocompatible polymers that are inert, safe, and do not compromise the bioactivity of the DNA will further enhance the applicability of these systems. A synthetic polymer use-

From: *Methods in Molecular Biology, vol. 245:*
Gene Delivery to Mammalian Cells: Vol. 1: Nonviral Gene Transfer Techniques
Edited by: W. C. Heiser © Humana Press Inc., Totowa, NJ

ful in this regard is poly(lactide-co-glycolide) (PLG), which has a history of safe use in humans *(12)*. Small molecule, peptide- and protein-sustained release systems that utilize PLG microparticles as a delivery vehicle are commercially available, including Lupron Depot® (TAP), Neutropin Depot® (Genentech), Zoladex® (Zeneca), Decapeptyl® (Ipsen Biotech), and Prostap SR® (Lederle).

Microencapsulation of DNA must not only address the criteria of a DNA delivery system as described earlier, but must include critical parameters that permit the production of defined size particles and the preservation of plasmid structure postformulation. Particles containing DNA can be large, as may be required for the expression of therapeutic proteins, or less than 10 µm in size so as to promote uptake by APC. Maintaining plasmid structure throughout the process results in a product with potentially enhanced biological activity and a longer shelf life. A successful procedure for DNA encapsulation will need to be economical at scale, allow for high drug loading, and be aseptic in nature, because terminal sterilization is not possible with this type of formulation. Techniques for the incorporation of plasmid DNA into PLG microparticles have been described *(13–16)*. Plasmid DNA released from PLG microparticles over time is functional and is expressed in cells *(17)*. Administration of the particles to mice elicits an immune response to the encoded antigen *(13,14,18,19,20)* and patients immunized with encapsulated plasmid DNA mount a significant immune response to the encoded antigen *(21,22)*.

In this chapter, detailed methods and protocols are provided for producing PLG-encapsulated DNA via a water/oil/water (w/o/w) encapsulation processsand for microparticle characterization. Protocols are also provided for measuring gene expression, and immune responses to the encoded antigen (antibody and T-cell responses). β-galactosidase (β-gal) was chosen as the reporter antigen for these studies because the plasmid and the tools to test β-gal expression and specific immune responses are commercially available.

2. Materials

2.1. Small-Scale Production of <10-µm Microparticles Containing Plasmid DNA (see Notes 1–3)

1. Silverson SL2T homogenizer equipped with 16-mm ID slotted mixing head (Silverson, East Longmeadow, MA).
2. Ring stand and clips to hold the homogenizer.
3. Freeze dryer (Labconoco Freeze dryer 4.5).
4. TE buffer: 10 mM Tris-HCl, pH 8.0, 1 mM ethylenediaminetetraacetic acid (EDTA), pH 8.0.
5. TE with 303 mM sucrose.
6. 1.2 mg plasmid DNA dissolved in 0.3 mL of TE/303 mM sucrose (*see* **Note 4**).

7. 0.2 g PLG (Resomer® RG502, 50:50 *random* D,L lactide-co-glycolide, ~12,000 Da; Boehringer Ingelheim, Germany).
8. Dichloromethane high-performance liquid chromatography ([HPLC]-grade, J.T. Baker, Phillipsburg, NJ).
9. Polyvinyl alcohol (PVA, mwt ~30-50,000, Sigma, St Louis, MA).
10. 1% PVA solution: 1% w/v PVA/303 m*M* sucrose in DI water (*see* **Note 5**).
11. 0.05% PVA solution: 0.05% (w/v) PVA/303 m*M* sucrose in DI water (*see* **Note 5**).

2.2. Lab-Scale Production of <10-μm Microparticles Containing Plasmid DNA (see Note 6)

1. Silverson SL2T homogenizer equipped with 16-mm ID homogenizing probe (Silverson).
2. Silverson L4RT homogenizer equipped with in-line mixer head.
3. Pump 1, MasterFlex L/S drive (Cat. no. 7554-90) equipped with Master-Flex EasyLoad pumping head (Cat. no. 75118-00, Upchurch Scientific, Seattle WA).
4. Pump 2, MasterFlex standard drive (Cat. no. 7520-00) equipped with Ea-syLoad II pumping head (Cat. no. 77200-50, Upchurch Scientific).
5. Five-mm ID viton tubing for pump 2 (Masterflex Cat. no. 6412-25, Up-church Scientific) (*see* **Note 7**).
6. One-mm ID viton tubing for pump 1 (Masterflex Cat. no. 6412-14, Up-church Scientific) (*see* **Note 7**).
7. Rind stand and clips to hold the homogenizers.
8. VWR adjustable jack.
9. Freeze dryer (Labconoco Freeze dryer 4.5).
10. 10.6 mg plasmid DNA dissolved in 1.6 mL TE/303 m*M* sucrose (*see* **Note 4**).
11. 1 g PLG (Resomer® RG502, 50:50 *random* D,L lactide-co-glycolide, ~12000 Da; Boehringer Ingelheim).
12. Dichloromethane (J.T. Baker, HPLC-grade)
13. PVA (mwt ~30–50,000, Sigma).
14. 1% PVA solution: 1% w/v PVA/303 m*M* sucrose in DI water (*see* **Note 5**).

2.3. Measuring DNA Encapsulation

1. TE buffer: 10 m*M* Tris-HCl, pH 8.0, 1 m*M* EDTA, pH 8.0.
2. Chloroform (J.T. Baker).

2.4. Measurement of DNA Supercoiling

See **Note 8** for a more quantitative measure of supercoiling.

1. 50X TAE buffer (Boston Bioproducts, Ashland MA).
2. Ethidium Bromide solution (Sigma)

3. 0.8% agarose gel made with TAE buffer (final concentration 1X) and Ethidium Bromide (5 µL/100 mL gel)
4. 10X agarose gel loading buffer (Boston Bioproducts)

2.5. Microparticle Sizing (see Note 9)

1. Multisizer II Coulter counter with a probe suitable for analyzing 1–10-µm sized particles (aperture 30 µm; Beckman Coulter, Inc., Palatine, IL).
2. Three-µm standard beads (Cat. no. 17143, Polysciences,Warrington, PA).
3. Isoton II diluent (Cat. no. PN 8546719, Beckman-Coulter).

2.6. In Vitro Release of DNA from Particles Over Time

1. Dulbecco's Phosphate-Buffered Saline Solution: (PBS, Cat. no. 59321-500M; JRH Biosciences, Kansas City, KS).

2.7. Analysis of β-Gal Specific Expression (see Note 10)

1. Oven.
2. Microtome.
3. Balb/c mice, females, 6–10 wk of age (The Jackson Laboratory, Bar Harbor, ME) (*see* **Note 11**).
4. Isoflurane (Abbott Lab, North Chicago, IL).
5. PBS solution (Cat. no. 59321-500M, JRH Biosciences).
6. 0.25% glutaraldehyde (J.T. Baker).
7. Xylene (Cat. no. X3P, Fisher Scientific).
8. 0.2% X-gal (5-bromo-4-chloro-3-indolyl-β-D-galactopyranoside; Promega, Madison, WI).
9. 10% neutral buffered formalin (Cat. no. 0133, Fisher Scientific, Pittsburgh, PA).
10. Hematoxylin Solution (Cat. no. HHS-32, Sigma).
11. Eosin Y solution (0.1% w/v, Cat. no. HT110-3-32, Sigma).
12. Permount (Cat. no. 1000-4, Sigma)

2.8. β-Gal Specific Enzyme-Linked Immunosorbent Assay (ELISA)

1. ELISA plate reader.
2. Balb/c mice, females, 6–10 wk of age (The Jackson Laboratory, Bar Harbor, ME) (*see* **Note 11**).
3. 96-well plates (Dynatech Laboratories, Chantilly, VA, Cat. no. 011-010-3455).
4. β-gal protein (Calbiochem Novabiochem, Pasadena, CA, Cat. no. 345788).
5. Normal mouse serum (NMS; *see* **Note 12**).
6. β-gal-specific monoclonal antibody (MAb) (Calbiochem Novabiochem, Cat. no. 0B02-100UG).

7. Horseradish peroxidase-conjugated antibodies specific for mouse IgG (heavy + light-chain specific) (Cat. no. 6175-05, Southern Biotechnology, Birmingham, AL).
8. ABTS substrate (Zymed, San Francisco, CA, Cat. no. 00-2011).
9. Dulbecco's PBS (Cat. no. 59321-500M, JRH Biosciences).
10. PBS/0.05% Tween-20 (PBST, Cat. no. P-3563, Sigma).
11. Bovine serum albumin (BSA, Sigma).

2.9. T-Cell Assays

1. Cell Harvester, filter mats, and bag sealer (Harvester Model no. MACH, Tomtec, Hamden, CT).
2. Liquid Scintillation Counter (Cat. no. 1205, Wallac, Atlanta, GA).
3. Centrifuge (Beckman Coulter Model G5-CR).
4. X ray source to irradiate cells (AXR Minishot 160 Model M160NH) (*see* **Note 13**)
5. Balb/c mice, females, 6–10 wk of age (The Jackson Laboratory) (*see* **Note 11**).
6. The synthetic peptide, TPHPARIGL, representing the naturally processed H-2 L^d restricted epitope spanning amino acids 876-884 of β-galactosidase (β-gal peptide; *see* **Note 14**).
7. The synthetic peptide, IPQSLDSWWTSL, the H-2 L^d epitope corresponding to residues 28–39 of hepatitis B surface Ag (HBV peptide) (*see* **Note 14**).
8. Fetal calf serum (FCS), (Cat. no. 12106-78P, JRH Bioscience).
9. RPMI-1640 (JRH Bioscience) supplemented with 2% or 10% FCS (RPMI-2 and RPMI-10), 2 mM L-glutamine (Invitrogen, Baltimore, MD) streptomycin (Invitrogen) at 100 g/mL, penicillin at 100 U/mL (Invitrogen) and 5.5×10^{-2} mM 2-mercaptoethanol (Cat. no. 21985-023, Invitrogen).
10. Trypan Blue (Invitrogen).
11. T-cell enrichment columns and buffers for CD3 or CD8 T cells (R&D Systems, Minneapolis, MN)
12. Interferon-γ (IFN-γ) ELISPOT kit, includes 10X wash buffer (Cat. no. EL485, R&D Systems).
13. 1X wash buffer (10X buffer is included in the ELISPOT kit, Cat. no. EL485, R&D Systems).
14. RPMI-10 containing 2.5 mg/mL Concanavalin A (ConA) (Cat. no. L7647, Sigma).
15. RPMI-10 containing 3 mg/mL Ovalbumin (OVA, Cat. no. A5503, Sigma).
16. RPMI-10 containing 3 mg/mL β-galactosidase (Cat. no. 345788, Calbiochem Novabiochem).
17. ^{3}H-TdR (Cat. no. 24066, ICN Biomedicals, Inc, Philadelphia, PA).

3. Methods

3.1. Small-Scale Production of <10-μm Microparticles Containing Plasmid DNA

1. Dissolve 0.2 g of PLG in 7 mL of dichloromethane (DCM) in a 17x 100-mm, 14 mL polypropylene tube at ambient temperature (*see* **Note 15**).
2. Pour 50 mL of 1.0% PVA solution into a 100-mL polypropylene beaker. Have a 250-mL polypropylene beaker containing 100 mL of 0.05% PVA with a 1.5-in stir bar ready on a stir plate.
3. Pour the dissolved PLG/DCM into a 50-mL polypropylene graduated cylinder that has been sawed off at the 35 mL mark.
4. Set the homogenizer to speed setting 10 (approx 10,000 rpm).
5. Pipet 1.2 mg of DNA in 300 μL TE/sucrose solution into the PLG/DCM and homogenize for 2 min at room temperature to create the water-oil emulsion. Turn off the homogenizer, remove the probe from the emulsion, and place the graduated cylinder in ice.
6. Immediately place the beaker containing the 50 mL of 1.0% PVA solution on an adjustable stand and raise the stand so that the probe of the homogenizer is in the solution and near, but not touching, the bottom of the beaker. Turn on the homogenizer (setting 10) and quickly pour in the water-oil emulsion. Homogenize for 1 min to create the water-oil-water emulsion. Begin timing the moment the DNA/PLGA emulsion is poured into the PVA.
7. Quickly pour the water-oil-water emulsion into the 100 mL of 0.05% PVA solution that is in the 250-mL polypropylene beaker on the magnetic stir-plate (this solution should be stirring during the addition of the emulsion).
8. Evaporate the organic solvent (dichloromethane) by stirring the particles at room temperature for 2 h under a chemical hood (*see* **Note 16**).
9. Pour the microparticle suspension into a 250-mL polypropylene centrifuge tube and centrifuge at 336*g* for 15 min at ambient temperature (*see* **Note 17**).
10. Remove the supernatant carefully with a 50-mL polypropylene disposable pipet. Do not remove all of the supernatant because this may result in a loss of microparticles. Approximately 1 mL of fluid should remain in the tube.
11. Gently resuspend the pellet in the residual fluid with a rounded spatula to avoid breaking the microparticles. Reconstitute with 40 mL of deionized (DI) water. Transfer the suspension to a 50-mL screw-capped tube. Wash the 250-mL tube with 10 mL of DI water and transfer that 10 mL into the 50-mL tube. Repeat the centrifugation process/reconstitution twice (*see* **Note 18**).
12. After the final wash, resuspend the pellet in 5 mL of DI water.
13. Cover the top of the 50-mL tube with a Kimwipe® and secure the Kimwipe with a rubber band.
14. Insert the tube into liquid nitrogen to freeze the microparticle suspension.

15. Place the tube into a 900-mL fast-freeze flask and dry for a minimum of 15 h under vacuum on a lab freeze dryer (<100 mm Hg).
16. After drying, remove the Kimwipe and cap the tube. Keep the cap loose and store in an airtight container with desiccant at −20°C (*see* **Note 19**).

3.2. Lab-Scale Production of <10-μm Microparticles Containing Plasmid DNA

1. Calibrate the Masterflex pumps. Calibrate pump 1 with 1-mm ID viton tubing at a flowrate of 46 mL/min and pump 2 with 5-mm ID viton tubing at a flow rate of 1 L/min.
2. Dissolve 1 g of PLG in 17 mL of DCM in a 50-mL sterile polypropylene centrifuge tube (*see* **Note 15**). Cap the tube to prevent solvent evaporation. Continue to invert the tube until all of the polymer dissolves.
3. Place a 50-mL polypropylene beaker in an ice tub and set the tub on an adjustable jack. Lower the tip of the Silverson homogenizer shaft (SL2T) Position to approximately 3 mm of the bottom of the beaker. Add ice to the ice tub just to the top of the beaker.
4. Install the in-line mixing assembly to the Silverson L4R homogenizer drive unit as directed in the manufacturer's instructions. Attach 5-mm viton tubing to the in-line mixer's inlet and outlet ports as shown in **Fig. 1.** Prime

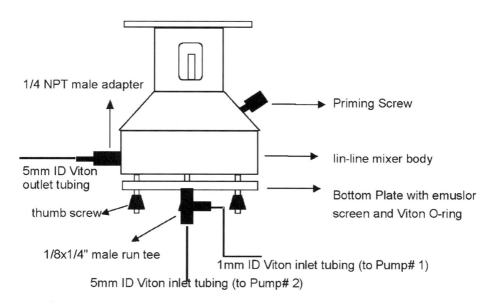

Fig. 1. Set up for production of w/o/w emulsion. A schematic of the in line mixer is depicted. Location of 1-mm and 5-mm viton tubing is indicated.

the in-line mixer with a 1% PVA solution. Insert the end of the 5-mm ID viton inlet tubing into a beaker containing 1 L of 1% PVA solution and the end of the 5-mm viton outlet tubing into a 4-L pyrex beaker.

5. Pour 18 mL of DCM into a 50-mL polypropylene graduated cylinder. Cover the graduated cylinder with foil.

6. Draw the DNA solution (10.6 mg dissolved in 1.6 mL TE/303 mM sucrose) into a 3-mL syringe with an 18-G needle.

7. Pour the PLGA/DCM solution from the 50 mL polypropylene tube into the 50-mL polypropylene beaker in the ice bucket and start the timer (T = 0:00:00, in h:mm:ss). Turn on the SL2T homogenizer to 10,000 rpm.

8. At T = 0:00:30, inject the DNA solution over a span of 5 s into the PLGA/ DCM solution. Continue homogenization until T = 0:04:00. At this time, quickly pour 18 mL of DCM from the graduated cylinder into the 50-mL beaker containing the emulsion. At T = 0:04:30, turn off the homogenizer. Hold the 50-mL beaker by hand. Slide out the jack and remove the beaker from under the homogenizing probe.

9. Place the 50-mL beaker on the bench next to pump 1. Turn the pump speed of pump 1 to 46 mL/min but do not turn on the pump. Place the inlet end of the 1-mm ID viton tubing into the 50 mL beaker.

10. At T = 0:05:30, turn on pump 2 and set the Silverson (L4R) homogenizing speed to 6000 rpm.

11. At T = 0:05:35, turn on pump 1.

12. When the emulsion in the 50-mL beaker is gone, wait 5 s, then turn off both pumps and the Silverson L4R. The w/o/w emulsion should be in the 4-L beaker.

13. Immediately place a 2.5-in magnetic bar into the 4-L beaker. Place the beaker on a magnetic stir plate and stir for 30 min at room temperature. After 30 min, heat the stirring solution to 37°C. Continue to monitor the solution so that it does not exceed this temperature.

14. After 2.5 h of stirring, dispense the milky dispersion evenly into four 250-mL polypropylene centrifuge tubes. Centrifuge the tubes at 336g for 10 min (*see* **Notes 16** and **17**). Gently pour off the supernatant.

15. Resuspend the particles in each tube with 5 mL of DI water using a spatula, then add an additional 35 mL of DI water to each tube and resuspend by gentle flicking with a rounded spatula. Do not vortex the particles (*see* **Note 18**). Transfer the particles from each 250-mL tube to a 50-mL screw-capped polypropylene centrifuge tube. Wash each of the 250-mL tubes with 10 mL of DI water and transfer to the 50-mL tubes.

16. Centrifuge the particles as in **step 14** and repeat the wash two more times. After the final wash, suspend the particles in each tube with 5 mL of water and cover the tubes with a Kimwipe. Secure the Kimwipe with a rubberband.

17. Insert the tubes into a bath of liquid nitrogen to freeze the suspensions.
18. Place the tubes in a 900-mL fast-freeze flask and dry for 15 h under vacuum on a lab freeze dryer (<100 mm Hg).
19. After drying, remove the Kimwipes and cover each of the tubes with a screw cap. Keep the caps loose and store in an airtight container (with desiccant) at −20°C (*see* **Note 19**).

3.3. Measuring DNA Encapsulation

1. Weigh approx 2.5 mg microparticles in a colorless microcentrifuge tube. Note the exact amount as it is needed to calculate the encapsulation efficiency.
2. Add 200 μL of TE buffer and resuspend microparticles by gently flicking the tube.
3. Add 500 μL of chloroform and rotate end-over-end for 90 min on a tube rotator at ambient temperature.
4. Centrifuge at 1400*g* for 5 min (*see* **Note 20**).
5. Remove 100 μL of the upper aqueous phase and transfer to a new tube containing 400 μL of TE buffer (dilution factor = 5).
6. Set the UV spectrophotometer to a wavelength of 260 nm and blank the spectrophotometer with 500 μL of TE buffer.
7. Pipet 500 μL of the diluted sample into the quartz cuvet and read absorbance. Reading should be between 0.1 and 0.5 (Beer-Lambert's range of linearity).

   Calculate the concentration of DNA in the microparticle sample (expect to recover approximately 2–4 μg DNA/mg of particles from the small scale batches and 4–7 μg DNA/mg of particles from the lab scale batches)

$$DNA\left[\frac{\mu g}{mg}\right] = \frac{Absorbance_{\lambda=260nm} \times dilution\ factor}{\varepsilon \times w \times b} \times 0.2\,ml$$

where $\varepsilon = 50\ \frac{\mu g \times cm}{Absorbance\ unit \times ml}$, w = weight of the microparticles in mg, and b = optical path length (1 cm).

3.4. Measurement of DNA Supercoiling

1. Extract the DNA from microparticles and quantify as in **Subheading 3.3**.
2. To a volume containing 200 ng of DNA, add an appropriate amount of 10X loading buffer to achieve a concentration of 1X loading buffer.
3. Prepare a sample containing 200 ng of unencapsulated control DNA with loading buffer.
4. Load the DNA into wells of a 0.8% agarose gel and electrophorese in 1X TAE until bromophenol blue has migrated two-thirds of the way down the gel.
5. Visualize on a UV lightbox and photograph if desired.

6. Estimate amount of supercoiling pre- and post-encapsulation by comparing the two samples (*see* **Note 21**).

3.5. Microparticle Sizing

1. Dispense 1 µL of 3 µm standard beads into a 50-mL polypropylene centrifuge tube containing 45 mL of Isoton II diluent and invert the tube five times.
2. Place the 50-mL polypropylene centrifuge tube underneath the aperture of the Coulter Multisizer II and adjust the platform height to immerse the aperture in the solution.
3. Press the "RESET" key to free any previous data stored in the instrument's memory. Turn the RESET/COUNT knob on the sampling stand to RESET. Wait 30 s.
4. Observe the coincidence meter on the screen. When the concentration is correct, the meter will read between 0% and 15%. Coinicidence is a measure of the percentage of counts that represent two or more particles passing through the aperture at a single time. If the coincidence is too low, add more particles and if it is too high, prepare the solution again starting with less sample.
5. Press the START key and after collecting at least 50,000 counts, press the STOP key.
6. Press "RESET" to clear data from instrument's memory.
7. Remove the tube containing particles and place a 100-mL beaker underneath the aperture to catch the wash waste. Flush the aperture with distilled water.
8. Add 40 mL of Isoton II diluent to a 50-mL polypropylene tube. Add a 1 µL sample to be tested (at ~50 mg PLG particles/mL), cap the tube, and invert the tube 5 times.
9. If visible particles are seen in the solution, continue to vortex until those particles vanish. If large particles continue to exist in the solution sonicate the sample for 1 min (*see* **Notes 17** and **18** regarding clumping).
10. Place the sample to be analyzed on the sample stand and adjust the platform height to immerse the aperture in the solution.
11. Follow **steps 4** and **5**.
12. Use the software provided with your instrument to print or store the results as average diameter (for number and volume) of the particles.

3.6. In Vitro Release of DNA from Particles Over Time

1. Weigh three 2.5-mg samples of microparticles containing DNA in 1.5-mL microcentrifuge tubes. Note the exact weight of the particles. Calculate the amount of total DNA in the weighed particles as described in **Subheading 3.3.**

2. Weigh three 2.5 mg samples of microparticles containing no DNA in 1.5-mL centrifuge tubes. It is necessary to perform the DNA release on these blank samples (*see* **Subheading 3.3**) because, at later time points, degradation of the PLG creates a byproduct that has an absorbance at 260 nm.
3. Add 1000 µL of PBS to each of the tubes.
4. Place tubes on an end-to-end rotator, and place the rotator in a 37° C incubator.
5. At each timepoint (1 h, 1 d, 3 d, 5 d, 7 d, 14 d, 21 d, etc.), remove the tubes from the incubator. Centrifuge the tubes at 5000*g* for 5 min.
6. Remove 800 µL of the supernatant from each of the pelleted samples and place the supernatant in a new tube.
7. Add 800 µL of PBS to each of the sample pellets. Reconstitute the pellets gently with a rounded spatula and replace the tubes in the incubator until the next time point.
8. Directly measure the Absorbance (260 nm) of the supernatant. Use PBS to zero the spectrophotometer.
9. Calculate the amount of DNA in the release buffer and then the % DNA released for that time interval:

$$(\text{DNA released} \div \text{Total DNA in tube}) \times 100, \text{ wherein:}$$

For the first time interval:

$$\text{DNA released (µg)} = (\text{Absorbance}_{\text{sample}} - \text{Absorbance}_{\text{blank}}) \div (\varepsilon \times b) \times 1 \text{ mL}$$

where $\varepsilon = 50$ (µg*cm/Absorbance*mL), and $b = 1$ cm for a 1 cm cuvet.

For the subsequent time intervals:

$$\text{DNA released (µg)} = \{[(\text{Absorbance}_{\text{sample}} - \text{Absorbance}_{\text{blank}}) \div (\varepsilon \times b)] \times 1 \text{ mL}\} - (0.2 \times \text{µg DNA released from previous time point})$$

3.7. Analysis of β-Gal Specific Expression

3.7.1. Immunizations

1. Suspend the microparticles in saline to achieve a dose of 30 µg DNA per animal in 50–200 µL. Minimizing the volume is preferable. Appropriate controls include saline-injected animals or blank microparticle injected animals, or most appropriately, vector (lacking the β-gal insert) containing microparticles.
2. Lightly sedate the animals with Isoflurane according to the institution's animal safety review committee accepted procedures.

3. Draw up the solution in a 3/10 cc insulin syringe with a 28-G1/2 needle. Inject half (up to 50 μL/muscle) of the formulation into each of the two tibialis anterior muscles and the other half (up to 50 μL/muscle) into each of the two gastrocnemius muscles of each animal. The tibialis is above the foot and below the knee joint (across the shin in comparison to a human) and the gastrocnemius is behind the tibialis.

3.7.2. Expression Analysis

1. On d 6 post-immunization, sacrifice the mice and douse the injected leg with 70% ethanol. Remove leg at hip by cutting with sharp, 12-cm scissors. Make an incision through the skin down the length of the leg, being careful not to cut muscle layer below. Peel away skin with forceps and remove. Identify the injected muscles (residual microparticles may still be visible) and excise the injected muscles. Rinse the muscles one time in 10 mL of PBS.
2. Fix the muscles by submerging in 3 mL of 0.25% glutaraldehyde at room temperature for 45 min.
3. Wash three times with 10 mL of PBS each time.
4. Stain the muscles by submerging in a 0.2% solution of X-gal at 37°C for 16 h with shaking. Protect from light.
5. Fix the stained muscles by submerging in 10% neutral buffered formalin for 24 h.
6. Wash the muscles three times in PBS.
7. Section the muscles at a thickness of 6 μm using a microtome and place on a glass slide. Dry the muscle sections overnight at 37°C.
8. Place the slides in a 60°C oven for 1 h.
9. Immerse the slides in xylene three times for 4 min each time.
10. Immerse the slides in 100% ethanol two times for 2 min each time.
11. Counterstain the slides by immersion in Hematoxylin solution for 1 min and then in Eosin Y solution for 20 s.
12. Dehydrate the slides by immersing 10 times in 95% ethanol, repeat two more times with a fresh solution of 95% ethanol. Continue by immersing the slides 10 times in 100% ethanol. Repeat with a fresh solution of 100% ethanol.
13. Clear the slides by immersing 10 times in two changes of xylene.
14. Place a drop of Permount onto tissue, and apply coverslip.
15. Examine under a microscope and photograph if desired.

3.8. β-Gal Specific ELISA

1. Immunize mice as described in **Subheading 3.7.1**.
2. At defined time points, bleed mice from the retro-orbital sinus using a 9-in Pasteur pipet. Approximately 60 μL of serum should be obtained from a single bleed in which 150 μL of blood was collected.

3. Transfer the blood to a microcentrifuge tube and centrifuge at 600*g* (*see* **Note 22**) for 15 min to separate the sera from the red blood cells. Remove the top serum layer and place into a new microcentrifuge tube for storage.

4. To each test well of a 96-well tissue-culture plate, add 50 µL of a solution of β-gal protein (2 µg/mL in PBS). Incubate the 96-well plate for 3 h at room temperature or overnight at 4°C.

5. Wash the plates two times with PBST. Fill the test wells with 150 µL of PBS containing 1% BSA and incubate at room temperature for 2 h.

6. Flick off the blocking buffer and wash two times with at least 200 µL of PBST per test well.

7. To the appropriate wells, add 100 µL of antibody diluted appropriately in PBS containing 1% BSA and incubate at room temperature for 1.5 h or at 4°C overnight.
   a. Negative control: NMS or antiserum.
   b. Positive control: β-gal specific MAb. Test five-fold dilutions of 1:500 to 1:62,500.
   c. Test samples: Dilute 1:100 to 1:12,500 at five-fold dilutions.

8. Wash plates four times with PBST. Fill the wells to the top each time.

9. Add 100 µL per test well of horseradish peroxidase (HRP)-conjugated antibodies specific for mouse IgG (H+L) diluted 1:5000 in PBS and incubate at room temperature for 3 h or at 4°C overnight.

10. Wash the plates four times with PBST. Fill the wells to the top each time.

11. Add 100 µL of ABTS substrate and incubate 20 min.

12. Add 50 µL of stop solution (1% sodium dodecyl sulfate [SDS] in distilled water) and measure the absorbance at 405 nm on the ELISA plate reader.

3.9. T-Cell Assays

*3.9.1. T-Cell Enrichment for Proliferation Assay (see **Note 23**).*

1. Immunize the animals following the protocol in **Subheading 3.7.1.**

2. Harvest spleens from immunized animals 2 wk post-injection. Place the freshly removed spleen in a 10 × 10-cm Petri dish containing 10 mL of RPMI-2.

3. Gently tease apart the spleen by crushing between two frosted, sterile glass slides to generate a single cell suspension and pipet cells into a 15-mL sterile, polypropylene centrifuge tube. Wash the Petri dish with additional RPMI-2 and add the wash to the tube.

4. Centrifuge for 10 min at 200*g* (*see* **Note 24**).

5. Decant the supernatant, disrupt the cell pellet by running the tube alongside test tube rack. Resuspend the cells in 1.5-mL of sterile distilled water and let sit for 8 s to lyse red blood cells. Quickly add 12 mL of RPMI-2.

6. Centrifuge for 10 min at 200g and then resuspend the cells in 1 mL of 1X column buffer.
7. For each spleen, prepare a mouse T-cell enrichment column by washing with 8 mL of 1X column buffer.
8. Load the 1 mL of cell suspension onto the top of the column and incubate at room temperature for 10 min.
9. Elute the cells from the column into a 15-mL sterile polypropylene centrifuge tube with 8 mL of 1X column buffer.
10. Centrifuge at 200g for 10 min. Resuspend the pellet in 2 mL of RPMI-10. Count the live cells using a hemocytometer and adjust the cell concentration to 2 × 10⁶/mL with RPMI-10.

3.9.2. Preparation of Antigen-Presenting Cells (APCs) for T-Cell Proliferation Assay (see **Note 25**)

1. Harvest the spleens from uninjected, control, age-matched, female syngeneic mice, prepare a single cell suspension, and lyse the red blood cells as described in **Subheading 3.9.1.**, **steps 4–6**.
2. Centrifuge at 200g for 10 min and resuspend the cells in 5 mL of RPMI-10. Count the cells in a hemocytometer using trypan blue to distinguish live from dead cells and adjust the cell concentration to 2 × 10⁷ cells/mL with RPMI-10.
3. Place tubes on ice and X-ray irradiate APCs with 3000 rad. Using an AXR Minishot 160 Model M160NH, a 3000 rad accumulated dose is delivered by irradiating cells for 18 min with a 150kV, 3mA X-ray beam at a distance of 5.7 in from the X-ray tube aperture.
4. Centrifuge the cells for 10 min at 200g. Resuspend in RPMI-10. Repeat the wash step one time.
5. Re-count the live cells as above and adjust the cell concentration to 2 × 10⁶ with RPMI-10.

3.9.3. T-Cell Proliferation Assay

1. Add 100 µl of effector cells (2 × 10⁵) prepared in **Subheading 3.9.1.** and 100 µL of APCs (2 × 10⁵) prepared in **Subheading 3.9.2.** to each well of a 96-well, flat-well, tissue-culture plate (each sample, mixture of APC and effector cell and stimulator (ConA, OVA, protein, or media), should be plated in triplicate)
2. To the test sample triplicate wells, add 50 µL of RPMI-10 containing 150 µg/mL of β-gal protein.
3. To the positive control triplicate wells, add 50 µL of RPMI-10 containing 12.5 µg/mL of ConA.
4. To the negative control triplicate wells, add 50 µL of RPMI-10 containing 150 µg/mL OVA.

5. To the unstimulated control triplicate wells, add 50 µL of RPMI-10 containing no antigen or mitogen.
6. Incubate the plates at 37°C in an incubator with 5% CO_2 for 72 h.
7. Pulse the cells in each well with tritiated thymidine by adding 50 µL of RPMI-10 containing 1 µCi of ^{3}H-TdR.
8. Continue to incubate plates at 37°C in an incubator with 5% CO_2 for 20 h.
9. Harvest the cells on a Cell Harvester according to the manufacturer's instructions. Place the filter mat in a scintillation bag, add 10 mL of scintillation fluid to the bag, ensure that the entire filter is wet with the fluid, and seal the bag with a bag sealer.
10. Count samples in a betaplate reader (*see* **Note 26**).

3.9.4. Preparation of CD8+ T Cells for Elispot Assay (see **Note 27**)

1. Harvest spleen cells 4–12 wk post-injection and prepare a cell suspension according to the protocol in **Subheading 3.9.1.**, **steps 2–6**. Adjust the concentration of cells to 2×10^8 spleen cells in 1 mL of 1X column buffer.
2. Mix 1 mL of cells with 1 mL of MAb cocktail designed for the purification of CD8 T-cell subsets. This cocktail is provided with the T-cell purification columns. Incubate at room temperature for 15 min.
3. Add 10 mL of 1X column buffer and centrifuge the cells at 200*g* (*see* **Note 24**) for 15 min. Repeat the wash one time and resuspend the pellet in 2 mL of 1X column buffer.
4. Wash the mouse T cell CD8 columns with 10 mL of 1X column buffer and apply the cells to the column. Incubate at room temperature for 10 min. Elute the cells with 10 mL of 1X column buffer.
5. Centrifuge the cells at 200*g* for 10 min. Resuspend the pellet in 2 mL of RPMI-10, count the live cells on a hemocytometer, and adjust the concentration to 2×10^6 cells/mL.

3.9.5. Preparation of Antigen-Presenting Cells (APCs) for the Elispot Assay (see **Note 28**)

1. Harvest the spleens from uninjected, control, age-matched, female syngeneic mice. Prepare a single cell suspension and lyse the red blood cells as described in **Subheading 3.9.2.**, **steps 2–6**, except following the last wash, resuspend the cells in 5 mL of RPMI-10. Count the cells in a hemocytometer and adjust the cell concentration to 2×10^7 cells/mL with RPMI-10.
2. Divide the APC cells equally into three tubes and treat as follows:
   a. To one-third of the cells, add β-gal peptide in RPMI-10 to final concentration 50 µg/mL.
   b. To the second tube (the negative control), add HBV peptide in RPMI-10 to a final concentration of 50 µg/mL.
   c. To the third tube, add nothing.

3. Incubate cells at 37°C in a 5% CO_2 incubator for 3 h. Gently resuspend the settled cells every 30 min.
4. Place the tubes on ice and X-ray irradiate APCs with 3000 rad. Using an AXR Minishot 160 Model M160NH, a 3000R accumulated dose is delivered by irradiating cells for 18 min with a 150kV, 3mA X-ray beam at a distance of 5.7 in from the X-ray tube aperture.
5. Centrifuge the cells for 10 min at 200g. Resuspend in RPMI-10 and repeat the wash step one additional time.
6. Re-count the live cells as in **step 1** and adjust the cell concentration to 2×10^6 in RPMI-10.

3.9.6. Elispot Assay

This assay is designed to enumerate T cells that secrete interferon-γ (IFN-γ) in response to antigen. It is performed with a kit from R&D Systems and all of the specialized reagents indicated below are included in the kit. Kits are available to test human and mouse cells.

1. Add 200 µL of RPMI-10 to each well of the Elispot plate and incubate for 30 min at room temperature. Alternatively, condition the Elispot plates with RPMI-10 according to the maunfacturer's instructions. Discard the media.
2. Immediately add 100 µL (2×10^5) each of effector cells and APCs to the wells. Note that each sample should be performed in triplicate. Each effector cell population should be incubated with each APC population (i.e., those pulsed with β-gal peptide/protein, negative control peptide/protein, or APC pulsed with nothing).
3. To the wells containing APCs pulsed with nothing, add 50 µL of a 25 µg/mL ConA solution to the wells containing effecter cells and APCs. This will serve as a positive control for T-cell activity.
4. Incubate the plate at 37°C in a 5% CO_2 incubator for 24 h.
5. Discard the culture medium and wash the wells by filling each well with 200 µL of 1X wash buffer and flicking off the wash buffer into the sink. Repeat for a total of four washes.
6. Add 100 µL of IFN-γ detection antibody into each well and incubate at 4°C overnight.
7. Discard the antibody and the wash wells by filling each well with 200 µL of 1X wash buffer and flicking off the wash buffer into the sink. Repeat for a total of four washes.
8. Add 100 µL of streptavidin-AP to each well and incubate for 2 h at room temperature.
9. Discard strepavidin-AP, wash wells by filling each well with 200 µL of 1X wash buffer and flicking off the wash buffer into the sink. Repeat for a total of four washes.

10. Add 100 µL of BCIP/NBT chromogen to each well and incubate for 1 h at room temperature (protect from light).
11. Discard the chromogen solution from the microplate and rinse each test well of the plate with distilled water thoroughly. Remove the flexible plastic support from the bottom of the microplate, rinse the bottom of the plate with distilled water, and blot it on a paper towel. Allow the microplate to dry completely under a hood.
12. Cells secreting IFN-γ are indicated by purple spots in each well. Spots can be enumerated by eye with a dissecting microscope, but this is tedious and subject to variation depending on an individual reader. Computerized plate reading can be easily and inexpensively performed by outside vendors. One such company is ZellNet Consulting, New York, NY (Website: www.zellnet.com; phone: 212-744-9526). Additional companies can be found by contacting R&D Systems customer service department.

4. Notes

1. Although a number of microencapsulation methods have been developed to date, the most commonly used method for the encapsulation of water-soluble drugs in synthetic biodegradable polymers is the water-oil-water (w/o/w) method *(12,23)*. In this method, the drug is dissolved in a buffered or plain aqueous solution, then added during homogenization or sonication to an organic solution containing the polymer. This mixing event promotes the formation of a fine emulsion in which the drug-containing water droplets are suspended in the organic oil in which polymer is dissolved. The w/o emulsion is added, again with intense mixing, to a large volume of an aqueous quench solution containing a surfactant such as poly(vinyl alcohol) to create the w/o/w emulsion. The resulting mixture is stirred to evaporate the solvent. The solvent is removed from the particles by initial extraction to saturation in the aqueous medium, then gradual evaporation. The rate of solvent removal depends upon the temperature of the quench water and the saturation solubility of the solvent in quench medium, among other parameters. The particles are concentrated, and dried or lyophilized. The final characteristics of the microparticles containing DNA are strongly influenced by several factors including the initial concentration of polymer, the homogenization speed, and the rate of solvent removal *(24)*.
2. Alternative methods that have been used to encapsulate plasmid DNA include spray drying *(25)* and phase coacervation *(26,27)*, but these will not be discussed here.
3. This scale is suitable for in vitro characterization of particles and small scale animal studies. Blank particles can be prepared following any of the protocols by simply omitting DNA from the solution. Expect to recover approx

150–180 mg dried particles with an encapsulation of 2–4 µg DNA/mg dried particles.

4. The plasmid should be endotoxin free. Good quality β-galactosidase plasmid with a certificate of analysis is available from Aldevron, LLC (Fargo, ND).

5. Make this solution fresh each time and filter it through a 0.2-µm filter.

6. This scale is suitable for in vitro characterization of particles and large-scale animal studies. Expect to recover 700–800 mg of dried microparticles with an encapsulation of 4–7 µg DNA/mg dried particles.

7. Viton tubing must be used as tubing made of other material will be dissolved by the dichloromethane.

8. This can be done quantitatively by high-performance liquid chromatography (HPLC) analysis *(28)*. In general this requires a HPLC system equipped with an autosampler, and an 8-µL flow-through photodiode array UV detector set at 260 nm. Columns include a reverse-phase HPLC column (TSK-GEL DNA-NPR, 4.6 mm $\times$ 7.5 mm, 2.5 µm (Toso-Haas, Montgomeryville, PA), and a guard column (TSK-GEL, DEAE-NPR, 4.6 mm $\times$ 5.0 mm, 5.0 µm). Column equilibration buffer is: 0.56 M NaCl, 50 mM Tris, pH 9.0. The DNA sample (extracted from microparticles as described in **Subheading 3.3.**) is loaded onto the column and the column is developed with a linear gradient of 0.56 M to 1.2 M NaCl in 50 mM Tris, pH 9.0, over approx 10 column volumes at a flow rate of 0.5 mL/min.

9. Sizing can also be done under a microscope. Place a drop of suspended particles on a sizing slide (1–100 µm scale) and determine the size of at least 200 particles. Calculations can be done to arrive at a rough estimate of the percentage of particles that fall within a given size range

10. Note that although the methods provided here have been optimized for detecting β-gal specific expression, the steps can be adapted for most reporter proteins. For more specific information regarding these assays and suggestions for modifications, an excellent reference is Current Protocols in Immunology *(29)*.

11. To measure expression, groups of 3–5 animals are suitable. For measurement of antibody responses, 8–10 per group is preferred. For measurement of T-cell responses, 3–5 per group is preferred. Note that the source of animals is not particularly important, as long as it is reliable and has been recognized as an approved vendor by the institution's animal care committee.

12. Use only serum from nonimmunized female BALB/c mice, 8–10 wk old that has been batched and pre-tested in the ELISA.

13. In the absence of an X-ray source, similar results can often be obtained by treatment of the APCs with mitomycin C *(29)* or a γ-irradiation source.

14. Peptides can be ordered from a variety of sources. We use Multiple Peptide Systems, (San Diego, CA). Allow a 4–6 wk turnaround time. Acquire peptides at a purity of more than 90% from the vendor.

15. Do not use polyethylene or polystyrene tubes because dichloromethane will dissolve both of these materials.

16. The solution can be stirred for a longer time if desired. Do not leave the particles longer than 6 h.

17. In a typical tissue-culture centrifuge (Beckman GS-6R), 336*g* is equivalent to 1500 rpm. This is a crucial step. If the particles are centrifuged too hard (i.e., too fast or too long), they will be difficult to resuspend and the final dried particles will be clumpy and difficult to analyze.

18. Ensure that the particles suspend well. If they are clumpy here, it is likely that they will be clumpy post-drying. One reason for clumpiness is that the dicholoromethane has not been completely removed. Try stirring longer than 2 h. It is also possible to warm the solution to about 30°C while stirring the particles to remove dichloromethane. However, do not start the warming process until an hour of stirring has occurred. It is also a good idea to weigh the empty 50-mL tubes (with the cap on). After drying the particles, replace the cap as directed, and re-weigh. The difference is the weight of the particles. This is easier than trying to scoop all the particles out of the tube to weigh them (if you are interested in total recovery weight).

19. Particles can be stored for up to 1 mo at −20°C.

20. 1400*g* is 14,000 rpm (maximum speed) in a standard microfuge

21. Plasmid DNA exists in three forms. Supercoiled DNA migrates fastest through the gels. Nicked isoforms (plasmid containing single-stranded breaks) migrate more slowly than supercoiled and linear plasmid migrates between the other two isoforms. The plasmid should be at least 50% supercoiled. Typically, little supercoiling should be lost when a comparison is done between the input DNA and the DNA extracted from the particles.

22. 600*g* is equivalent to 6000 rpm in a standard microfuge.

23. Note that this procedure will enrich for CD3+ T cells. It is also possible to isolate CD4+ or CD8+ T cells using columns and methods available for this purpose from R&D Systems.

24. 200*g* is 1000 rpm in a Beckman G5-CR tissue-culture centrifuge.

25. This process prepares APCs from the splenic population. This includes B-cells, macrophages, and DCs. Do not perform a T-cell enrichment step with these spleen cells.

26. Results are presented as stimulation index (SI): the mean of the cpm of triplicate samples in stimulated cultures (those containing OVA, ConA, or β-gal protein) divided by the mean of the cpm of triplicate samples in wells containing only media (unstimulated). The SI obtained with ConA should be very high (average results produce an SI of 10–20) and the SI obtained with OVA should be very low (average results produce an SI <1.5). Test sample SI values should be greater than 4.

27. This assay can be used to determine T-cell responses to β-gal specific peptides. The addition of specific MHC class I restricted T-cell epitopes will permit detection of T cells that are typically CD8+ T cells associated with killing functions. The addition of longer peptides will permit the detection of both class II (typically CD4+ T-helper cells) and class I restricted responses. Additional responses can be detected by using T cells enriched for CD4+ or CD3+ markers with specific T-cell columns and kits available from R&D Systems. Note that the purification protocol provided is a summary of that available in the R&D Systems kit. Specific questions regarding the use of these columns may be addressed by reading this literature.

28. This process prepares APCs from the splenic population. This includes B-cells, macrophages, and DCs. Do not perform a T-cell enrichment step with these spleen cells. These cells are pulsed with a class I restricted test peptide (derived from β-gal), or a negative control peptide (derived from HBV).

References

1. Gurunathan, S., Klinman, D. M., and Seder, R. A. (2000). DNA vaccines: immunology, application, and optimization. *Ann. Rev. Immunol.* **18,** 927–974.

2. Wang, R. Doolan, D. L., Le, T. P., Hedstrom, R. C., Coonan, K. M., Charoenvit, Y., et al. (1998) Induction of antigen-specific cytotoxic T lymphocytes in humans by a malaria DNA vaccine. *Science* **282,** 476–480.

3. Calarota, S., Bratt, G., Nordlund, S., Hinkula, J., Leandersson, A. C., Sandstrom, E., and Wahren, B. (1998) Cellular cytotoxic response induced by DNA vaccination in HIV-1-infected patients. *The Lancet* **351,** 1320–1325.

4. Calarota, S., Leandersson, A. C., Bratt, G., Hinkula, J., Klinman, D. M., Weinhold, K. J., et al. (1999) Immune responses in asymptomatic HIV-1-infected patients after HIV-DNA immunization followed by highly active antiretroviral treatment. *J. Immunol.* **163,** 2330–2338.

5. Boyer, J. D., Chattergoon, M. A., Ugen, K. E., Shah, A., Bennett, M., Cohen, A., et al. (1999) Enhancement of cellular immune response in HIV-1 seropositive individuals: a DNA-based trial. *Clin. Immunol.* **90,** 100–107.

6. Boyer, J. D., Cohen, A. D., Vogt, S., Schumann, K., Nath, B., Ahn, L., et al. (2000) Vaccination of seronegative volunteers with a human immunodeficiency virus type 1 env/rev DNA vaccine induces antigen-specific proliferation and lymphocyte production of β-chemokines. *J. Infect. Dis.* **181,** 476–483.

7. Roy, M. J., Wu, M. S., Barr, L. J., Fuller, J. T., Tussey, L. G., Speller, S., et al. (2000) Induction of antigen-specific CD8+ T cells, and protective levels of antibody in humans by particle-mediated administration of a hepatitis B virus DNA vaccine. *Vaccine* **19,** 764–778.

8. Mincheff, M., Tchakarov, S., Zoubak, S., Loukinov, D., Botev, C., Altankova, I., et al. (2000) Naked DNA and adenoviral immunizations for immunotherapy of prostate cancer: a phase I/II clinical trial. *Europ. Urol.* **38,** 208–217.

9. Kanke, M., Geissler, R. G., Powell, D., Kaplan, A., and DeLuca, P. P. (1988) Interaction of microspheres with blood constituents. III. macrophage phagocytosis of various types of polymeric drug carriers. *J. Parenteral Sci. Technol.* **42,** 157–165.

10. Tabata, Y. and Ikada, Y. (1990) Phagocytosis of polymer microspheres by macrophages. *Adv. Polymer Sci.* **94,** 107–141.

11. Lunsford, L., McKeever, U., Ecktein, V., and Hedley, M. L. (2000) Tissue distribution and persistence in mice of plasmid DNA encapsulated in a PLGA-based microsphere delivery vehicle. *J. Drug Target.* **8,** 39–50.

12. Okada, H. and Toguchi, H. (1995) Biodegradable microspheres in drug delivery. *Crit. Rev. Ther. Drug Carrier Sys.* **12,** 1–99.

13. Jones, D. H., Corris, S., McDonald, S., Clegg, J., and Farrar, G. (1997) Poly (DL-lactide-co-glycolide)-encapsulated plasmid DNA elicits systemic and mucosal antibody responses to encoded protein after oral administration. *Vaccine* **15,** 814–817.

14. Hedley, M. L., Curley, J., and Urban, R. (1998) Microspheres containing plasmid-encoded antigens elicit cytotoxic T-cell responses. *Nature Med.* **4,** 365–368.

15. Wang, D., Robinson, D. R., Kwon, G. S., and Samuel, J. (1999) Encapsulation of plasmid DNA in biodegradable poly(D,L-lactic-co-glycolic acid) microspheres as a novel approach for immunogene delivery. *J. Controlled Rel.* **57,** 9–18.

16. Ando, S., Putnam, D., Pack, D. W., and Langer, R. (1999) PLGA microspheres containing plasmid DNA: preservation of supercoiled DNA via cryopreparation and carbohydrate stabilization. *J. Pharmaceut. Sci.* **88,** 126–130.

17. Hao, T., McKeever, U. and Hedley, M. L. (2000) Biological potency of encapsulated plasmid DNA. *J. Controlled Rel.* **69,** 249–259.

18. Chen, S.C., Jones, D. H., Fynan, E. F., Farrar, G. H., Clegg, J.C., Greenberg, H.B., and Herrmann, J.E. (1998) Protective immunity induced by oral immunization with a rotavirus DNA vaccine encapsulated in microparticles. *J. Virol.* **72,** 5757–5761.

19. Jones, D. H., Clegg, J., and Farrar, G. H. (1998) Oral delivery of micro-encapsulated DNA vaccines, in *Developmental Biological Standards*, vol. 92 (Brown, F. and Haaheim, L., eds.), pp. 149–155.

20. Herrmann, J. E., Chen, S. C., Jones, D. H., Tinsley-Bown, A., Fynan, E. F., Greenberg, H. B., and Farrar, G. H. (1999) Immune responses and protection obtained by oral immunization with rotavirus VP4 and VP7 DNA vaccines encapsulated in microparticles. *Virology* **259,** 148–153.

21. Klencke, B., Matijevic, M., Urban, R. G., Lathey, J. L., Hedley, M. L., Berry, M., Thatcher, J., Weinberg, V., Wilson, J., Darragh, T., Jay, N., Da Costa, M., Palefsky, J. M. (2002) Encapsulated plasmid DNA treatment for human papillomavirus 16-associated anal dysplasia: a Phase I study of ZYC101. *Clin. Cancer Res.* **8,** 1028–1037.

22. Sheets, E. E., Urban, R. G., Crum, C. P., Hedley, M. L., Politch, J. A., Gold, M. A., Muderspach, L. I., Cole, G. A., Crowley-Nowick, P. A. (2003) Immunotherapy of human cervical high-grade cervical intraepithelial neoplasia with microparticle-delivered human papillomavirus 16 E7 plasmid DNA. *Am. J. Obstet. Gynecol.* **188,** 916–926.

23. Lewis, D. (1990) Controlled release of bioactive agents from lactide/glycolide polymers, in *Biodegradable Polymers as Drug Delivery Systems* (Chasin, M. and Langer, R., eds.) Marcel Dekker, New York, pp. 1–41.

24. Hsu, Y. Y., Hao, T., and Hedley, M. L. (1999) Comparisons of process parameters for microencapsulation of plasmid DNA in poly (D, L-lactic-co-glycolic) acid microspheres. *J. Drug Target.* **7,** 313–323.

25. Walter, E., Moelling, K., Pavlovik, J., and Merke, H. (1999) Microencapsulation of DNA using poly(DL-lactide-co-glycolide) stability issues and release characteristics. *J. Controlled Rel.* **61,** 361–374.

26. Leong, K. W. Mao, H. Q., Truong-Le, V. L., Roy, K., Walsh, S.M., August, J.T. (1998) DNA-polycation nanospheres as non-viral gene delivery vehicles. *J. Controlled Rel.* **53,** 183–193.

27. Truong-Le, V. and August, T. (1998) Controlled gene delivery by DNA-gelatin nanospheres. *Hum. Gene Ther.* **9,** 1709–1717.

28. Bergan, D., Galbraith, T., and Sloane, D. L. (2000) Gene transfer in vitro and in vivo by cationic lipids is not significantly affected by levels of supercoiling of a reporter plasmid. *Pharmaceut. Res.* **17,** 967–973.

29. Colligan, J. E, Kruisbeek, A. M., Margulies, D.H., Shevach, E.M., and Strober, W. (eds.) (1992) *Current Protocols in Immunology.* Wiley Interscience, New York, NY, pp 3.0.1–3.18.10.

23

DNA Delivery to Cells in Culture Using Ultrasound

Thomas P. McCreery, Robert H. Sweitzer, and Evan C. Unger

1. Introduction

Effective gene delivery to mammalian cells is a necessary step in the development of gene medicines and therapy. A variety of techniques have been developed to enhance the transfection of exogenous DNA into mammalian cells. These techniques include viral, nonviral, and energy-based systems. The most common energy-based system is electroporation. There are drawbacks to electroporation owing to the spacing required for the electrodes. This necessitates the use of special electroporation cuvets to actually carry out the transfection. We have conducted experiments that demonstrate the use of ultrasound to enhance gene delivery into mammalian cells (1). The process of SonoPorationSM involves exposing cells to ultrasound energy. The ultrasound energy enhances gene expression in two ways. First, it creates pores in the cell membrane. This has been demonstrated by uptake of macromolecules and in photomicrographs by Dr. Tachibana's group in Japan (personal communication). Second, it upregulates DNA repair and the stress response genes in the cell (1). This upregulation has been measured using reverse transcription polymerase chain reaction (RT-PCR).

Gene-expression experiments have been carried out in a variety of mammalian cell lines. The most marked enhancement of gene expression has been observed in cell lines that are typically difficult to transfect. Ultrasound energy has been used in concert with liposomal DNA delivery systems, naked DNA, and peptide-based delivery systems.

From: *Methods in Molecular Biology, vol. 245:*
Gene Delivery to Mammalian Cells: Vol. 1: Nonviral Gene Transfer Techniques
Edited by: W. C. Heiser © Humana Press Inc., Totowa, NJ

The major steps in conducting DNA delivery to cells with ultrasound are similar to conventional expression systems. The cells are plated into a 6-well culture dish and grown to 80% confluence. The transfection mixture is added to the wells. At this point the system is treated with ultrasound. Ultrasound energy is delivered using a transducer to conduct sound through the culture medium into the cells. Mammalian cells differ in their tolerance to ultrasound energy. Some cells are quite robust, whereas others are sensitive. It is best to conduct a small pilot study to determine the sensitivity of the cell line being transfected. The cells are then grown normally and assayed as appropriate to the characteristics of the delivered gene.

2. Materials
2.1. Ultrasound Device, Plasmid, and Assay Kits

1. SonoPorator® 100 Ultrasound with a 1 MHz transducer (RichMar, Inola, OK).
2. pCMVCAT plasmid (Invitrogen, Carlsbad, CA).
3. Cell Proliferation Kit II (XTT) (Roche Diagnostics GmbH, Mannheim, Germany).
4. CAT Elisa Kit (Roche Diagnostics GmbH).

2.2. Lipids, Chemicals, and Tissue-Culture Materials

1. Six-well tissue-culture plate (VWR International, Seattle, WA).
2. Dipalmitoyl ethyl phosphocholine (DPEPC, Avanti Polar Lipids, Alabaster, AL).
3. Dioleoyl phosphoethanolamine (DOPE, Avanti Polar Lipids): Combine 1:1 with DPEPC at a final lipid concentration of 1 mg/mL in HEPES-buffered saline).
4. HEPES Buffered Saline: 10 mM HEPES, pH 7.8, at 37°C, 150 mM NaCl; autoclave and store at room temperature.
5. Culture Medium: Dulbecco's Modified Eagle's Medium (MEM) supplemented with 10% fetal calf serum (FCS).

3. Methods
3.1. Ultasound Tolerance Test

The tolerance test is designed to give a working level of ultrasound that is tolerated by a particular cell type. For best results, the cells should be exposed to the maximum amount of ultrasound that does not cause cell death. There are many factors that can create undesirable effects during the ultrasound process. The main factors that need to be regulated are the power and the time of exposure.

1. Plate mammalian cells at $1-4 \times 10^5$ cells per well in 6-well plates with 4 mL of culture medium in each well. Incubate at 37°C for 24–48 h or until the cells are 80% confluent.
2. Apply ultrasound to the cells at powers of 0.5, 1, 1.5, and 2 W/cm^2 (*see* **Note 1**). Just touch the ultrasound transducer to the surface of the media.
3. Vary the time of ultrasound treatment from 5 s to 1 min per well (*see* **Note 2**).
4. Assay cell viability immediately following application of ultrasound using the XTT assay. Perform the assay following the manufacturer's protocol.

3.2. Transfection

The transfection method used in these experiments can utilize liposome-encapsulated, peptide-based, or naked DNA. The transfection procedure described in this section utilizes a liposomal delivery system.

1. Plate cells in 4 mL of culture medium in 6-well plates and incubate until they are 80% confluent.
2. Form the Liposome:DNA complex using a 6:1 ratio in HEPES-buffered saline (e.g., 6 µL of DPEPC:DOPE [1:1] plus 1 µg of plasmid). Adjust the final volume of complex to allow for 200 µL for each test well. Incubate the lipid/DNA mixture at room temperature for 45 min.
3. Add 200 µL of the Liposome:DNA complex directly to the culture media in the 6-well plates and slowly mix with a pipet. Incubate the cells for 1 h at 37°C and then treat with ultrasound; adjust the power and time based on tolerance data.
4. Surface-sterilize the transducer using 70% ETOH and then apply ultrasound to the cells. Carry out the experiments in a biological safety cabinet to minimize the possibility of contaminating the cells.
5. Incubate the cells at 37°C for 24–72 h and then assay for gene expression using the CAT Elisa Kit according to the manufacturer's instructions.

4. Notes

1. The transducer only needs to touch the surface of the media to allow for ultrasound transmission. It is not necessary for the transducer surface to contact the cells (**Fig. 1**).

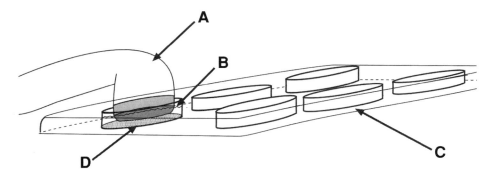

Fig. 1. Application of ultrasound to cells. (**A**) transducer; (**B**) culture medium; (**C**) 6-well plate; (**D**) cells.

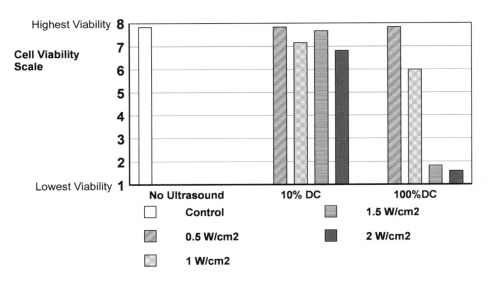

Fig. 2. Effect of differing levels of ultrasound energy on cell viability. DC, duty cycle or the percent of time that ultrasound transmission occurs.

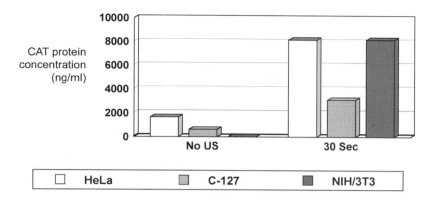

Fig. 3. Effect of ultrasound energy on transfection efficiency in three cell lines.

2. The timing device on the machine is set to protect the transducer. If a user turned on the machine and left the transducer powered up for this amount of time, it would not be damaged. It is best to use an external timer to time the experiments.
3. These techniques have been applied to several mammalian cell lines *(1)*. **Figure 2** shows the effect of ultrasound energy on HeLa cells. **Figure 3** shows the effect of differing levels of ultrasound on gene expression.

References

1. Unger, E. C., McCreery, T. P., and Sweitzer, R. H. (1997) Ultrasound enhances gene expression of liposomal transfection. *Invest. Radiol.* **32**, 723–772.

24

DNA Delivery to Cells In Vivo by Ultrasound

Thomas P. McCreery, Robert H. Sweitzer, Evan C. Unger,
and Sean Sullivan

1. Introduction

In vivo gene delivery has proven to be much more difficult than in vitro delivery. Many techniques have worked in vitro only to fail in vivo. Techniques such as electroporation require invasive and painful procedures to deliver genes. Ultrasound-mediated delivery has the potential to help overcome these difficult and painful aspects with a simple noninvasive procedure.

Ultrasound energy can be used to enhance the expression of DNA delivered intravenously or injected directly into tissue. The experiments described in this chapter were carried out to show that ultrasound can enhance delivery of both naked DNA or DNA complexed to cationic lipids. The procedure consists of preparing the transfection mixture, administering it either intravenously or intratumorally, and then applying ultrasound.

2. Materials

2.1. Ultrasound Device, Plasmid, Cells, and Assay Kits

1. SonoPorator® 100 Ultrasound with a 1 MHz transducer (RichMar, Inola, OK).
2. pCMVCAT (Invitrogen, Carlsbad, CA).
3. CAT ELISA Kit (Roche Diagnostics GmbH, Mannheim, Germany).
4. IL-12 ELISA Kit (Roche Diagnostics GmbH).
5. LS180 (human colon adenocarcinoma; ATCC, Manassas, VA).

From: *Methods in Molecular Biology, vol. 245:*
Gene Delivery to Mammalian Cells: Vol. 1: Nonviral Gene Transfer Techniques
Edited by: W. C. Heiser © Humana Press Inc., Totowa, NJ

2.2. Lipids, Chemicals, and Equipment

1. Dipalmitoyl ethyl phosphocholine (DPEPC, Avanti Polar Lipids, Alabaster, AL).
2. Dioleoyl phosphoethanolamine (DOPE, Avanti Polar Lipids): combine 1:1 with DPEPC at a final lipid concentration of 1 mg/mL in HEPES-buffered saline.
3. N-[(1-(2-3-dioleyloxy) propyl)]-N-N-N-trimethylammonium chloride:cholesterol (4:1 mol/mol ratio, DOTMA:CHOL, Avanti Polar Lipids).
4. HEPES-buffered saline: 10 mM HEPES, pH 7.8, at 37°C, 150 mM NaCl; autoclave and store at room temperature.
5. Tissue homogenizer (Model PT1300D, Kinematica AG, Italy)

3. Methods

The transfection method used in these experiments can utilize either naked DNA or liposome-encapsulated DNA. The experiments in this study were conducted using liposomal-based delivery systems.

3.1. Liposomal Transfection to Muscle (4–5)

1. Prepare DPEPC:DOPE liposomes using a 1:1 (w/w) ratio of DPEPC and DOPE according to the manufacture's instructions.
2. Form the Liposome:DNA complex using a 6:1 ratio in HEPES-buffered saline (e.g., 30 µL of DOPE:DPEPC (1:1) plus 5 µg of plasmid). Use 30 µg of lipid and 5 µg of plasmid per mouse. Incubate the Liposome:DNA mixture at room temperature for 45 min.
3. Inject 100 µL of the lipid/DNA mixture intravenously into the tail vein of the mouse.
4. Immediately apply ultrasound to the hind limb (or target area) of the mouse at a power of 1 W/cm^2 for 2 min (*see* **Notes 1** and **2**).
5. Harvest tissues 72 h after ultrasound treatment.
6. Using a tissue homogenizer, homogenize the target tissues in Cell Lysis Buffer (CAT ELISA Kit) for 1 h at 4°C.
7. Centrifuge at 1000g for 15 min at 4°C to remove tissues from buffer.
8. Assay CAT activity in 200 µL of supernatant using the CAT ELISA Kit according to the manufacturer's instructions.

3.2. Liposomal Transfection in Tumor Cells (1–3, Figs. 1–4)

1. Inject 5×10^5 (100uL) of LS180 colon adenocarcinoma cells intradermally in the lower back of nude mice.

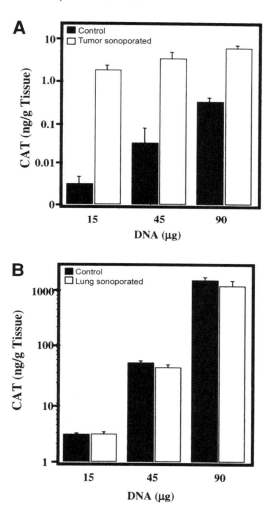

Fig. 1. Effect of ultrasound treatment on cationic liposome-mediated systemic gene transfer to sc SCCVII tumor-bearing mice. A 15, 45, or 90 µg plasmid dose of DOTMA:CHOL formulated CAT plasmid was administered by tail vein to SCCVII tumor-bearing mice. Tumors were sonoporated at 1.5 W/cm^2 for 5 min immediately after plasmid injection. CAT expression levels in tumor (**A**) and lung (**B**) were measured 18–20 h after DNA administration. The data is expressed as mean + SD, $n = 5$.

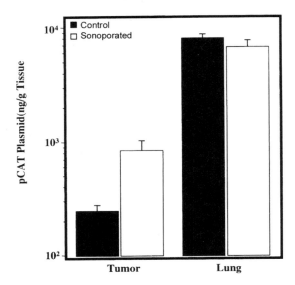

Fig. 2. Effect of ultrasound on plasmid uptake by SCCVII primary tumors following tail vein administration of transfection complexes. A 90 µg plasmid dose of transfection complex was administered to SCCVII tumor-bearing mice by tail vein injection and tissues were collected 1 h later for quantitation of CAT plasmid (mean + SD, $n = 5$) by qPCR.

2. Allow the tumors to develop for 7–10 d to reach an adequate size for transfection.
3. Form the Liposome:DNA complexes as in **Subheading 3.1.1., step 2**, but at a 3:1 ratio using either pCMVCAT or pCMV-IL-12.
4. Inject 100 µL of Liposome:DNA mixture into the tail vein of the mouse.
5. Immediately after injection apply ultrasound to the tumor at 1.5 W/cm² for 5 min (*see* **Notes 1** and **2**).
6. Harvest tumors and lungs 18–20 h after ultrasound.
7. Using a tissue homogenizer, grind up target tissues, and suspend in cell lysis buffer (CAT ELISA Kit) for 1 h at 4°C.
8. Centrifuge at 1000*g* for 15 min at 4°C to remove tissues from buffer.
9. Assay CAT activity in 200 µL of supernatant using the CAT ELISA Kit according to the manufacturer's instructions.

4. Notes

1. The area of the animal being treated with ultrasound must be shaved to allow for good ultrasound contact. The transducer needs ultrasound gel to conduct the energy into the tissue.

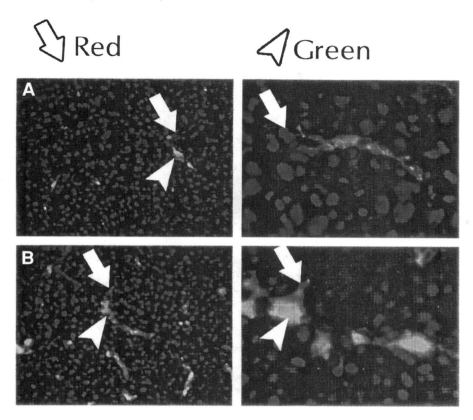

Fig. 3. Intratumoral distribution and endothelial localization of intravenously administered fluorescent-labeled plasmid. Psoralen-fluorescein-labeled pCMV-CAT complexed with DOTMA:CHOL at 90 µg dose was administered into sc tumor-bearing mice by tail-vein injection. Tumors were sonoporated at 1.5 W/cm² for 5 min immediately after plasmid injection. Control (**A**) and ultrasound treated (**B**) tumors were collected 15 min after plasmid injection, tissue cryosections (5 µm) were prepared for CD31 endothelial immunostaining and examined by fluorescence microscopy for fluorescent plasmid (green) and endothelial cell marker (red); magnification ×20 (left) and ×60 (right).

2. The timing device on the machine is set to protect the transducer. If a user turned on the machine and left the transducer powered up for this amount of time, it would not be damaged. It is best to use an external timer to time the experiments.

3. Ultrasound-mediated gene delivery has been conducted in muscle and tumor tissue (*1–5*).

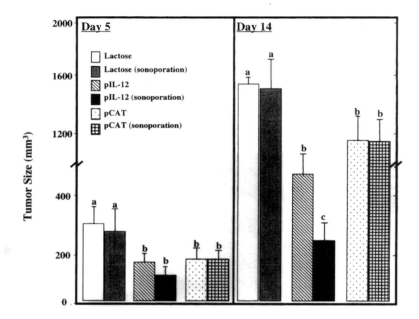

Fig. 4. Effect of ultrasound treatment on inhibition of tumor growth following systemic administration of IL-12 transfection complexes. SCCVII tumor-bearing mice were
sonoporated for 2 min at 1.5 W/cm^2 followed by a single tail-vein injection of
DOTMA:CHOL transfection complexes containing 10 µg IL-12 plasmid or CAT plasmid. Control animals received 10% lactose. Tumor size was measured 14 d after plasmid
injection. Bars with different superscripts are statistically different ($p < 0.05$, mean +
SD, $n = 5$) as determined by one-way analysis of variance (ANOVA) and Duncan's multiple range test.

References

1. Anwer, K., Kao, G., Proctor, B., Anscombe, I., Florack, V., Earls, R., et al. (2000)
   Ultrasound enhancement of cationic lipid-mediated gene transfer to primary tumors
   following systemic administration. *Gene Ther.* **7,** 1833–1839.
2. Anwer, K., Kao, G., Proctor, B., Rolland, A., and Sullivan, S. (2000) Optimization
   of cationic lipid/DNA complexes for systemic gene transfer to tumor lesions. *J.
   Drug Target.* **8,** 125–135.
3. Anwer, K., Meaney, C., Kao, G., Hussain, N., Shelvin., R, Earls., R. M., et al.
   (2000) Cationic lipid-based delivery system for systemic cancer gene therapy. *Cancer Gene Ther.* **7,** 1156–1164.
4. Unger, E. C., Hersh, E., Vannan, M., Matsunaga, T. O., and McCreery, T. (2001)
   Local drug and gene delivery through microbubbles. *Prog. Cardiovasc. Dis.* **44,**
   45–54.
5. Unger, E. C., McCreery, T. P., and Sweitzer, R. H. (1997) Ultrasound enhances gene
   expression of liposomal transfection. *Invest. Radiol.* 32, 723–727.

Index

A

Accell™ gene gun, 185
Alanine aminotransferase, in serum, 260
Anesthesia, 140, 192, 220, 221, 222, 230,
    232, 239, 240, 253, 255, 257, 258,
    259, 275
Aurintricarboxylic acid, 224

B

β-galactosidase, 17, 223
    assay of, 275
    expression of, 249, 261, 262
Brain slices, preparation of, 200

C

Calcium phosphate, 4, 25
    BES method, 26, 28
Cationic lipids (*see also* liposomes,
        synthetic vectors), 5, 95, 289, 294
    charge ratio, 6
    helper lipids, 6, 8
    preparation of aqueous lipid
        suspensions, 100
    preparation of dry lipid films, 99
    preparation of dry lipid films
        fluorescently labeled films, 112
    preparation of, 98
    structure, 6
    use in gene therapy, 95
Cationic polymers, 6
    dendrimers, 7
    PEI, 6
Chloramphenicol acetyltransferase (CAT),
        17, 211
    assay of, 70
    codon-optimized, 102
    expression of, 211, 290, 294
Chloroquine, augmentation of transfection, 74
Co-lipid, see helper lipid
CpG motifs, 96
    removal from plasmids, 101
Cytomegalovirus Immediate Early 1 gene
        (CMV IE1), 212

D

DEAE-Dextran, 5, 25, 148
    augmentation of transfection, 74
    standard method, 27
Dendrimer, 67
    properties, 67
    Starburst PAMAM, 67
Dendrimer-DNA complex
    charge ratio, 72
    preparation of, 69, 72
DNA vaccines, 147, 186, 265

E

Electroporation, 156, 207, 215, 227
    advantages of, 227
    analysis of, 210
    electrodes, 223, 233, 234, 242
    in vitro, 156
    in vivo, 156
    instrumentation, 230, 233
    intradermal delivery, 216, 221, 224
    optimization, 210, 211, 224,
    of skeletal muscle, 227, 234
    of skin, 215
    theory of, 207, 215, 228
    transdermal delivery, 216, 220, 224
    of tumors, 237
Electrotransfer, *see* Electroporation
Elispot assay
    Antigen-presenting cells
        preparation of, 279
    CD8+ T cells
        preparation of, 279
    procedure, 280
Enhanced green fluorescent protein, *see*
        Green fluorescent protein

F

Flow cytometry, 58, 63
Fusogenic peptides (*see* peptide-mediated
        delivery)

299